Defense Innovation Handbook

Systems Innovation Series

Series Editor
Adedeji B. Badiru
Air Force Institute of Technology (AFIT) – Dayton, Ohio

PUBLISHED TITLES

Project Management Simplified: A Step-by-Step Process, *Barbara Karten*

A Six Sigma Approach to Sustainability: Continual Improvement for Social Responsibility, *Holly A. Duckworth & Andrea Hoffmeier*

Project Management for Research: A Guide for Graduate Students, *Adedeji B. Badiru, Christina Rusnock, & Vhance V. Valencia*

Handbook of Construction Management: Scope, Schedule, and Cost Control, *Abdul Razzak Rumane*

Essentials of Engineering Leadership and Innovation, *Pamela McCauley*

Project Feasibility: Tools for Uncovering Points of Vulnerability, *Olivier Mesly*

Project Management for the Oil and Gas Industry: A World System Approach, *Adedeji B. Badiru & Samuel O. Osisanya*

Learning Curves: Theory, Models, and Applications, *Mohamad Y. Jaber*

Handbook of Emergency Response: A Human Factors and Systems Engineering Approach, *Adedeji B. Badiru & LeeAnn Racz*

Profit Improvement through Supplier Enhancement, *Ralph R. Pawlak*

Work Design: A Systematic Approach, *Adedeji B. Badiru & Sharon C. Bommer*

Additive Manufacturing Handbook: Product Development for the Defense Industry, *Adedeji B. Badiru, Vhance V. Valencia, & David Liu*

Global Engineering: Design, Decision Making, and Communication, *Carlos Acosta, V. Jorge Leon, Charles Conrad, & Cesar O. Malave*

Design for Profitability: Guidelines to Cost Effectively Manage the Development Process of Complex Products, *Salah Ahmed Mohamed Elmoselhy*

Handbook of Measurements: Benchmarks for Systems Accuracy and Precision, *Adedeji B. Badiru, LeeAnn Racz*

Introduction to Industrial Engineering, Second Edition, *Avraham Shtub, Yuval Cohen*

Company Success in Manufacturing Organizations: A Holistic Systems Approach, *Ana M. Ferreras, Lesia L. Crumpton-Young*

Handbook of Industrial and Systems Engineering, Second Edition, *Adedeji B. Badiru*

Quality Management in Construction Projects, Second Edition, *Abdul Razzak Rumane*

Quality Tools for Managing Construction Projects, *Abdul Razzak Rumane*

Industrial Project Management: Concepts, Tools, and Techniques, *Adedeji B. Badiru, Abidemi Badiru, & Adetokunboh Badiru*

Productivity Theory for Industrial Engineering, *Ryspek Usubamatov*

Defense Innovation Handbook: Guidelines, Strategies, and Techniques, *Adedeji B. Badiru, Cassie B. Barlow*

For more information about this series, please visit: https://www.crcpress.com/Systems-Innovation-Book-Series/book-series/CRCSYSINNOV

Defense Innovation Handbook
Guidelines, Strategies, and Techniques

Edited by
Adedeji B. Badiru
Cassie B. Barlow

CRC Press
Taylor & Francis Group
Boca Raton London New York

CRC Press is an imprint of the
Taylor & Francis Group, an **informa** business

CRC Press
Taylor & Francis Group
6000 Broken Sound Parkway NW, Suite 300
Boca Raton, FL 33487-2742

First issued in paperback 2020

ISBN-13: 978-1-138-05067-9 (hbk)
ISBN-13: 978-0-367-78091-3 (pbk)

Library of Congress Cataloging-in-Publication Data

Names: Badiru, Adedeji Bodunde, 1952- editor. | Barlow, Cassie B., editor.
Title: Defense innovation handbook : guidelines, strategies, and techniques /
[edited by] Adedeji B. Badiru and Cassie B. Barlow.
Description: Boca Raton : Taylor & Francis, a CRC title, part of the Taylor &
Francis imprint, a member of the Taylor & Francis Group, the academic
division of T&F Informa, plc, 2018. | Series: Systems innovation series |
Includes bibliographical references.
Identifiers: LCCN 2018027135 | ISBN 9781138050679 (hardback : acid-free paper)
| ISBN 9781315168623 (ebook)
Subjects: LCSH: Defense industries--Technological innovations.
Classification: LCC HD9743.A2 D37 2018 | DDC 623.4068/4--dc23
LC record available at https://lccn.loc.gov/2018027135

Visit the Taylor & Francis Web site at
http://www.taylorandfrancis.com

and the CRC Press Web site at
http://www.crcpress.com

Dedication

To Innovation, in all its diverse ramifications

Contents

Foreword

I was asked by my good friends and colleagues Dr. Adedeji B. Badiru and Dr. Cassie B. Barlow who are contributing authors, compilers, and collaborating editors on this *Defense Innovation Handbook* you're reading to consider writing a Foreword "on the importance of Innovation to the DoD." I thought about it for a bit and sent Dr. Barlow a note back stating I wrestle with the term "innovation" in the DoD and in the tech community at large—much overused, not understood, too much money dumped in its name down the drain .... soooooooooooo not sure I'm your "Foreword" writer...... To which Dr. Barlow replied, "we actually completely agree that the term 'innovation' is overused and not well understood. That is exactly the reason why we wanted to pull the handbook together." If you're reading this, you see I acquiesced and decided with the above exchange to put a few thoughts together and share in this Foreword.

Dr. Badiru and Dr. Barlow are two of the finest, most giving and passionate people I know who work diligently every day to advance the end-state capabilities of our Department of Defense. From sharing their vast knowledge in the academic environment with students of all ages; to being staunch contributors to K-12 STEM programs; to leading regional workforce development in support of the Aerospace enterprise; and, most recently, seeing a chasm between the needs of the Department of Defense and how that "chasm" supposedly is being addressed in the name of "innovation" ..... yet knowing that there is a fundamental disconnect in the term, the risk levels and willingness of our DoD to truly allow "innovation" in fact to occur, decided to undertake the development of this handbook.

My personal belief is that the term "innovation" is overused in our technological communities. I find it to be a "label" placed on what perhaps is a good, maybe even solid, technical concept to fundamentally sell the idea to decision makers, who themselves would never stand in the way of something "innovative" and end up supporting an idea as innovative, but the idea actually provides no value. After all, who would want to be seen as standing in the way of innovation? Sadly, most don't seek the basic understandings of what is "innovative" in the context of what is being proposed and sold to them. In most simplistic terms I believe a run through of the Heilmeier Catechism would quickly separate the wheat from the chaff and help leaders and the Department to truly know whether a concept or idea is technologically sound, adds value and is truly "innovative."

The Catechism: (1) In the most simplistic terms what are you trying to do and why? (2) How is it done today and what are the limits of current practice? (3) What is new in your approach and why do you think it will be successful? Who cares? (4) If you are successful, what difference will it make? (5) What are the risks? (6) How much will it cost? (7) How long will it take? (8) What are the mid-term and final "exams" to check for success?

So, you've read this far and are possibly thinking, okay all fine and dandy, Joe, I passed all the gates laid out above... now what? Several years ago, many of us in the Dayton, Ohio region invited Dr. John Kao to come work with a large cross section of the Dayton Regional leadership. We spent a few days looking at exactly the issue of innovation and what had to happen to bring to fruition the world's truly innovative capabilities, such as powered flight; refrigeration; all electric starting, ignition and lighting for automobiles; cellophane tape; the step ladder; and traffic lights; to name a few.

Along with the innovation came a solid 100 years of prosperity in the industries that spawned the automotive, military, and civil aviation products which provided our USAF unparalleled air, space and cyber superiority. So, what was it that caused these advancements to occur? Several things in my mind. First it was a full-contact, hands-on sport, the people behind these great technological advancements had calluses and dirty fingernails, and worked shoulder-to-shoulder to solve the wicked problems of their day. No PowerPoint slides or Keynote involved. As we studied this with Dr. Kao, we understood that the inventors and innovators whose shoulders we stand on worked tirelessly to "convert possibilities into value." They were not focused on the world's definition of value around wealth-based tangible assets, but value in terms of the capability to continuously realize the future they wanted. It wasn't about money, it was about enhancing the quality of life, advancing the body of knowledge, advancing capabilities, especially in the DoD, where we, to this day, seek to have an unfair technological advantage over our near-peer adversaries. Scientists, engineers, inventors, and lay people saw a future and went after it.

So, were they Innovators?

I submit to you—yes, based on the following unwrapping of innovation by Dr. Kao. "Innovation a (noun) + a (verb)—The portfolio of financial, intellectual, organization and human capabilities that enable a society's journey to its desired future." Innovation occurs across a value chain; innovation then is an idea followed by a reasoned complementary or juxtaposed action; Science and Art; Engineering and Design; Incremental and Disruptive; Inside Out and Outside In; Public and Private; Left Brain and Right Brain; Analytics and Values; Facts and Possibilities; Risk Taking and Prudence; Inspiration and Planning; Closure and Treasure Hunting. The sum then of innovation is agility, foresight, enlightened leadership, risk appetite, and the appetite for experimentation.

As you step forward to "convert possibilities into value" heed the words of this Foreword and the compendium of thought leadership that is contained in this DoD Handbook. Thank you for your Service to this great Nation, be it in uniform, as a civilian, an academician, or a defense contractor. Above all else, remember our US Military, Allied and Coalition partners require the best of the best, 210% reliable in every off-nominal condition you can't even imagine—so, yes, be innovative, be realistic and yet please be humbled enough to understand that though you may have thought up the greatest innovative thing since sliced bread, it may have no place in the DoD inventory.

Joe Sciabica, SES Retired
President, Universal Technology Corporation
Dayton, Ohio

Preface

"Innovation" is one of the most recognized and most used words, not only in the defense enterprise, but also in many science and technology realms. Indeed, it is also frequently cited in business and industry. When people talk of innovation, their term of reference is usually technological developments. But innovation goes well beyond the technical realm. Innovation in process and strategies is just as important and relevant as technological developments. Many times, process and business innovations are even more important than technological innovations because the manifestation of technology can be realized only through effective processes and strategies. The *Defense Innovation Handbook: Guidelines, Strategies, and Techniques* represents a monumental collection of diverse views of innovation, from technological requirements to process and managerial requirements. Specific themes addressed by the 23 chapters in the handbook include "Innovation for national defense," "Definitional analysis of innovation," "The aerospace and defense industry in Southwest Ohio: A model for workforce-driven economic development," "Other transactions: Increasing importance in the Department of Defense," "Commercial technologies in the Department of Defense: Technology evolution and implications for acquisition professionals," "A system and statistical engineering enabled approach for process innovation," "Building resilient systems via innovative human systems integration," "Innovative model for situation awareness in dynamic defense systems," "Globalization and defense manufacturing," "Is your organization ready for innovation?" "Human monitoring systems for health, fitness and performance augmentation," "Enhancing innovation: Methods, cultural aspects, ideation approaches, and box busters," "Self-jamming behavior: Joint interoperability, root causes, and thoughts on solutions," "4D Weather Cubes and defense applications," "Innovative approach to infrastructure resilience: A case study of evaluating Department of Defense sites for small modular reactors," "Three innovations for defense acquisition reform," "Strategy and military technology: The three offsets," "Prescription for an affordable full spectrum defense and innovation policy," "Anatomy of arms races and technological innovation," "Innovation dynamics in the defense space sector," "Innovative applications of polymer materials for 3D printing," "Innovation project management," and "Innovation in systems framework for intelligence operations." With this collection of diverse and thought-provoking chapters, all readers will find this handbook to be a useful reference at home, work, industry, education, and business environments. Please join us in innovative thought!

Acknowledgments

We acknowledge Ms. Anna E. Maloney for her extraordinary contributions to this monumental handbook. Not only did she serve as a co-author of one of the chapters, but she also provided superior editorial and administrative support for the book project from the beginning to the end. Her insightful suggestions and technical finessing of the chapters greatly increased the overall quality of the handbook. Without her consistent dedication to the project, the handbook would not have been completed on time.

Editors

Dr. Adedeji B. Badiru is a Professor of Systems Engineering at the Air Force Institute of Technology (AFIT). He is a registered professional engineer and a fellow of the Institute of Industrial Engineers as well as a fellow of the Nigerian Academy of Engineering. He earned a BS in Industrial Engineering, MS in Mathematics, and MS in Industrial Engineering from Tennessee Technological University, and PhD in Industrial Engineering from the University of Central Florida. He is the author of several books and technical journal articles. Prof. Badiru has served as a consultant to several organizations around the world including Russia, Mexico, Taiwan, Nigeria, and Ghana. He has conducted customized training workshops for numerous organizations including Sony, AT&T, Seagate Technology, US Air Force, Oklahoma Gas & Electric, Oklahoma Asphalt Pavement Association, Hitachi, Nigeria National Petroleum Corporation, and ExxonMobil. He holds a leadership certificate from the University Tennessee Leadership Institute. Prof. Badiru has served as a Technical Project Reviewer, curriculum reviewer, and proposal reviewer for several organizations including The Third-World Network of Scientific Organizations, National Science Foundation, National Research Council, and the American Council on Education. He is on the editorial and review boards of several technical journals and book publishers. Prof. Badiru has also served as an Industrial Development Consultant to the United Nations Development Program. He is also a Program Evaluator for ABET. In 2011, Prof. Badiru led a research team to develop analytical models for Systems Engineering Research Efficiency (SEER) for the Air Force acquisitions integration office at the Pentagon. He has led a multi-year composite manufacturing collaborative research between the Air Force Institute of Technology and Wyle Aerospace Company. Prof. Badiru has diverse areas of avocation. His professional accomplishments are coupled with his passion for writing about everyday events and interpersonal issues, especially those dealing with social responsibility. Outside of the academic realm, he writes motivational poems, editorials, and newspaper commentaries; he also engages in paintings and crafts.

Dr. Cassie B. Barlow is the Chief Operating Officer at the Southwestern Ohio Council for Higher Education. Her focus is on developing the defense workforce of the next generation. Previously, she was the 88th Air Base Wing and Installation Commander, Wright-Patterson AFB, Ohio, where she was in command of one of the largest air base wings in the Air Force with more than 5,000 Air Force military and civilian employees. Dr. Barlow was commissioned in 1988 as a distinguished graduate of Georgetown University, Washington, DC. She has served in a variety of positions in the information management, behavioral science and personnel career fields at squadron, Wing, direct reporting unit, major command, Air Force and combatant command levels. Dr. Barlow was selected by the Air Force Institute of Technology to attend Rice University to earn a doctorate in Organizational Psychology. After graduation, she served as a behavioral scientist at the Air Force Research Laboratory and the Air Force Academy. She was then selected to lead the analysis team for the Chief of Staff of the Air Force Developing Aerospace Leaders Program. Dr. Barlow commanded the 355th Mission Support Squadron at Davis Monthan Air Force Base, Arizona, and the 48th Mission Support Group at Royal Air Force Lakenheath. She was also the Director of Manpower and Personnel of the North American Aerospace Defense Command and United States Northern Command, headquartered at Peterson Air Force Base, Colorado. Dr. Barlow attended the Air Command and Staff College and the Industrial College of the Armed Forces.

Contributors

Darryl Ahner
OSD Scientific Test and Analysis
Techniques Center of Excellence
Air Force Institute of Technology
Dayton, Ohio

Adedeji B. Badiru
Air Force Institute of Technology
Dayton, Ohio

Bud Baker
Wright State University
Dayton, Ohio

Cassie B. Barlow
SOCHE
Dayton, Ohio

Sally J. F. Baron
US Air Force Academy
Colorado Springs, Colorado

Randy J. Belles
Oak Ridge National Laboratory
Oak Ridge, Tennessee

Budhendra L. Bhaduri
Oak Ridge National Laboratory
Oak Ridge, Tennessee

Kimberly Bigelow
University of Dayton
Dayton, Ohio

Sarah Burke
OSD Scientific Test and Analysis
Techniques Center of Excellence
Air Force Institute of Technology
Dayton, Ohio

Jarred L. Burley
Air Force Institute of Technology
Dayton, Ohio

Cory A. Cooper
US Air Force Academy
Colorado Springs, Colorado

Amanda Delaney
University of Dayton
Dayton, Ohio

Mark M. Derriso
Air Force Institute of Technology
Dayton, Ohio

Ed Downs
ProTerf Training
Miami, Florida

Brannon J. Elmore
Air Force Institute of Technology
Dayton, Ohio

Mica R. Endsley
SA Technologies, LLC
Santa Clara, California

Steven T. Fiorino
Air Force Institute of Technology
Dayton, Ohio

Daniel D. Jensen
US Air Force Academy
Colorado Springs, Colorado

Christine Schubert Kabban
Air Force Institute of Technology
Dayton, Ohio

Bandana Kar
Oak Ridge National Laboratory
Oak Ridge, Tennessee

Ibrahim Katampe
Central State University
Wilberforce, Ohio

Kevin J. Keefer
Air Force Institute of Technology
Dayton, Ohio

Anna E. Maloney
Air Force Institute of Technology
Dayton, Ohio

Gary T. Mays
Oak Ridge National Laboratory
Oak Ridge, Tennessee

Jan P. Muczyk
Air Force Institute of Technology
Dayton, Ohio

Olufemi A. Omitaomu
Oak Ridge National Laboratory
Oak Ridge, Tennessee

Michael P. Poore
Oak Ridge National Laboratory
Oak Ridge, Tennessee

Aaron Ramert
OSD Scientific Test and Analysis
Techniques Center of Excellence
Air Force Institute of Technology
Dayton, Ohio

Matthew Richards
The Boeing Company
El Segundo, California

Kristy Rochon
SOCHE
Dayton, Ohio

Jaclyn E. Schmidt
Air Force Institute of Technology
Dayton, Ohio

David E. Shahady
Air Force Research Laboratory
Wright-Patterson Air Force Base, Ohio

Zoe Szajnfarber
The George Washington University
Washington, District of Columbia

Alfred E. Thal, Jr.
Air Force Institute of Technology
Dayton, Ohio

Claude D. Vance
United States Air Force
    Materiel Command
Wright-Patterson Air Force Base, Ohio

Noah R. Van Zandt
Air Force Research Laboratory
Albuquerque, New Mexico

Annalisa Weigel
Fairmont Consulting Group
Boston, Massachusetts

Roy L. Wood
Northeastern State University
Broken Arrow, Oklahoma

and

Defense Acquisition University
Fort Belvoir, Virginia

Stephen R. Woodall
President and CEO
Strategic Synthesis, Ltd. (LLC)
Fairfax Station, Virginia

chapter one

Innovation for national defense

Adedeji B. Badiru

Contents

> Innovation: Give it to someone who can make something of it or keep it to yourself and make nothing of it.
>
> **Adedeji Badiru, 2018**

Collaboration is the essence of actualizing innovation for practical applications as opined by the quote above.

Introduction

Evolution, revolution, and innovation have defined human existence for millennia. From the Ice Age to the Stone Age, the Bronze Age, the Iron Age, and the modern age, innovation, rudimentary as it may be in many cases, has determined how humans move from one stage to the next. Innovation is the lifeline of national development. This handbook presents a collection of chapters that provide techniques and methodologies for achieving the transfer of defense-targeted science and technology development for general industrial applications. Experts from national defense institutions, government laboratories, business, and industry contributed chapters to the handbook. The handbook provides a lasting guidance for nations, communities, and businesses expecting to embark upon science and technology transfer to industry under the auspices of national defense pursuits. We don't often make a connection between a viable industrial base and a robust national defense. The fact is that a vibrant base of industrial activities can promote and protect national defense pursuits, particularly where economic vitality is concerned.

There is a need for a good utility framework for this handbook because of the globalization of modern industries desirous of capitalizing on technical developments in the defense industry for the purpose of developing new consumer products. Many nations are interested in embarking on rapid prototyping of new technologies from their defense organizations for the advancement of their nations. Guidelines, strategies, and techniques are needed to actualize their aspirations. Allied nations often conduct joint defense exercises, the coalitions from which can advance their respective local industries. Some good examples of how national defense products enhance general consumer products include the following:

1. There are several consumer products that originated from initial defense focus, such as the microwave oven and the global positioning system (GPS).
2. R&D personnel from defense organizations often end up working in general business and industry, where their expertise is needed through consumer technology transfer processes.
3. The International Space Station combines the efforts of cooperating nations, thus paving the way for potential advancement of tech-transfer industries at the national level.
4. Many formerly classified defense-related developments have been declassified, thereby necessitating the need for tech transfer strategies to industry.

The overall conclusion is that a strong national defense program fuels a strong industrial base. Every country, even the poorest ones, must be engaged in national defense pursuits, which are predicated on innovation, both soft and hard. Not all innovation is of a technical breed. Soft innovation may pertain only to the processes and managerial principles for managing and deploying innovation. Hard innovation may relate to technical and technology-based developments that enhance the focus on national defense.

Digital revolution and innovation

The digital environment has created new opportunity for new innovation developments both in technology and in operational processes. For example, in the digital emergence of 3D printing (additive manufacturing), the lead editor offers the following operational quotes:

"Little thoughts make up big ideas."

—Adedeji Badiru

"Big components are made in little layers of material."

—Adedeji Badiru

Manufacturing is rapidly shifting from manual labor to digital labor. The digital revolution has landed on the doorstep of conventional manufacturing. What was once limited to the realm of laboratory research has now been transformed, through innovation, to the platform of practicality and reality. For decades, manufacturing had languished within the same old framework of mold-and-cast type of product development. This traditional approach has made manufacturing subject to the inability to respond quickly and adaptively to new product requirements. With the advent of direct digital manufacturing (aka 3D printing or additive manufacturing), product designers and developers now have a mechanism to respond to the requirements for new intricately designed and delivered products, often at the immediate point of need. The defense sector is well positioned to leverage the capabilities of this new digital innovation for designing and making products. The emerging proliferation of 3D printing in business and industry has made it imperative that a structural forum be organized to guide the path of full utilization of innovative developments in digital manufacturing. The conventional product development environment is vastly different from what 3D printing will require. Hitherto, individuals and organizations have been jumping on the 3D printing bandwagon without strategic consideration of downstream and upstream aspects of "printed" products. This handbook forum offers a structured platform of enabling innovation in the defense sector. Both technical and management issues related to this new wave of innovation are addressed in the handbook. The expected benefits of innovation dialogue and exchanges include a better alignment of product technology with future developments and the need to secure, maintain, and advance national defense. Specifically, readers of this handbook will learn about the systems engineering aspect of 3D printing to achieve a faster translation of innovation into real products as well as operational effectiveness, raw material efficiency, higher return on manufacturing investment, rapid and focused product deployment, technology transfer potentials, manufacturing flexibility, and anywhere-anytime agility for product generation.

With the additional emergence of virtual reality (VR), augmented reality (AR), and mixed reality (MR), the platform of innovation for the defense industry is growing rapidly. These emerging technologies can be leveraged to provide cost-effective development of new products. The best way to accomplish this is to mix innovation and collaboration.

Central role of innovation

The central role of innovation in national defense is evidenced by the fact that "Drive Innovation" is one of the top five priorities announced by the US Air Force in August 2017. The priorities, released by USAF secretary Heather Wilson, are

1. Restore readiness
2. Cost-effectively modernize
3. Drive innovation
4. Develop exceptional leaders
5. Strengthen alliances

Figure 1.1 shows the cross-linkages of the five priorities and how innovation has a central role. We cannot restore readiness without employing new innovative tools and techniques. We cannot cost-effectively modernize without developing and utilizing radically innovative quantitative and qualitative methodologies. We cannot develop exceptional leaders without directing efforts at new, innovative, and specialized education, including advanced education. We cannot strengthen alliances without innovative partnering strategies. In a systems approach, a system is defined as the collection of interrelated elements whose collective output is higher than the sum of the individual outputs of the elements. As a specific tool, the DEJI® (Design, Evaluation, Justification, and Integration) model of systems engineering is unique and innovative because it explicitly calls for a *justification* and *integration* of actions, which requires a more rational decision process during the *design* and *evaluation* stages. The model facilitates a recursive *design-evaluate-justify-integrate* process for enhancing

Priority Cross-Linkages ⇨	Restore Readiness	Cost-effectively modernize	**Drive Innovation**	Develop Exceptional Leaders	Strengthen Alliances
Restore Readiness		*	⊘	*	*
Cost-effectively Modernize	*		⊘	*	*
Drive Innovation	⊘	⊘		⊘	⊘
Develop Exceptional Leaders	*	*	⊘		*
Strengthen Alliances	*	*	⊘	*	

Figure 1.1 Central role of innovation in air force priority cross-linkages. ⊘ Innovation Alignment across priorities and * Inter-priority alignment.

operations. The design stage is essentially the decision stage, which must be evaluated and justified before moving to the implementation stage. The typical implementation stage must be pursued with respect to how well the decision (i.e., design) integrates into the prevailing infrastructure and resource base of the organizations involved. Thus, the model covers the broad spectrum of people, process, and technology in national defense pursuits. Some of the analytical tools used in the DEJI model include state-space modeling, simulation, systems value modeling, learning curve analysis, workload analysis, cognitive modeling, and hierarchical decision transformation. The DEJI model is further discussed later in this chapter. Based on a systems approach, priorities are best pursued from a system of systems perspective. In this regard, multifaceted collaboration approaches must be embraced.

Multifaceted collaboration for innovation

Innovation is best pursued via multifaceted collaboration. No one entity has all the answers. Together, innovation is stronger. Figure 1.2 shows a framework for academia-government-industry collaboration that can be leveraged for the pursuit and sustainment of innovation.

In executing the desired multifaceted collaboration for innovation, some of the technical topics of interest include, but are not limited to, the following:

- Hypersonic weaponry
- Stealth technology
- Autonomous systems
- Mobile radar platforms
- Directed energy systems
- Laser warning systems
- Cognitive radio networks
- Human performance systems
- Quantum computing
- Neuromorphic computing
- Additive manufacturing
- CUBESAT (Cube Satellite), a type of miniaturized satellite for space research that is made up of multiples of 10×10×10 cm cubic units (U-Class Spacecraft) and conventional satellite technology
- Artificial intelligence and machine learning

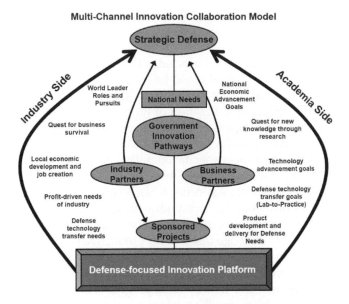

Figure 1.2 Framework for multifaceted defense-focused innovation.

There are several complementary and auxiliary topics affiliated with the earlier list. In addition to the technological aspects of innovation, there are also issues of human-centric innovation as well as process innovations, such as logistics and integrated supply chain, to get the mission done in contested environments. Thus, the span of innovation is quite expansive.

Innovation transfer paths

Just as we may have technology transfer paths, so can we have innovation transfer paths.

Innovation transfer is not just about the hardware, technology, or technical components of a system. It can involve a combination of several components, including software (computer-based) and peopleware. Thus, this chapter addresses the transfer of innovation knowledge as well as the transfer of innovation skills.

Due to its many interfaces, the area of technology adoption and implementation is a prime candidate for the application of project planning and control techniques. Technology managers, engineers, and analysts should make an effort to take advantage of the effectiveness of project management tools. This applies the various project management techniques available to the problem of innovation transfer. The project management approach is presented within the context of innovation adoption and implementation for national defense. The Triple C model of Communication, Cooperation, and Coordination is applied as an effective tool for ensuring the acceptance of new innovation products.

Characteristics of innovation transfer

To transfer innovation, like any technology transfer, we must know what constitutes innovation. A working definition of innovation will enable us to determine how best to transfer it. A basic question that should be asked is: What is innovation?

Innovation can mean different things to different audiences. Innovation can be defined as follows:

Innovation is a combination of physical and nonphysical processes that make use of the latest available knowledge, skills, technology, etc. to achieve business, service, or organizational goals.

Innovation is a specialized body of knowledge that can be applied to achieve a mission or purpose. The knowledge concerned could be in the form of methods, processes, techniques, tools, machines, materials, and procedures. Technology design, development, and effective use is driven by effective utilization of human resources and effective management systems. Technological progress is the result obtained when the provision of technology is used in an effective and efficient manner to improve productivity, reduce waste, improve human satisfaction, or meet specific operational needs.

Innovation all by itself is useless. However, when the right innovation is put to the right use, with effective supporting management systems, it can be very effective in achieving organizational goals. Innovation implementation starts with an idea and ends with a productive process. Innovative progress is said to have occurred when the outputs of innovation, in the form of information, instrument, or knowledge that is used productively and effectively in industrial operations, leads to a lowering of costs of production, better product quality, higher levels of output (from the same amount of inputs), and better alignment with mission requirements. The information and knowledge involved in innovation progress includes those which improve the performance of management, labor, and the total resources expended for a given activity.

Innovation progress plays a vital role in improving overall national defense. Experience in the developed countries, such as the United States, show that in the period 1870–1957, 90% of the rise in real output per man-hour can be attributed to technological progress fueled by innovation. It is conceivable that a higher proportion of increases in per capita income is accounted for by technological change. Changes occur through improvements in the efficiency in the use of existing technology; that is, through learning and through the adaptation of other technologies, some of which may involve different collections of technological equipment. The challenge to developing countries is how to develop infrastructure that promotes, uses, adapts, and advances technological knowledge.

Most of the developing nations today face serious challenges arising not only from the worldwide imbalance of dwindling revenue from industrial products and oil, but also from major changes in a world economy that is characterized by competition, imports, and exports of not only oil, but also of basic technology, weapon systems, and electronics. If technology utilization is not given the right attention in all sectors of the national economy, the much-desired national defense cannot occur or cannot be sustained. If innovation is stymied, the ability of a nation to compete in the world market will, consequently, be stymied, with potential adverse implication for national defense.

The important characteristics or attributes of a new technology may include productivity improvement, improved quality, cost savings, flexibility, reliability, and safety. An integrated evaluation must be performed to ensure that a proposed technology is justified both economically and technically. The scope and goals of the proposed technology must be established right from the beginning of the project. Table 1.1 summarizes some of the common "ilities" characteristics of innovation transfer assessment.

An assessment of a technology transfer opportunity will entail a comparison of unit-level objectives with the overall organizational goals in the following areas.

1. *Marketing and outreach strategy*: This should identify the customers of the proposed technology. It should also address items such as market cost of proposed product, assessment of competition, and market share. Import and export considerations should be a key component of the marketing strategy.

Table 1.1 The "ilities" assessment of innovation

Characteristics	Definitions, questions, and implications
Adaptability	Can the technology be adapted to fit the needs of the organization? Can the organization adapt to the requirements of the technology?
Affordability	Can the organization afford the technology in terms of first-cost, installation cost, sustainment cost, and other incidentals?
Capability	What are the capabilities of the technology with respect to what the organization needs? Can the technology meet the current and emerging needs of the organization?
Compatibility	Is the technology compatible with existing software and hardware?
Configurability	Can the technology be configured for the existing physical infrastructure available within the organization?
Dependability	Is the technology dependable enough to produce the outputs expected?
Desirability	Is the particular technology desirable for the prevailing operating environment of the organization? Are there environmental issues and/or social concerns related the technology?
Expandability	Can the technology be expanded to fit the changing needs of the organization?
Flexibility	Does the technology have flexible characteristics to accomplish alternate production requirements?
Interchangeability	Can the technology be interchanged with currently available tools and equipment in the organization? In case of operational problems, can the technology be interchanged with something else?
Maintainability	Does the organization have the wherewithal to maintain the technology?
Manageability	Does the organization have adequate management infrastructure to acquire and use the technology?
Re-configurability	When operating conditions change or organizational infrastructure change, can the technology be re-configured to meet new needs?
Reliability	Is the technology reliable in terms of technical, physical, and/or scientific characteristics?
Stability	Is the technology mature and stable enough to warrant an investment within the current operating scenario?
Sustainability	Is the organization committed enough to sustain the technology for the long haul? Is the design of the technology sound and proven to be sustainable?
Volatility	Is the technology devoid of volatile developments? Is the source of the technology devoid of political upheavals and/or social unrests?

2. *Industry growth and long-range expectations*: This should address short-range expectations, long-range expectations, future competitiveness, future capability, and prevailing size and strength of the industry that will use the proposed technology.
3. *National defense benefit*: Any prospective technology must be evaluated in terms of direct and indirect benefits to be generated by the technology. These may include product price versus value, increased international trade, improved standard of living, cleaner environment, safer work place, and higher productivity.
4. *Economic feasibility*: An analysis of how the technology will contribute to profitability should consider past performance of the technology, incremental benefits of the new technology versus conventional technology, and value added by the new technology.
5. *Capital investment*: Comprehensive economic analysis should play a significant role in the technology assessment process. This may cover an evaluation of fixed and sunk

costs, cost of obsolescence, maintenance requirements, recurring costs, installation cost, space requirement cost, capital substitution options, return on investment, tax implications, cost of capital, and other concurrent projects.

6. *Innovation resource requirements*: The utilization of resources (human resources and equipment) in the pre-technology and post-technology phases of industrialization should be assessed. This may be based on material input-output flows, high value of equipment versus productivity improvement, required inputs for the technology, expected output of the technology, and utilization of technical and nontechnical personnel.

7. *Innovation technology stability*: Uncertainty is a reality in technology adoption efforts. Uncertainty will need to be assessed for the initial investment, return on investment, payback period, public reactions, environmental impact, and volatility of the technology.

8. *National defense improvement*: An analysis of how the technology may contribute to national productivity may be verified by studying industrial throughput, efficiency of production processes, utilization of raw materials, equipment maintenance, absenteeism, learning rate, and design-to-production cycle.

Embracing new innovation

Opportunity lost can be a recurring risk in industry. When new innovation knocks, it should be embraced. A good case example of opportunity lost and innovation ignored is the case of digital photography first developed (and ignored) at Kodak in the mid-1970s. Kodak ignored the new innovation, perhaps because it conflicted with their traditional market model. In 1998, Kodak had 170,000 employees and sold 85% of all photo paper worldwide. Within just a few years, Kodak's business model disappeared and the company went out of its traditional business. Had Kodak aggressively embraced and leveraged the new digital photography in 1975, the future of the company might have taken a different positive and profitable path. If innovation is not timely embraced and capitalized on, what happened at Kodak can happen to many other companies in the prevailing digital engineering and manufacturing environment, particularly those dealing with artificial intelligence, health, autonomous and electric cars, (Science, Technology, Engineering, Mathematics [STEM]) education, 3D printing, agriculture, and knowledge-based jobs.

Fortunately, new industrial and service technologies have been gaining more attention in recent years. This is due to the high rate at which new productivity improvement technologies are being developed. The fast pace of new technologies has created difficult implementation and management problems for many organizations. New technology can be successfully implemented only if it is viewed as a system whose various components must be evaluated within an integrated managerial framework. Such a framework is provided by a project management approach. A multitude of new technologies has emerged in recent years. It is important to consider the peculiar characteristics of a new technology before establishing adoption and implementation strategies. The justification for the adoption of a new technology is usually a combination of several factors rather than a single characteristic of the technology. The potential of a specific technology to contribute to industrial development goals must be carefully assessed. The technology assessment process should explicitly address the following questions:

What is expected from the new technology?
Where and when will the new technology be used?
How is the new technology similar to or different from existing technologies?

What is the availability of technical personnel to support the new technology?
What administrative support is needed for the new technology?
Who will use the new technology?
How will the new technology be used?
Why is the technology needed?

The development, transfer, adoption, utilization, and management of technology is a problem that is faced in one form or another by business, industry, and government establishments. Some of the specific problems in technology transfer and management include the following:

- Controlling technological change
- Integrating technology objectives
- Shortening the technology transfer time
- Identifying a suitable target for technology transfer
- Coordinating the research and implementation interface
- Formally assessing current and proposed technologies
- Developing accurate performance measures for technology
- Determining the scope or boundary of technology transfer
- Managing the process of entering or exiting a technology
- Understanding the specific capability of a chosen technology
- Estimating the risk and capital requirements of a technology

Integrated managerial efforts should be directed at solving the problems stated earlier. A managerial revolution is needed in order to cope with the ongoing technological revolution. The revolution can be initiated by modernizing the long-standing and obsolete management culture relating to technology transfer. Some of the managerial functions that will need to be addressed when developing a technology transfer strategy include the following:

1. Development of an innovation and technology transfer plan
2. Assessment of technological risk
3. Assignment/reassignment of personnel to implement the technology transfer
4. Establishment of a transfer manager and a technology transfer office; in many cases, transfer failures occur because no individual has been given the responsibility to ensure the success of technology transfer
5. Identification and allocation of the resources required for technology transfer
6. Setting of guidelines for technology transfer; for example,
   a. Specification of phases (Development, Testing, Transfer, etc.)
   b. Specification of requirements for interphase coordination
   c. Identification of training requirements
   d. Establishment and implementation of performance measurements
7. Identification of key factors (both qualitative and quantitative) associated with technology transfer and management
8. Investigation of how the factors interact and development of the hierarchy of importance for the factors
9. Formulation of a loop system model that considers the forward and backward chains of actions needed to effectively transfer and manage a given technology
10. Tracking of the outcome of the technology transfer

Technological developments in many industries appear in scattered, narrow, and isolated areas within a few selected fields. This makes it so technology efforts are rarely coordinated, thereby hampering the benefits of technology. The optimization of technology utilization is, thus, very difficult. To overcome this problem and establish the basis for effective technology transfer and management, an integrated approach must be followed. An integrated approach will be applicable to technology or innovation transfer between any two organizations, whether public or private.

Some nations concentrate on the acquisition of bigger, better, and faster technology. But little attention is given to how to manage and coordinate the operations of the technology once it arrives. When technology fails, it is not necessarily because the technology is deficient. Rather, it is often the communication, cooperation, and coordination functions of technology management that are deficient. Technology encompasses factors and attributes beyond mere hardware, software, and peopleware, which refers to people issues affecting the utilization of technology. This may involve social-economic and cultural issues of using certain technologies or innovative techniques. Consequently, innovation transfer involves more than the physical transfer of hardware and software. Several flaws exist in the common practices of technology transfer and management. These flaws include the following:

- *Poor fit*: This relates to an inadequate assessment of the needs of the organization receiving the technology. The target of the transfer may not have the capability to properly absorb the technology.
- *Premature transfer of technology*: This is particularly acute for emerging technologies that are prone to frequent developmental changes.
- *Lack of focus*: In the attempt to get a bigger share of the market or gain an early lead in the technological race, organizations frequently force technology in many incompatible directions.
- *Intractable implementation problems*: Once a new technology is in place, it may be difficult to locate sources of problems that have their roots in the technology transfer phase itself.
- *Lack of transfer precedents*: Very few precedents are available related to the management of brand new technology. Managers are, thus, often unprepared for their new technology management responsibilities.
- *Stuck on technology*: Unworkable technologies sometimes continue to be recycled needlessly in the attempt to find the "right" usage.
- *Lack of foresight*: Due to the nonexistence of a technology transfer model, managers may not have a basis against which they can evaluate future expectations.
- *Insensitivity to external events*: Some external events that may affect the success of technology transfer include trade barriers, taxes, and political changes.
- *Improper allocation of resources*: There are usually not enough resources available to allocate to technology alternatives. Thus, a technology transfer priority must be developed.

The following steps provide specific guidelines for pursuing the implementation of manufacturing technology transfer:

1. Find a suitable application
2. Commit to an appropriate technology
3. Perform economic justification

4. Secure management support for the chosen technology
5. Design the technology implementation to be compatible with existing operations
6. Formulate the project management approach to be used
7. Prepare the receiving organization for the technology change
8. Install the technology
9. Maintain the technology
10. Periodically review the performance of the technology based on prevailing goals

Innovation transfer modes

The transfer of technology can be achieved in various forms. Project management provides an effective means of ensuring proper transfer of technology. Three technology transfer modes are presented here to illustrate basic strategies for getting one technological product from one point (technology source) to another point (technology sink). A conceptual integrated model of the interaction between the technology source and sink is presented in Figure 1.3.

Innovation and technology application centers may be established to serve as a unified point for linking technology sources with interested targets. The center will facilitate interactions between business establishments, academic institutions, and government agencies to identify important technology needs. With reference to Figure 1.3, technology can be transferred in one or a combination of the following strategies:

1. Transfer of complete technological products: In this case, a fully developed product is transferred from a source to a target. Very little product development effort is carried out at the receiving point. However, information about the operations of the product is fed back to the source so that necessary product enhancements can be pursued. So, the technology recipient generates product information which facilitates further improvement at the technology source. This is the easiest mode of technology transfer and the most tempting. Developing nations are particularly prone to this type of transfer. Care must be exercised to ensure that this type of technology transfer does

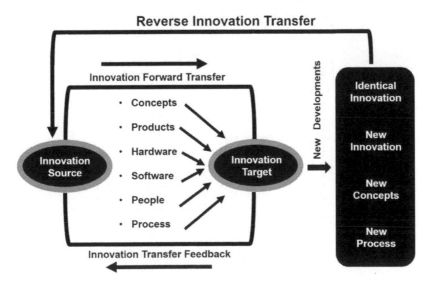

Figure 1.3 Technology transfer modes.

not degenerate into "machine transfer." It should be recognized that machines alone do not constitute technology.

2. Transfer of technology procedures and guidelines: In this technology transfer mode, procedures (e.g., blueprints) and guidelines are transferred from a source to a target. The technology blueprints are implemented locally to generate the desired services and products. The use of local raw materials and manpower is encouraged for the local production. Under this mode, the implementation of the transferred technology procedures can generate new operating procedures that can be fed back to enhance the original technology. With this symbiotic arrangement, a loop system is created whereby both the transferring and the receiving organizations derive useful benefits.

3. Transfer of technology concepts, theories, and ideas: This strategy involves the transfer of the basic concepts, theories, and ideas behind a given technology. The transferred elements can then be enhanced, modified, or customized within local constraints to generate new technological products. The local modifications and enhancements have the potential to generate an identical technology, a new related technology, or a new set of technology concepts, theories, and ideas. These derived products may then be transferred back to the original technology source as new technological enhancements. Figure 1.4 presents a specific cycle for local adaptation and modification of technology. An academic institution is a good potential source for the transfer of technology concepts, theories, and ideas.

It is very important to determine the mode in which technology will be transferred for defense purposes. There must be a concerted effort by people to make the transferred technology work within local infrastructure and constraints. Local innovation, patriotism, dedication, and willingness to adapt technology will be required to make technology transfer successful. It will be difficult for a nation to achieve national defense through total dependence on transplanted technology. Local adaptation will always be necessary.

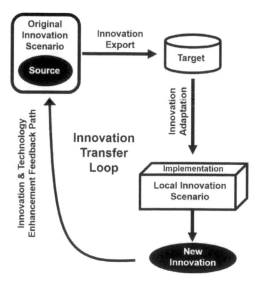

Figure 1.4 Local adaptation and enhancement of technology.

Innovation change-over strategies

One good innovation begets another. Thus, change-over arrangements are essential for a smooth transition between stages of innovation. Any development project will require changing from one form of technology to another. The implementation of a new technology to replace an existing (or a nonexistent) technology can be approached through one of several options. Some options are more suitable than others for certain types of technologies. The most commonly used technology change-over strategies include the following:

Parallel change-over: In this case, the existing technology and the new technology operate concurrently until there is confidence that the new technology is satisfactory.

Direct change-over: In this approach, the old technology is removed totally and the new technology takes over. This method is recommended only when there is no existing technology or when both technologies cannot be kept operational due to incompatibility or cost considerations.

Phased change-over: In this incremental change-over method, modules of the new technology are gradually introduced one at a time using either direct or parallel change-over.

Pilot change-over: In this case, the new technology is fully implemented on a pilot basis in a selected department within the organization.

Post-implementation evaluation

The new technology should be evaluated only after it has reached a steady-state performance level. This helps to avoid the bias that may be present at the transient stage due to personnel anxiety, lack of experience, or resistance to change. The system should be evaluated for the following aspects:

- Sensitivity to data errors
- Quality and productivity
- Utilization level
- Response time
- Effectiveness

Innovation systems integration

With the increasing shortages of resources, more emphasis should be placed on the sharing of resources. Technology resource sharing can involve physical equipment, facilities, technical information, ideas, and related items. The integration of technologies facilitates the sharing of resources. Technology integration is a major effort in technology adoption and implementation. Technology integration is required for proper product coordination. Integration facilitates the coordination of diverse technical and managerial efforts to enhance organizational functions, reduce cost, improve productivity, and increase the utilization of resources. Technology integration ensures that all performance goals are satisfied with a minimal expenditure of time and resources. It may require the adjustment of functions to permit sharing of resources, development of new policies to accommodate product integration, or realignment of managerial responsibilities. It can affect both hardware and software components of an organization. Important factors in technology integration include the following:

- Unique characteristics of each component in the integrated technologies
- Relative priorities of each component in the integrated technologies
- How the components complement one another
- Physical and data interfaces between the components
- Internal and external factors that may influence the integrated technologies
- How the performance of the integrated system will be measured

Role of government in innovation transfer

The malignant policies and operating characteristics of some of the governments in under-developed countries have contributed to stunted growth of technology in those parts of the world. The governments in most developing countries control the industrial and public sectors of the economy. People either work for the government or serve as agents or contractors for the government. The few industrial firms that are privately owned depend on government contracts to survive. Consequently, the nature of the government can directly determine the nature of industrial technological progress.

The operating characteristics of most of the governments perpetuate inefficiency, corruption, and bureaucratic bungles. This has led to a decline in labor and capital productivity in the industrial sectors. Using the Pareto distribution, it can be estimated that in most government-operated companies, there are eight administrative workers for every two production workers. This creates a non-productive environment that is skewed towards hyper-bureaucracy. The government of a nation pursuing industrial development must formulate and maintain an economic stabilization policy. The objective should be to minimize the sacrifice of economic growth in the short run and while maximizing long-term economic growth. To support industrial technology transfer efforts, it is essential that a conducive national policy be developed.

More emphasis should be placed on industry diversification, training of the work force, supporting financial structure for emerging firms, and implementing policies that encourage productivity in a competitive economic environment. Appropriate foreign exchange allocation, tax exemptions, bank loans for emerging businesses, and government-guaranteed low-interest loans for potential industrial entrepreneurs are some of the favorable policies to spur growth and development of the industrial sector.

Improper trade and domestic policies have adversely affected industrialization in many countries. Excessive regulations that cause bottlenecks in industrial enterprises are not uncommon. The regulations can take the form of licensing, safety requirements, manufacturing value-added quota requirements, capital contribution by multinational firms, and high domestic production protection. Although regulations are needed for industrial operations, excessive controls lead to low returns from the industrial sectors. For example, stringent regulations on foreign exchange allocation and control have led to the closure of industrial plants in some countries. The firms that cannot acquire essential raw materials, commodities, tools, equipment, and new technology from abroad due to foreign exchange restrictions are forced to close and lay off workers.

Price controls for commodities are used very often by developing countries especially when inflation rates for essential items are high. The disadvantages involved in price control of industrial goods include restrictions of the free competitive power of available goods in relation to demand and supply, encouragement of inefficiency, promotion of dual markets, distortion of cost relationships, and increases in administrative costs involved in producing goods and services.

NASA examples of innovation and technology transfer

One way that a government can help facilitate industrial technology transfer involves the establishment of technology transfer centers within appropriate government agencies. A good example of this approach can be seen in the government-sponsored technology transfer program by the US National Aeronautics and Space Administration (NASA). In the Space Act of 1958, the US Congress charged NASA with a responsibility to provide for the widest practical and appropriate dissemination of information concerning its activities and the results achieved from those activities. With this technology transfer responsibility, technology developed in the United States' space program is available for use by the nation's business and industry sector.

In order to accomplish technology transfer to industry, NASA established the Technology Utilization Program (TUP) in 1962. The Technology Utilization Program uses several avenues to disseminate information on NASA technology. The avenues include the following:

- Complete, clear, and practical documentation is required for new technology developed by NASA and its contractors. This is available to industry through several publications produced by NASA. An example is the monthly *Tech Briefs*, which outlines technology innovations. This is a source of prompt technology information for industry.
- Industrial Application Centers (IAC) were developed to serve as repositories for vast computerized data on technical knowledge. The IACs are located at academic institutions around the country. All the centers have access to a large data base containing millions of NASA documents. With this data base, industry can have access to the latest technological information quickly. The funding for the centers is obtained through joint contributions from several sources including NASA, the sponsoring institutions, and state government subsidies. Thus, the centers can provide their services at very reasonable rates.
- NASA operates a Computer Software Management and Information Center (COSMIC) to disseminate computer programs developed through NASA projects. COSMIC, which is located at a university, has a library of thousands of computer programs. The center publishes an annual index of available software.

In addition to the specific mechanisms discussed earlier, NASA undertakes Application Engineering Projects. Through these projects, NASA collaborates with industry to modify aerospace technology for use in industrial applications. To manage the application projects, NASA established a Technology Application Team (TAT), consisting of scientists and engineers from several disciplines. The team interacts with NASA field centers, industry, universities, and government agencies. The major mission of the team interactions is to define important technology needs and identify possible solutions within NASA. NASA Application Engineering Projects are usually developed in a five-phase approach with go or no-go decisions made by NASA and industry at the completion of each phase. The five phases are outlined in the following:

1. NASA and the Technology Applications Team meet with industry associations, manufacturers, university researchers, and public-sector agencies to identify important technology problems that might be solved by aerospace technology.

2. After a problem is selected, it is documented and distributed to the Technology Utilization Officer at each of NASA's field centers. The officer in turn distributes the description of the problem to the appropriate scientists and engineers at the center. Potential solutions are forwarded to the team for review. The solutions are then screened by the problem originator to assess the chances for technical and commercial success.

3. Next is the development of partnerships and a project plan to pursue the implementation of the proposed solution. NASA joins forces with private companies and other organizations to develop an Application Engineering Project. Industry participation is encouraged through a variety of mechanisms such as simple letters of agreement or joint endeavor contracts. The financial and technical responsibilities of each organization are specified and agreed upon.

4. At this point, NASA's primary role is to provide technical assistance to facilitate utilization of the technology. The costs for these projects are usually shared by NASA and the participating companies. The proprietary information provided by the companies and their rights to new discoveries are protected by NASA.

5. The final phase involves the commercialization of the product. With the success of commercialization, the project would have widespread impact. Usually, the final product development, field testing, and marketing are managed by private companies without further involvement from NASA.

Through this well-coordinated government-sponsored technology transfer program, NASA has made significant contributions to US industry, thereby providing an anchor for national defense pursuits. The results of NASA's technology transfer abound in numerous consumer products either in subtle forms or in clearly identifiable forms. Food preservation techniques constitute one area of NASA's technology transfer that has had a significant positive impact on the society. Although the specific organization and operation of the NASA technology transfer programs have changed in name or in deed over the years, the basic descriptions outlined earlier remain a viable template for how to facilitate manufacturing technology transfer. In a similar government-backed strategy, the US Air Force Research Laboratory (AFRL) also has very structured programs for transferring non-classified technology to the industrial sector. It is believed that a project management approach can help in facilitating success with innovation and technology transfer efforts.

PICK'ing the right innovation

It is important to pick the right innovation to adopt and adapt. The question of which innovation is appropriate to transfer in or transfer out is relevant for technology transfer considerations. While several methods of technology selection are available, this book recommends methods that combine qualitative and quantitative factors. The Analytic Hierarchy Process (AHP) is one such method. Another useful, but less publicized, is the PICK chart. The PICK chart was originally developed by Lockheed Martin to identify and prioritize improvement opportunities in the company's process improvement applications. The technique is just one of the several decision tools available in process improvement endeavors. It is a very effective technology selection tool used to categorize ideas and opportunities. The purpose is to qualitatively help identify the most useful ideas. A 2 × 2 grid is normally drawn on a white board or large flip-chart. Ideas that were written on sticky notes by team members are placed on the grid based on a group assessment of the payoff relative the level of difficulty. The PICK acronym comes from the labels for each of the quadrants of the grid: **P**ossible (easy, low payoff), **I**mplement (easy, high payoff), **C**hallenge (hard, high payoff), and **K**ill (hard, low payoff). The PICK chart quadrants are summarized as follows:

Possible (easy, low payoff)	→ Third quadrant
Implement (easy, high payoff)	→ Second quadrant
Challenge (hard, high payoff)	→ First quadrant
Kill (hard, low payoff).	→ Fourth quadrant

The primary purpose is to help identify the most useful ideas, especially those that can be accomplished immediately with little difficulty. These are called "Just-Do-Its." The general layout of the PICK chart grid is shown in Figure 1.5. The PICK process is normally done subjectively by a team of decision makers under a group decision process. This can lead to bias and protracted debate of where each item belongs. It is desired to improve the efficacy of the process by introducing some quantitative analysis. Badiru and Thomas (2013) present a methodology to achieve a quantification of the PICK selection process. The PICK chart is often criticized for its subjective rankings and lack of quantitative analysis. The approach presented by Badiru and Thomas (2013) alleviates such concerns by normalizing and quantifying the process of integrating the subjective rakings by those involved in the group PICK process. Human decision is inherently subjective. All we can do is to develop techniques to mollify the subjective inputs rather than compounding them with subjective summarization.

PICK chart quantification methodology

The placement of items into one of the four categories in a PICK chart is done through expert ratings, which are often subjective and non-quantitative. In order to put some quantitative basis to the PICK chart analysis, Badiru and Thomas (2013) present the methodology

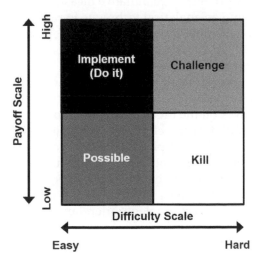

Figure 1.5 Basic layout of the PICK chart.

of dual numeric scaling on the impact and difficulty axes. Suppose each technology is ranked on a scale of one to ten and plotted accordingly on the PICK chart. Then, each project can be evaluated on a binomial pairing of the respective rating on each scale. Note that a high rating along the x axes is desirable while a high rating along the y axis is not desirable. Thus, a composite rating involving x and y must account for the adverse effect of high values of y. A simple approach is to define $y' = (11-y)$, which is then used in the composite evaluation. If there are more factors involved in the overall project selection scenario, the other factors can take on their own lettered labeling (e.g., a, b, c, z, etc.). Then, each project will have an n-factor assessment vector. In its simplest form, this approach will generate a rating such as the following:

$$PICK_{R,i}(x,y') = x + y'$$

where:

$PICK_{R,i}(x, y')$ is the PICK rating of project i ($i = 1, 2, 3,...., n$)
x is the rating along the impact axis ($1 \le x \le 10$)
y is the rating along the difficulty axis ($1 \le y \le 10$)
y' is the $(11-y)$

If $x + y'$ is the evaluative basis, then each technology's composite rating will range from 2 to 20, 2 being the minimum and 20 being the maximum possible. If $(x)(y)$ is the evaluative basis, then each project's composite rating will range from 1 to 100. In general, any desired functional form may be adopted for the composite evaluation. Another possible functional form is:

$$PICK_{R,i}(x,y'') = f(x,y'')$$

$$= (x+y'')^2,$$

where y'' is defined as needed to account for the converse impact of the axes of difficulty. The previous methodology provides a quantitative measure for translating the entries in a conventional PICK chart into an analytical technique to rank the technology alternatives, thereby reducing the level of subjectivity in the final decision. The methodology can be extended to cover cases where a technology has the potential to create negative impacts, which may impede organizational advancement.

The quantification approach facilitates a more rigorous analytical technique compared to traditional subjective approaches. One concern is that although quantifying the placement of alternatives on the PICK chart may improve the granularity of relative locations on the chart, it still does not eliminate the subjectivity of how the alternatives are assigned to quadrants in the first place. This is a recognized feature of many decision tools. This can be mitigated by the use of additional techniques that aid decision makers to refine their choices. The analytic hierarchy process (AHP) could be useful for this purpose. Quantifying subjectivity is a continuing challenge in decision analysis. The PICK chart quantification methodology offers an improvement over the conventional approach.

Although the PICK chart has been used extensively in industry, there are few published examples in the open literature. The quantification approach presented by Badiru and Thomas (2013) may expand interest and applications of the PICK chart among technology researchers and practitioners. The steps for implementing a PICK chart are summarized in the following:

Step 1: On a chart, place the subject question. The question needs to be asked and answered by the team at different stages to be sure that the data that is collected is relevant.

Step 2: Put each component of the data on a different note like a post-it or small cards. These notes should be arranged on the left side of the chart.

Step 3: Each team member must read all notes individually and consider the importance of each. The team member should decide whether the element should or should not remain a fraction of the significant sample. The notes are then removed and moved to the other side of the chart. Now, the data is condensed enough to be processed for a particular purpose by means of tools that allow groups to reach a consensus on priorities of subjective and qualitative data.

Step 4: Apply the quantification methodology presented earlier to normalize the qualitative inputs of the team.

DEJI model for innovation integration

In the Foreword of this handbook, Joe Sciabica suggested taking any innovation pursuit through of the *Heilmeier Catechism*, which helps separate the wheat from the chaff when assessing the value of new innovation. How do we know if a concept or idea is technologically sound, adds value, and is truly "innovative"? The stages of *Heilmeier Catechism* are:

1. In the most simplistic terms what are you trying to do and why?
2. How is it done today and what are the limits of current practice?
3. What is new in your approach and why do you think it will be successful? Who cares?
4. If you are successful, what difference will it make?
5. What are the risks?
6. How much will it cost?
7. How long will it take?
8. What are the mid-term and final "exams" to check for success?

It is believed that the stages espoused by the DEJI model of systems engineering align well with fulfilling the requirements of the catechism with respect to design, evaluation, justification, and integration. It is the requirement for explicit integration that makes the DEJI model effective and applicable to all spheres of human endeavor. If a new product or process cannot be sustainably integrated into normal practice and prevailing pattern of operation, the new innovation would be out of alignment and would not add value.

Technology is at the intersection of efficiency, effectiveness, and productivity. Efficiency provides the framework for quality in terms of resources and inputs required to achieve the desired level of quality. Effectiveness comes into play with respect to the application of product quality to meet specific needs and requirements of an organization. Productivity is an essential factor in the pursuit of quality as it relates to the throughput of a production system. To achieve the desired levels of quality, efficiency, effectiveness, and productivity, a new technology integration framework must be adopted. This section presents a technology integration model for design, evaluation, justification, and integration (DEJI) based on the product development application presented by Badiru (2012). The model is relevant for

Table 1.2 DEJI model for technology integration

DEJI model	Characteristics	Tools & techniques
Design	Define goals	Parametric assessment
	Set performance metrics	Project state transition
	Identify milestones	Value stream analysis
Evaluate	Measure parameters	Pareto distribution
	Assess attributes	Life cycle analysis
	Benchmark results	Risk assessment
Justify	Assess economics	Benefit-cost ratio
	Assess technical output	Payback period
	Align with goals	Present value
Integrate	Embed in normal operation	SMART concept
	Verify symbiosis	Process improvement
	Leverage synergy	Quality control

Figure 1.6 DEJI systems model for innovation integration.

research and development efforts in industrial development and technology applications. The DEJI model encourages the practice of building quality into a product right from the beginning so that the product or technology integration stage can be more successful. The essence of the model is summarized in Table 1.2. Figure 1.6 shows the graphical framework for the model.

Design for innovation implementation

The design of quality in product development should be structured to follow point-to-point transformations. A good technique to accomplish this is the use of state-space transformation, with which we can track the evolution of a product from the concept stage to a final product stage. For the purpose of product quality design, the following definitions are applicable:

Product state: A state is a set of conditions that describe the product at a specified point in time. The *state* of a product refers to a performance characteristic of the product which relates input to output such that a knowledge of the input function over time and the state of the product at time $t = t_0$ determines the expected output for $t \geq t_0$. This is particularly important for assessing where the product stands in the context of new technological developments and the prevailing operating environment.

Product state space: A product *state-space* is the set of all possible states of the product lifecycle. State-space representation can solve product design problems by moving from an initial state to another state, and eventually to the desired end-goal state. The movement from state to state is achieved by means of actions. A goal is a description of an intended state that has not yet been achieved. The process of solving a product problem involves finding a sequence of actions that represents a solution path from the initial state to the goal state. A state-space model consists of state variables that describe the prevailing condition of the product. The state variables are related to inputs by mathematical relationships. Examples of potential product state variables include schedule, output quality, cost, due date, resource, resource utilization, operational efficiency, productivity throughput, and technology alignment. For a product described by a system of components, the state-space representation can follow the quantitative metric in the following:

$$Z = f(z, x); \; Y = g(z, x)$$

where f and g are vector-valued functions. The variable Y is the output vector while the variable x denotes the inputs. The state vector Z is an intermediate vector relating x to y. In generic terms, a product is transformed from one state to another by a driving function that produces a transitional relationship given by:

$$S_s = f(x \mid S_p) + e,$$

where:
S_s = subsequent state
x = state variable
S_p = the preceding state
e = error component

The function f is composed of a given action (or a set of actions) applied to the product. Each intermediate state may represent a significant milestone in the project. Thus, a descriptive state-space model facilitates an analysis of what actions to apply in order to achieve the next desired product state. The state-space representation can be expanded to cover several components within the technology integration framework. Hierarchical linking of product elements provides an expanded transformation structure. The product state can be expanded in accordance with implicit requirements. These requirements might include grouping design elements, linking precedence requirements (both technical and procedural), adapting to new technology developments, following required communication links, and accomplishing reporting requirements. The actions to be taken at each state depend on the prevailing product conditions. The nature of subsequent alternate states depends on what actions are implemented. Sometimes there are multiple paths that can lead to the desired end result. At other times, there exists only one unique path to the desired objective. In conventional practice, the characteristics of the future states can only be recognized after the fact, thus making it impossible to develop adaptive plans. In the implementation of the **DEJI** model, adaptive plans can be achieved because the events occurring within and outside the product state boundaries can be taken into account. If we describe a product by P state variables s_i, then the composite state of the product at any given time can be represented by a vector S containing P elements. That is,

$$S = \{s_1, s_2, ..., s_P\}$$

The components of the state vector could represent either quantitative or qualitative variables (e.g., cost, energy, color, time). We can visualize every state vector as a point in the state-space of the product. The representation is unique since every state vector corresponds to one and only one point in the state-space. Suppose we have a set of actions (transformation agents) that we can apply to the product information so as to change it from one state to another within the project state-space. The transformation will change a state vector into another state vector. A transformation may be a change in raw material or a change in design approach. The number of transformations available for a product characteristic may be finite or unlimited. We can construct trajectories that describe the potential states of a product evolution as we apply successive transformations with respect to technology forecasts. Each transformation may be repeated as many times as needed. Given an initial state S_0, the sequence of state vectors is represented by the following:

$$S_n = T_n(S_{n-1})$$

The state-by-state transformations are then represented as $S_1 = T_1(S_0)$; $S_2 = T_2(S_1)$; $S_3 = T_3(S_2)$; …; $S_n = T_n(S_{n-1})$. The final State, S_n, depends on the initial state S and the effects of the actions applied.

Evaluation of innovation

A product can be evaluated on the basis of cost, quality, schedule, and meeting requirements. There are many quantitative metrics that can be used in evaluating a product at this stage. Learning curve productivity is one relevant technique that can be used because it offers an evaluation basis of a product with respect to the concept of growth and decay. The half-life extension (Badiru, 2012) of the basic learning is directly applicable because the half-life of the technologies going into a product can be considered. In today's technology-based operations, retention of learning may be threatened by fast-paced shifts in operating requirements. Thus, it is of interest to evaluate the half-life properties of new technologies as they impact the overall product quality. Information about the half-life can tell us something about the sustainability of learning-induced technology performance. This is particularly useful for designing products whose life cycles stretch into the future in a high-tech environment.

Justification of innovation tool

We need to justify a program on the basis of quantitative value assessment. The Systems Value Model (SVM) is a good quantitative technique that can be used here for project justification on the basis of value. The model provides a heuristic decision aid for comparing project alternatives. It is presented here again for the present context. Value is represented as a deterministic vector function that indicates the value of tangible and intangible attributes that characterize the project. It is represented as $V = f(A_1, A_2, ..., A_p)$, where V is the assessed value and the A values are quantitative measures or attributes. Examples of product attributes are quality, throughput, manufacturability, capability, modularity, reliability, interchangeability, efficiency, and cost performance. Attributes are considered to be a combined function of factors. Examples of product factors are market share, flexibility, user acceptance, capacity utilization, safety, and design functionality. Factors are themselves considered to be composed of indicators. Examples of indicators are debt ratio, acquisition volume, product responsiveness, substitutability, lead time, learning curve, and scrap volume. By combining the earlier definitions, a composite measure of the operational value of a product can be quantitatively assessed. In addition to the quantifiable factors, attributes, and indicators that impinge upon overall project value, the human-based subtle factors should also be included in assessing overall project value.

Integration of innovation

Without being integrated, a system will be in isolation and it may be worthless. We must integrate all the elements of a system on the basis of alignment of functional goals. The overlap of systems for integration purposes can conceptually be viewed as projection integrals by considering areas bounded by the common elements of subsystems. Quantitative metrics can be applied at this stage for effective assessment of the technology state. Trade-off analysis is essential in technology integration. Pertinent questions include the following:

What level of trade-offs on the level of technology are tolerable?
What is the incremental cost of more technology?
What is the marginal value of more technology?
What is the adverse impact of a decrease in technology utilization?

What is the integration of technology over time? In this respect, an integral of the form in the following may be suitable for further research:

$$I = \int_{t_1}^{t_2} f(q)dq$$

where:
   I is the integrated value of quality
   $f(q)$ is the functional definition of quality
   t_1 is the initial time
   t_2 is the final time within the planning horizon.

Guidelines and important questions relevant for technology integration are presented in the following:

- What are the unique characteristics of each component in the integrated system?
- How do the characteristics complement one another?
- What physical interfaces exist among the components?
- What data/information interfaces exist among the components?
- What ideological differences exist among the components?
- What are the data flow requirements for the components?
- What internal and external factors are expected to influence the integrated system?
- What are the relative priorities assigned to each component of the integrated system?
- What are the strengths and weaknesses of the integrated system?
- What resources are needed to keep the integrated system operating satisfactorily?
- Which organizational unit has primary responsibility for the integrated system?

The recommended approach of the DEJI model will facilitate a better alignment of product technology with future development and needs. The stages of the model require research for each new product with respect to design, evaluation, justification, and integration. Existing analytical tools and techniques can be used at each stage of the model.

Conclusion

Technology transfer is a great avenue to advancing industrialization. This chapter has presented a variety of principles, tools, techniques, and strategies useful for managing technology transfer. Of particular emphasis in the chapter are the management aspects of technology transfer. The technical characteristics of the technology of interest are often well understood. What is often lacking is an appreciation of the technology management requirements for achieving a successful technology transfer. This chapter presents the management aspects of manufacturing technology transfer.

References

Badiru, A. B. (2012, Fall), Application of the DEJI model for aerospace product integration, *Journal of Aviation and Aerospace Perspectives (JAAP)*, 2(2):20–34.

Badiru, A. B. and M. Thomas (2013), Quantification of the PICK chart for process improvement decisions, *Journal of Enterprise Transformation*, 3(1):1–15.

chapter two

Definitional analysis of innovation*

Adedeji B. Badiru

Contents

Introduction

A definitional analysis of innovation is essential for getting the intended full benefit of this chapter. Joe Sciabica, in the FOREWORD, reminded us of a dictionary definition of innovation:

> "Innovation a (noun) + a (verb) - The portfolio of financial, intellectual, organization and human capabilities that enable a society's journey to its desired future."

This definition of innovation conveys the multifaceted operational meaning of the word. This may help readers to put everything into the proper perspective. Innovation is widely heralded as essential for successful competition in the increasingly global economy. However, to enhance innovation in education, organizations and countries require transformative thinking. National thought leaders and organizations such as the National Academy of Engineering are supporting projects to explore this relationship. The Educate to Innovate (ETI) project was designed to explore the issue regarding teaching innovation and the expected outcome, entrepreneurship [1].

---

* This chapter is adapted and modified with copyright permission from Chapter 2 (The role of Creativity and Innovation in Leadership) of McCauley, Pamela (2017), **Essentials of Engineering Leadership and Innovation**, CRC Press/Taylor & Francis, Boca Raton, Florida.

During the 1950s and 1960s, Sputnik and the space race stimulated a generation of Americans to follow education and careers in science and technology. Half a century later, American students are now graded 22nd and 21st among their peers all over the world in science and math, respectively. Students in the United States, formerly a leader in science, technology, engineering, and mathematics (STEM), are now outperformed by students from Slovenia, Hungary, and Estonia, among others [2].

In 1983, the National Commission on Excellence in Education published "A Nation at Risk," a nationwide study that highlighted the intolerable state of the American education system:

> Our nation is at risk. Our once unchallenged preeminence in commerce, industry, science, and technological innovation is being overtaken by competitors throughout the world. This report is concerned with only one of the many causes and dimensions of the problem, but it is the one that undergirds American prosperity, security, and civility. What was unimaginable a generation ago has begun to occur—others are matching and surpassing our educational attainments. If an unfriendly foreign power had attempted to impose on America the mediocre educational performance that exists today, we might well have viewed it as an act of war. [3]

More than two decades afterward, in 2010, the National Academies of Science, Engineering, and Medicine published *Rising above the Gathering Storm, Revisited: Rapidly Approaching Category 5*, which built on the findings of its 2005 "Gathering Storm" report. Notably, the report warns that

> "Today, for the first time in history, America's younger generation is less well-educated than its parents" [4].

In an effort to respond to the faltering academic status of American students and in a quest to elevate them "from the middle to the top of the pack in science and math," the Obama Administration announced its ETI initiative in November 2009 [5].

President Barack Obama's ETI campaign is publicized as a joint effort between the federal government, the private sector, and the nonprofit and research communities to raise the standing of American students in science and math through dedication of time and money, and volunteering. The program attempts to enhance STEM literacy, improve teaching quality, and develop educational and career opportunities for America's youth.

At the time the program was first declared in November 2009, the participating organizations offered a financial and in-kind commitment of more than $260 million. Taxpayer commitments for the federal government's portion of ETI add to that total. In addition, five public-private partnerships were announced, as well as commitments by key societal and private sector leaders to muster funds for STEM education, innovation, and awareness [6]. These partnerships and commitments are:

- Time Warner Cable's Connect a Million Minds (CAMM), which pledges to connect children to after-school STEM programs and activities in their area.
- Discovery Communications' "Be the Future" will broadcast dedicated science programming to more than 99 million homes and offer interactive science education to approximately 60,000 schools.

- Sesame Street's "Early STEM: Literacy" commits to a two-year focus on STEM subjects.
- National Lab Day will promote hands-on learning with 100,000 teachers and 10 million students over the next four years and foster communities of collaboration between volunteers, students, and educators in STEM education. These initiatives will then culminate in a nationally recognized day centered on science activities.
- The National STEM Video Game Challenge promotes the design and creation of STEM-related video games.
- The annual White House Science Fair will bring the winners of science fairs from across the nation to the White House to showcase their STEM creations and innovation.
- Sally Ride, the first female astronaut, Craig Barrett, the former Intel chairman, Ursula Burns, CEO of XEROX, and Glenn Britt, CEO of Eastman Kodak, committed to fostering interest and support for STEM: education among American corporations and philanthropists [6].

In January 2010, President Obama announced the continuation of the program, stressing the half-billion-dollar monetary obligation from the administration's partners. This development includes an additional commitment of $250 million in financial and in-kind support, and a pledge by 75 of the nation's biggest public universities to train 10,000 new teachers by 2015. The program expansion also incorporated additional public-private partnerships anticipated to aid the training of new STEM educators, together with the launch of Intel's Science and Math Teachers Initiative and the PBS Innovative Educators Challenge, as well as the expansion of the National Math and Science Initiative's UTeach program and Woodrow Wilson Teaching Fellowships in math and science. In addition, the president called on 200,000 federal government staff working in the fields of Science and Engineering (S&E) to volunteer to work with educators in order to foster enhanced STEM education [6].

A STEM-educated workforce is very important for the protection and the wealth of the United States as industry and government increasingly demand exceedingly trained STEM professionals to vie in the international market and look to science and technology to help stay one step ahead of national security threats. The United States must not permit itself to be outcompeted in science, technology, engineering, and mathematics. While the Obama Administration's ETI enterprise is projected to raise the United States "from the middle to the top of the pack in science and math," this one-size-fits-all federal approach fails to cure the primary problems of educational performance and does not stop the permeable pipeline in the American education system.

The evolution of innovation

The principles associated with innovation can be applied to organizations, individuals, and product development. These three categories of innovation can also be applied simultaneously to create a culture where individuals are continually seeking to be innovative and create enhanced product outcomes. The meaning of innovation has evolved with US Federal funding agencies as well. For example, consider the National Science Foundation (NSF), one of the premier research funding agencies in the United States that funds 24 percent of all federally supported basic research conducted by colleges and universities in the United States each year [7].

For many years NSF largely focused on funding only basic research rather than funding applied research and technology transition. Now the NSF's funding goals are extending beyond basic research to support various aspects of groundbreaking applied research and the transition of research outcomes into useful products, services, and technologies. There's a good reason for this change in focus. Historically, it was thought that it could take up to 50 years for the knowledge learned from basic research to be applied to products and services. However, as the pace of change itself continues to increase, the speed of technology and new development has compressed the time it takes to move basic research from reaction to knowledge to actual application. The NSF reflects this shift quite powerfully in its desire to now fund more applied research. The quick transition of the NSF's innovation core and its desire to swiftly convert new knowledge into new products and services is solid evidence of change.

Discussion of 1-corp program and related National Science Foundation initiatives

America's affluence grew in part from the capability to profit economically on groundbreaking developments from science and engineering research. At the same time, a well-informed, imaginative labor force has maintained the country's international leadership in significant areas of technology. These essential discoveries and competent labor force resulted from substantial, incessant investment in science and engineering. A strong capability for leveraging essential science discoveries into influential engines of innovation is necessary to maintain our competitive edge in the future. The NSF supports fundamental research and education in science and engineering. NSF's dual role, distinctive among government agencies, results in new knowledge and paraphernalia as well as a competent ground-breaking workforce. These corresponding building blocks of innovation have led to innovatory high-tech advances and completely new industries. Through this program, NSF seeks to hasten the improvement of new technologies, products, and processes that arise from elementary study. NSF investments will advantageously strengthen the innovation ecosystem [8] by addressing the challenge built into the early stages of the innovation process. This solicitation will support partnerships that are designed to triumph over scores of obstacles in the path of innovation.

Program description

The objectives of this program are to encourage translation of fundamental research, to facilitate collaboration between the academic world and business, and to train students to comprehend innovation and entrepreneurship. The rationale of the NSF I-Corps program is to spot NSF-funded researchers who will obtain extra support—in the form of mentoring and funding—to hasten the conversion of knowledge derived from essential research into up-and-coming products and services that can attract successive third-party funding.

About the National Science Foundation

The NSF is an autonomous federal agency created by the National Science Foundation Act of 1950, as amended (42 USC 1861–1875). The act states the function of the NSF is "to promote the progress of science; [and] to advance the national health, prosperity, and welfare by supporting research and education in all fields of science and engineering" [7]. NSF funds research and learning in most fields of science and engineering through grants and cooperative agreements to more than 2000 colleges, universities, K-12 school systems, businesses, informal science organizations, and other research organizations all over the United States. The foundation accounts for about one-fourth of federal support to educational institutions for essential research. NSF receives in the region of 40,000 proposals each year for study, learning, and training projects, of which roughly 11,000 are funded. In addition, the foundation receives thousands of applications for graduate and post-doctoral fellowships. The agency operates no laboratories itself but does support national research centers, user facilities, certain oceanographic vessels, and Arctic and Antarctic research stations. The foundation furthermore supports joint research between universities and industry, US participation in global scientific and engineering efforts, and educational activities at every academic level [7].

The role of creativity and innovation has changed our nation because now we are pushing more to see these new developments converted into new products and services, and the driving factor in accomplishing this is leadership. There is even more accountability in terms of wanting to understand what has been done with research funding for over the past several years. Generally, Americans convey extremely favorable attitudes toward science and technology (S&T). In 2001, overpowering majorities of NSF survey respondents agreed with the following statements:

- "Science and technology are making our lives healthier, easier, and more comfortable" (86 percent agreed and 11 percent disagreed).
- "Most scientists want to work on things that will make life better for the average person" (89 percent agreed and 9 percent disagreed).
- With the application of science and technology, work will become more interesting" (72 percent agreed and 23 percent disagreed).
- "Because of science and technology, there will be more opportunities for the next generation" (85 percent agreed and 14 percent disagreed) [9].

In addition, Americans give the impression of having more positive attitudes toward S&T than their counterparts in the United Kingdom and Japan [10].

Despite these positive indicators, a sizable segment, although not a majority, of the public has some reservations concerning science, especially technology. For example, in 2001, approximately SO percent of NSF survey respondents agreed with the following statement: "We depend too much on science and not enough on faith" (46 percent disagreed). In addition, 38 percent agreed with the statement: "Science makes our way of life change too fast" (59 percent disagreed) [11].

Public attitudes toward federal funding of scientific research

All indicators point to general support for government funding of essential research. In 2001, 81 percent of NSF survey respondents agreed with the following statement: "Even if it brings no immediate benefits, scientific research that advances the frontiers of knowledge is necessary and should be supported by the Federal Government" [12]. The level of agreement with this statement has consistently been in the 80 percent range. In 2000, 72 percent of U.K. residents agreed with the statement, as did 80 percent of Japanese residents (in 1995).

These differences in the measure of public support worldwide for basic research are notable. This may be attributed to the increased expectations in terms of transitioning science to technology and innovations. The result is people expect basic research to more readily provide benefits to society and in fact, in 2001, 16 percent disagreed with the statement completely. This suggests that they expected immediate benefits from basic research and this trend of expectation has continued.

Although there is strong evidence that the public supports the government's investment in basic research, few Americans are able to name the two agencies that provide most of the federal funds for this type of research. In a recent survey, only 5 percent identified the National Institutes of Health (NIH) as the agency that "funds most of the taxpayer-supported medical research performed in the United States," and only 3 percent named NSF as "the government agency that funds most of the basic research and educational programming in the sciences, mathematics and engineering" [13].

In addition, those with more positive attitudes toward S&T were more likely to express support for government funding of basic research. In 2001, 93 percent of those who scored 75 or higher on the Index of Scientific Promise agreed that the federal government should fund basic scientific research compared with only 68 percent of those with relatively low index scores [13].

In 2001, only 14 percent of NSF survey respondents thought the government was spending too much on scientific research; 36 percent thought the government was not spending enough, a percentage that has grown steadily since 1990, when 30 percent chose that answer [14]. Men are more than likely than women to say the government is spending too little in support of scientific research (40 percent versus 33 percent in 2001).

To put the response to this item in perspective, at least 65 percent of those surveyed thought the government was not spending enough on other programs, including programs to improve health care, help senior citizens, improve education, and reduce pollution. Only the issues of space exploration and national defense received less support for increased spending than scientific research.

In 2001, 48 percent of those surveyed thought spending on space exploration was excessive, the highest percentage for any item in the survey—and nearly double the number of those who felt that the government was spending too much on national defense [15]. In contrast, the latter fell steadily, from 40 percent in 1990 to 25 percent in 2001.

Definitions of innovation

Definitions of innovation differ but the general thread among these definitions is that innovations present a new or better product, service, or resource that adds "value" to those seeking it. The ETI study conducted 60 interviews that revealed common characteristics of innovators. A prevailing aspect of innovation is team interaction or team activities. For these teams to be effective, they are often managed by a technical person with detailed knowledge of the proposed innovation. In these situations, it is imperative that the team leader understands how to inspire, motivate, and lead the team as they move toward a useful innovation. When innovators were asked to describe characteristics of innovations or innovative products, the following characteristics emerged:

- *Innovation provides societal value*: The interviewees felt powerfully that innovations must offer societal value. The innovation must be supportive to society. It's great if one makes an invention, but it's even better if the invention can be used to develop individual lives. Part of the importance of an innovation is connected to timely adoption. It should be helpful in the near future. In truth, unless an innovation is in fact used by society, it cannot be called an innovation: R. Graham Cooks [16] warned innovators against believing that all they do is collectively meaningful or useful. In other words, if you feel that you have some fondness for innovation, then the big danger is that you'll convince yourself even in cases where the work is trivial or doesn't have the implications that you hoped it would have. Robert Dennard [16] agreed: "Lots of inventions aren't innovations. I have 62 patents and only one or two are actually being used, and if it's not used, it's really not innovating very much. So, innovation's a breakthrough, something that's really useful and it doesn't have to be patentable, even."

- *Innovation is an Improvement*: Innovations are naturally seen as "something new." Nevertheless, all the interviewees and workshop participants accentuated that innovations are improvements, not necessarily just new. Laurie Dean Baird [16] gives details of her approach to telling the difference in the value of an innovation. "If I look at something that is new and ask 'Is this innovative?' then I ask how was this problem solved before? What was the industry standard and how is this different?' And if the answer is that, in addition to being new (the problem or solution), it takes the hassle out of something (i.e., it improves life), then it is innovative."

> "I don't see innovation being the introduction of something [that is just] new," Tim Cook says. "There are many things new every day, and I wouldn't say they all are innovative. I think to be innovative, something has to be better than the predecessor product, materially better, not just a small percentage better" [16].

In terms of the level of improvement, innovations can be transformational, for instance, creating large-scale changes in the way technology is used or thought about. Mary Lou Jepsen [16] said, "I think of innovation as doing some transformative work in an area or in a combination of areas that trail blazes in a way that people recognize has moved the ball forward... in a way that is a leap."

But it is not compulsory that every innovation be pioneering or radically change the world. Bernard Meyerson [16] referred to "continuous innovators." "The danger is there are other types of innovators that are just as necessary, what I call the continuous innovators. These are the guys who come to work every day and make it 5 to 10 percent better, and there's a terrible undervaluation of that."

- *Innovation occurs at the interfaces of different disciplines*: Innovators in all the areas represented, that is, academia, large companies, small businesses, and the arts, acknowledged that innovation occurs at the edge of disciplines and necessitates the synthesis of knowledge from dissimilar fields. Yo-Yo Ma [16] captured this aspect using the concept of the edge effect from ecology: "If you think about where new ideas can come from, you need proximity to density, and if you're at the edge of something you see both sides; you already see over the wall. You could be part of one ecosystem, but you actually are constantly interacting with another ecosystem, and so you see the possibility of what another ecosystem can bring. And... if the center uses the knowledge at the edge, the center does benefit."

- *Teamwork is important to the process of innovation*: Innovation is the effect of joint effort, a point frequently made by the innovators. And it relies on the work of the team as a whole, not the work of one key innovator and other Supporters." Ivan Seidenbergt [16] observed: "I get comfort in knowing that life is cumulative, innovation is cumulative, and it's not individual. Let's take some of the greatest examples: Let's start with the example everybody's using right now, and I knew him well. Steve Jobs was a genius, but he didn't invent the computer. He didn't invent anything that went into the iPhone, but he made it all work together... so what did he invent? Take another example: Bill Gates had enough common sense and enough vision to know that PCs couldn't talk to each other, so he built operating systems to make them talk to each other, but along the way, they didn't work very well when they first came out with them. They (Jobs and Gates) needed a full team and with their superior insights and innovative spirit, they made something bigger than any one person could have made. So, all I'm getting at is that there's really no one innovator who can innovate all alone. I can't think of any one person that gets it all right. Is there anybody? Is there anybody in the literature that gets it right the whole time?"

• *Innovation is part of an invention-value continuum*: Innovation is part of a field between invention and worth. Innovators may start with a discovery and then innovate to generate value from it, or start with a problem and solve it innovatively. Innovation was portrayed as the use of inventions to real-world needs. Innovation can also be driven by the impression of marketability or attempt to solve a problem. As Robert Fischelr [16] said, "Sometimes we see an invention and then we can apply it to another thing, but that doesn't happen very often. Most times, we hear about something and it occurs to us that the way they're doing it is not good, and so we innovate a better way."

Analysis of the 60 innovators' observations disclosed that innovation is an enhanced product, process, or service that profits society in a timely and, sometimes, transformational manner. It is a team activity at the meeting point of diverse fields, bringing as one diverse ideas, skills, and/or methods to result in the production of value.

Types of innovation

Innovation applications are commonly applied to a product, a process, or a service. To additionally comprehend how this is done, let's reflect on three categories of innovation:

Product innovation

Product innovation is about making valuable changes to material products. Interrelated terms that are frequently used interchangeably comprise product design, research and development, and new product development (NPD). All of these terms proffer a particular viewpoint on the degree of alteration to products. Well-known organizations characteristically have a collection of products that must be incrementally enhanced or adjusted as problems are recognized in service or as new requirements emerge. It is imperative that they also work on add-ons to the product families. One of the major actions of the product design team is the work it carries out on next-generation products or new models of products. They might also work on designing far-reaching new products or new core products that enlarge the portfolio considerably and frequently involve drastically new processes to produce them. These new core products idyllically present the organization with the possibility of major increases in revenue and growth, which can also create the potential of short-term monopoly in the market.

The product development process for next-generation and new core products, according to Cooper, follows a familiar cycle in most organizations:

1. Ideation
2. Preliminary investigation
3. Detailed investigation

4. Development
5. Testing and validation
6. Market launch and full production [17]

All of these steps involve communication with customers, who might take part in idea creation and element recognition. Key performance criteria in the design process revolve around the following:

1. Time to market
2. Product cost
3. Customer benefit delivery
4. Development costs [18]

These standards can be traded off against one another. For instance, development costs can be traded against time to market, customer benefits can be traded against product costs, and so on. Three blueprint systems have become known as providing a management system for efficient product innovation: phase review, stage gate, and product and cycle-time excellence (PACE).

1. *Phase review*: This technique splits the product development life cycle into a sequence of different phases. Every phase encompasses a body of work that, once finished and evaluated, is dispensed over to the next phase. No consideration is paid to what may or may not occur in the succeeding phases, principally for the lack of knowledge or exclusive focus on the job in the existing phase. The phase review technique is a chronological rather than a simultaneous product design method, that is, each phase is accomplished and concluded before the commencement of the next phase.
2. *Stage gate*: This technique is a simultaneous product design procedure that follows a prearranged life cycle from idea creation to market commencement [17]. The stages in this technique are first and foremost cross-functional. Stage gates appear at the end of each stage, where a design evaluation takes place. Each stage gate evaluates the decided deliverables for completion at the conclusion of the stage, a checklist of the standard agreed for each stage, and a choice about how to advance from a particular stage.
3. *PACE*: This method is concerned mainly with enhancing product improvement strategies [18]. The technique connects product strategy with the general strategy and goal of the organization. A key element is positioning of the voice of the customer all through the product design procedure. Strategies are divided into six product strategic thrusts: expansion, innovation, strategic balance, platform strategy, product line strategy, and competitive strategy. Product innovation methods and processes are one element in an organization's mission to create value for customers.

Process innovation

Process innovation can be observed as the launching of a new or considerably enhanced method for the construction or delivery of production that append value to the organization. The term *process* refers to an interconnected set of actions designed to convert inputs into a specific result for the customer. It implies a strong prominence on how work is done within an organization rather than what an organization does [19].

Processes recount every operational action by which value is presented to the end client, such as the purchase of raw materials, production, logistics, and after-sales service. The process innovation in the 1970s and 1980s gave Japanese manufacturing a viable advantage that permitted them to take over some international markets with cars and electronic goods. Likewise, process innovation has permitted organizations such as Dell and Zara to achieve competitive advantage by offering higher-quality products, delivered faster and more proficiently to the market than by the competitors. By focusing on the resources by which they transform inputs, such as raw materials, into results, such as products, organizations have achieved efficiencies and have added importance to their production. Process innovation permits some organizations to contend by having a further proficient value chain than their rivals have.

Process innovation has resulted in organizational enhancement such as lower stock levels; quicker, additional flexible production processes; and more responsive logistics. Organizations can develop the competence and value of their processes with a huge array of diverse enablers. Even though the use of these enablers is dependent on the organizational framework, many present the possibilities for improved process performance. The application of technology such as robotics, enterprise resource planning systems, and sensor technologies can change the process by decreasing the price or variation of its output, improving safety, or decreasing the throughput time of the process.

Service innovation

Service innovation is concerned with making changes to intangible products. Services are frequently linked with work, play, and recreation. Examples of these types of service consist of education, banking, government, recreation, entertainment, hospitals, and retail stores. In the past decade, an enormous amount of knowledge-based services has been accessible through websites. These services involve intangible products, have a high quantity of customer dealings, and are typically set in motion on demand by the customer. Defining a service can be to some extent problematic. Some define service as a sequence of overlapping value-creating activities.

Others define service in terms of performance, where customer and provider coproduce value. There are three categories of service operations:

1. Quasi-manufacturing (e.g., warehouses, testing labs, recycling)
2. Mixed services (e.g., banks, insurance, realtors)
3. Pure services (e.g., hospitals, schools, retail)

Services can without a doubt involve products that form a comprehensive part of the product life cycle, from preliminary sales to end-of-life recycling and clearance. Service business in areas such as finance, food, education, transportation, health, and government make up most organizations in any economy.

These organizations as well require innovation incessantly so that they can enhance levels of service to their customers. A key characteristic of a service is a very high level of communication with the end user or customer. The customer is often not capable of separating the service from the person delivering the service and so will make quality postulation based on impressions of the service, the group delivering the service, and any product delivered as part of the service. An additional feature of some service organizations is that their product may be perishable; consequently, the product must be consumed as soon as possible following purchase. Consequently, the timing of the delivery and customer opinion of quality are vital to success.

The notion of service quality is of particular significance. Service quality is a function of numerous factors including the uniqueness of offerings, intangibilities such as customized customer contact or perishable manufacture, and a continued capacity for innovations of the service. Another important driver of service innovation comes from the possibilities afforded by the new information technology podium, predominantly the Internet. The Internet is a priceless resource on which new service associations between organizations and their customers are being developed every day.

Innovation and entrepreneurship

If innovation is successful, the expected outcome is the transitioning of these new products, processes, or services into useful products that people are willing to pay for in the United States and globally. Although innovation and entrepreneurship are related, many caution against focusing too much on entrepreneurship in the initial stages of the creative aspect of innovation. This perspective believes that entrepreneurship should be a natural outcome of entrepreneurship but should not be the initial focus.

It is really important to lead with "innovation" and have it evolve into "entrepreneurship" because innovation is the large end of the funnel that appeals to and actually requires participation by a much broader audience. Non-business, non-engineering, and non-STEM people are every bit as important to include in that innovation process because the process is not as rich and has inferior outcomes without that diversity [20]. In order to see this type of innovation systematically realized, engineering leaders must understand principles that should be integrated into the creative process to produce effective innovations.

The terms *entrepreneurship* and *innovation* are over and over again used interchangeably; nevertheless, this is deceptive. Innovation is frequently the starting point from which an entrepreneurial business is built for the reason of the competitive advantage it offers. On the contrary, the act of entrepreneurship is simply one means of bringing an innovation result to the marketplace. Technology entrepreneurs regularly decide to build a startup company for a technological innovation. This will offer financial and skill-based resources that will take advantage of the chance to grow and commercialize the innovation. Once the entrepreneur has set up a business, the focal point shifts in the direction of its sustainability, and the best way to attain this is through managerial innovation. Nonetheless, innovation can be conveyed to the market by ways other than entrepreneurial startups; it can also be subjugated through well-known organizations and deliberate alliances between organizations.

Case study: Charles Dow

In 1896, Charles Dow created the Dow Jones Industrial Average in order to provide a snapshot of the US economy through the stock market There were 12 companies on Dow's original list: American Cotton Oil, American Sugar, American Tobacco, Chicago Gas, Distilling & Cattle Feeding, General Electric (GE), Laclede Gas, National Lead, North American, Tennessee Coal and Iron, US Leather pfd., and US Rubber. Of all of those companies, which were financial leaders at the turn of the 20th century, there is only one you might recognize that is still in business today: General Electric.

What is the key to GE's century-long tenure? Product innovation. According to business researchers Heath Downie and Adela J. McMurray, "The consistency of GE's commitment to product innovation was made possible by the steadiness of the company's leadership." Even during the Great Depression, GE found a way to allocate diminishing financial resources to its research and development initiatives.

Today, GE has taken their commitment to innovation even further, crowdsourcing both internally and externally to drive advancements in several industries. In fact, GE has an Open Innovation Manifesto, in which they state:

We believe openness leads to inventiveness and usefulness. We also believe it's impossible for any organization to have all the best ideas, and we strive to collaborate with experts and entrepreneurs everywhere who share our passion to solve some of the world's most pressing issues [...] We'll never stop experimenting, collaborating and learning-we'll get smarter as we go, and the Global Brain will evolve and grow with us.

GE has a hand in advancing just about every engineering industry you can think of such as aviation, software, consumer goods, water and wastewater, power and energy, transportation, and healthcare, to name a few. Named "America's Most Admired Company" in a poll conducted by Fortune magazine and one of "The World's Most Respected Companies" in polls by Barron's and the Financial Times, the quality work GE has done for the planet has not gone unnoticed, and their leadership is extremely dedicated to quality and innovation. Take, for instance, Deb Frodl, global executive director of GE Ecomagination. Ecomagination is a business initiative designed by GE to develop innovative solutions to environmental challenges while driving economic growth. In an interview with Cleantech Group, Frodl said:

> Innovation is the foundation for Ecomagination and we have really developed a lot of solutions that solve complex problems for a multitude of industries. [...] Ecomagination has really been the catalyst within GE to step outside and get those ideas and that outside innovation moving forward.

In 2016, GE announced it would be relocating its corporate headquarters to Boston, Massachusetts, in part to enable GE to place additional emphasis on digital industrial innovation. This is further proof of the company's commitment to innovation and its leaders' push to improve access to a more innovative workforce and relocate to a better environment for innovation.

References

1. National Academy of Engineering. *Educate to Innovate*. Washington, DC: National Academies Press, 2015.
2. US Department of Education, National Center for Education Statistics. *Highlights from PISA 2006: Performance of US 15-Year-Old Students in Science and Mathematics Literacy in an International Context*, 2007. Available at: https://nces.ed.gov/pubs2008/2008016.pdf, accessed October 12, 2016.
3. National Commission on Excellence in Education. *A Nation at Risk*, 1983, https://en.wikipedia.org/wiki/A_Nation_at_Risk, accessed March 25, 2018.
4. The National Academies Press. *Rising above the Gathering Storm, Revisited: Rapidly Approaching Category 5*. 2010. Available at https://www.nap.edu/catalog/11463/rising-above-the-gathering-storm-energizing-and-employing-america-for, accessed March 25, 2018.
5. The White House. *Remarks by the President at the National Academy of Sciences Annual Meeting*. The White House. 2009. Available at: https://obamawhitehouse.archives.gov/the-press-office/2013/04/29/remarks-president-150th-anniversary-national-academy-sciences, accessed March 25, 2018.

6. The White House. *President Obama Launches "Educate to Innovate" Campaign for Excellence in Science*. Technology, Engineering & Math (Stem) Education, Available at https://obamawhitehouse.archives.gov/the-press-office/president-obama-launches-educate-innovate-campaign-excellence-science-technology-en, accessed March 25, 2018.

7. National Science Foundation. Fact Sheet, January 7, 2016. P.l. Available at: https://www.nsf.gov/news/news-factsheets.jsp, accessed March 25, 2018.

8. NSF Directorate for Engineering version. *The Role of the National Science Foundation in the Innovation Ecosystem*. Available at https://www.nsf.gov/eng/iip/innovation.pdf, accessed March 25, 2018.

9. Dunlap, R. E., and Lydia, S. Gallup Organization. Science & Engineering Indicators-200, *Only One in Four Americans Are Anxious About the Environment*. Princeton, NJ: Poll Release, 2001. Available at: http://news.gallup.com/poll/1801/only-one-four-americans-anxious-about-environment.aspx, accessed March 25, 2018.

10. Prime Minister's Office. *Public Opinion Survey of Relations of Science and Technology to Society*. Survey taken in February 1995, and Public Opinion Survey of Future Science and Technology. 2001. Survey taken in October 1998. *S&T Today* 14(6): 12. Tokyo, Japan: Japan Foundation of Public Communication on Science and Technology, Tokyo, Japan.

11. Gaskell, G., and Bauer, M. W. (Eds.). Biotechnology 1996–2000. East Lansing, Michigan: National Museum of Science and Industry (U.K.) and Michigan State University Press, 2016.

12. Research! America. Poll Data Booklet. Vol. 2. Research! America, Alexandria, VA: Research! America, 2001.

13. *National Science Foundation Survey of Public Attitudes toward and Understanding of Science and Technology*. Arlington, VA: National Science Foundation, 2001.

14. Pew Research Center for the People and the Press. America's Place in the World II: More Comfort with Post-Cold War, 1997. Available at http://www.people-press.org/1997/10/10/americas-place-in-the-world-ii/, accessed March 25, 2018.

15. Carlson, O. K. Public Views NASA Positively, but Generally Disinterested in Increasing Its Budget. Gallup News Service. Poll Analyses, 2001. Available at http://news.gallup.com/poll/1927/public-views-nasa-positively-generally-disinterested-increasin.aspx, accessed March 25, 2018.

16. Bement, A., Dutta, D., and Patil, L. *Educate to Innovate: Factors that Influence Innovation: Based on Input from Innovators and Stakeholders*. Washington, DC: National Academies Press, 2015.

17. Cooper, R. G., and Kleinschmidt, E. J. New product performance: What distinguishes the star products? *Australian Journal of Management*, 2000; 25(1):17–46.

18. McGrath, M. *Setting the Pace in Product Development: A Guide to Product and Cycle-Tune Excellence*. Revised Edition. Stoneham, MA: Butterworth: Heinemann, 1996.

19. Davenport, T. H. *Process Innovation: Reengineering Work through Information Technology*. Boston, MA: Harvard Business School Press, 1993.

20. National Academy of Engineering. What is innovation? In: *Educate to Innovate: Factors That Influence Innovation: Based on Input from Innovators and Stakeholders. Academy of Management*, 1993; 7(2); Washington, DC: The National Academies Press, 2015. doi:10.17226/21698.

chapter three

The aerospace and defense industry in Southwest Ohio

A model for workforce-driven economic development

Cassie B. Barlow and Kristy Rochon

Contents

Executive summary

This chapter came to life through two aerospace and defense industry studies completed by the authors in Ohio in 2011 and 2016. In these studies, the authors investigated in-demand industries in Ohio and the need to focus on workforce development to fully support the industries. Both the 2011 and the 2016 studies defined and highlighted details of the aerospace and defense industry, while assessing the current and future growth of the Industry. Building on their previous two studies, in this chapter the authors represent the aerospace and defense industry in Ohio with distinct clusters that best define the industry: Aerospace Manufacturing, Research and Development (R&D), Federal and Military, and Aviation.

The chapter delves into the details of each distinct cluster, specifically focusing on workforce supply and demand. In addition, the aging workforce and demographic shifts are examined.

Data suggests that the higher education system in Ohio is not necessarily underproducing STEM graduates in the context of the patterns that support the defined clusters analyzed in this report. The analysis found Ohio's percent of degree holders has experienced small, steady increases over the past six years. The rate of increase has been just above the US average, but Ohio still lags behind the nation in the percent of the population holding bachelor's degrees and graduate or professional degrees. Overall, the State's growth rate across demographics was comparable with the overall population growth over the same period, a 1 percent growth. This modest growth is directly tied to the largest industry sectors in the region. It may also be attributable to the economic development model employed in the region. Tying economic growth to the growth of the regional military installation will lead to growth rates which are relative to the installation's growth.

There are workforce challenges in the aerospace and defense industry which cannot be ignored. While Ohio's colleges and universities are generally keeping pace with the required STEM workforce in R&D, Aerospace Manufacturing demands are outpacing training and education, which is further stressed by the replacement rates—retirements and other workforce churn. Nationally, more than 53 percent of the industry workforce supporting Aerospace Manufacturing are over 45 years old with a median average age of 44.5. At the same time, only 26 percent of the workforce is under the age of 35, meaning that retirements will continue to drive higher workforce demand.

Aerospace Manufacturing directly supports more than 20,000 jobs in Ohio, ranking it eighth in the country. Ohio's growth is relatively flat in this Industry cluster in Ohio. Aerospace Manufacturing in Northeast Ohio is projected to decline, while Southwest Ohio is predicted to grow. The Aerospace Manufacturing workforce requires some of the highest-skilled manufacturing workers. Almost 20 percent of its workforce are engineers.

Ohio ranks 15th in the country in the number of R&D jobs, with a projected growth of 25 percent in the next decade. This cluster is the only one of those analyzed with significant projected growth. Economic growth can be realized through dedicated efforts to leverage this research for commercial applications. Analysis of R&D expenditure data suggests Ohio has a disproportionate emphasis on companies and individuals contracted to conduct this R&D compared to the top 10 states in R&D expenditures. More analysis is needed to understand why this is occurring in Ohio and who owns the intellectual property being produced because of this structure. Ohio ranks 14th in the US for the number of civilian and military jobs. Like most of the other states employing large numbers of civilian and military workers, Ohio's federal jobs are projected to decrease. This aligns with the decrease of federal budgets and expenditures.

To best address the workforce demand within the aerospace and defense industry in Southwest Ohio, the authors recommend building a robust ecosystem with participation across the region. The ecosystem requires a broad audience with maximum participation from all parts of the ecosystem. Strong leadership is required and can come from any of the stakeholders. The ecosystem will be successful if there is a strategic plan that is regularly monitored and measured.

This chapter could serve as a step-by-step model for any region of the country that desires to focus on the workforce and economic development of an in-demand industry.

Introduction

The aerospace and defense industry is critical to national defense and therefore a very unique industry in the world market. It is also a very large industry with many complexities. The aerospace and defense industry serves both the commercial and the defense markets with a vast array of products. The Industry also includes some very large employers, like Boeing and Lockheed Martin as well as thousands of small companies. The industry is smaller than it once was, but is still quite large and has strategic importance in the US.

The end of the cold war meant decreased spending in Aerospace and Defense, which led to diversification of the Industry. There were many new commercial applications discovered for aerospace and defense products. This also led to the Department of Defense spending a much smaller percentage of their budget on research and development. When the cold war ended, the US went through a recession which led to the Industry cutting workforce and consolidating operations. Examples of these consolidations include Northrup and Grumman and Lockheed and Martin. Because of the recession and a reduction in spending, Aerospace and Defense companies started to focus sales on foreign governments. This new strategy worked until the Asian economic crisis in the late 1990s. This crisis resulted in another hit to the market bottom-line which led to more reductions for companies across the Industry. After 9/11, the Industry was again impacted, but this time by a large infusion of new dollars that were part of an investment into a new conflict in the Middle East. The aerospace and defense industry is forecasted to continue to grow.

Because of the complexity of the aerospace and defense industry and the vast nature of the market, the Industry is the world's leading producer of technology and supports one of the largest high-skill and high-wage workforce in the country. In 2015, the US aerospace and defense industry supported almost 2.4 million jobs (AIA, Dec 2016). The primary share of jobs within the aerospace and defense industry are in manufacturing (917,000 jobs) and then Information and Professional Services (465,000) (AIA, Dec 2016). Growth areas within the Industry are foreign sales, and research and development. An evolving demand and continued growth within the aerospace and defense industry as well as an aging workforce make the next ten years within the Industry a challenge. In this chapter, the authors will focus on workforce demand and how a region can build a robust ecosystem to react to needs within an Industry.

Defining the industry

To thoroughly examine workforce supply and demand within the aerospace and defense industry, the authors chose to scope their work in terms of a region within Ohio, specifically, the Southwest region. To represent the existing demand within the aerospace and defense industry, the authors divided the Industry into four key areas:

- Aerospace Manufacturing
- Research and Development (including Information Technology related to Aerospace)
- Federal and Military
- Aviation

Aerospace manufacturing

Although aircraft are not manufactured in Ohio, many Aerospace Manufacturing suppliers are in Ohio. For instance, GE Aviation manufactures jet engines and other components in Southwest Ohio. In 2014, Boeing reported that it had 375 suppliers and vendors in Ohio, accounting for $11.4 billion in purchases that supported up to 385,000 jobs in Ohio (Boeing, 2016).

Aerospace Manufacturing is an important component of Ohio's economy. Ohio supports direct Aerospace Manufacturing jobs. However, this cluster does not define all of

Aerospace Manufacturing since occupations such as metalworking, sensors, and composites may serve automotive, medical, and other kinds of manufacturing in addition to the aerospace industry. The direct Aerospace Manufacturing jobs are defined with the industry codes in Figure 3.1.

Code	Description
333314	Optical instrument and lens manufacturing
334511	Search, detection, navigation, guidance, aeronautical, and nautical system and instrument manufacturing
334515	Instrument manufacturing for measuring and testing electricity and electrical signals
335931	Current-carrying wiring device manufacturing
336411	Aircraft manufacturing
336412	Aircraft engine and engine parts manufacturing
336413	Other aircraft parts and auxiliary equipment manufacturing
336414	Guided missile and space vehicle manufacturing
336415	Guided missile and space vehicle propulsion unit and propulsion unit parts manufacturing
336419	Other guided missile and space vehicle parts and auxiliary equipment manufacturing

Figure 3.1 Aerospace manufacturing occupations.

Research and development

The Aerospace R&D cluster includes more than just the research companies but also includes testing, engineering services, and software and computer design, including the following occupational codes (Figure 3.2).

Code	Description
511210	Software publishers
541330	Engineering services
541380	Testing laboratories
541512	Computer systems design services
541690	Other scientific and technical consulting services
541712	Research and development in the physical, engineering, and life sciences (except biotechnology)

Figure 3.2 Research and development occupations.

Federal/military

Federal jobs are a substantial part of Ohio's Aerospace Industry. Wright-Patterson Air Force Base (WPAFB) in Dayton, Ohio is the largest single-site employer in the state, employing more than 27,000 civilians and military. Ohio is also home to a robust National Guard and to Defense Finance and Accounting Service and Defense Logistics Agency. NASA Glenn in Cleveland is an important part of Ohio's Aerospace Industry and is a large federal employer (Figure 3.3).

Code	Description
901199	Federal Government, Civilian, Excluding Postal Service
901200	Federal Government, Military

Figure 3.3 Federal/military occupations.

Aviation

While Aviation is directly linked to Aerospace, from a workforce development perspective, it is a demand that follows the market, not business need. As consumers spend more on commercial flights and as businesses become more nationally and internationally connected the demand for pilots and other support staff will naturally grow. As this happens, commercial airlines will look for a pool of qualified pilots. That search typically starts with Department of Defense pilots and for this reason, the growth in commercial pilots is worth benchmarking (Figure 3.4).

Code	Description
481111	Scheduled Passenger Air Transportation
481112	Scheduled Freight Air Transportation
481211	Nonscheduled Chartered Passenger Air Transportation
481212	Nonscheduled Chartered Freight Air Transportation
481219	Other Nonscheduled Air Transportation
488111	Air Traffic Control
488119	Other Airport Operations
488190	Other Support Activities for Air Transportation
611512	Flight Training

Figure 3.4 Aviation occupations.

National workforce trends

The aging workforce in the US has been a popular topic among economists and the government for decades. There is good reason for this concern and for discussion of this topic. The aging baby boomers, recession of the late 2000s as well as a rising life expectancy have changed the face of our current and future workforce (Bureau of Labor Statistics, 2017). Internationally, the number of older people is expected to exceed the number of children for the first time by 2047. Although the retirement age seems to be increasing, there is still a need to replace this generation with a new workforce as they make the retirement decision. The companies that figure out how best to use this aging workforce to their betterment, for example in training and mentoring of the new workforce, will be the most successful companies in the world. This generation has extensive knowledge and skills that can continue to contribute to companies across the country. At the same time, strategic workforce planning is critical as the baby boomers leave the workforce.

Nationally, more than half of the workforce in the industries supporting Aerospace Manufacturing are over 45 years old with only a quarter below the age of 35. This composition is slightly better for the R&D industry with 43 percent over the age of 45 and 32 percent below the age of 35. This is a metric to watch as the workforce between 35 and 44 years old are critical for the long-term growth and stability of the industry. Industry experience has been a common theme in Aerospace and Defense workforce shortage studies. The higher education system can develop strategies to increase competition rates in desired degrees; however, industry must take responsibility for developing their workforce to meet future experience demands (Figure 3.5).

In addition to an aging workforce, there are other demographic changes occurring in the workforce that must be part of an organizational strategic workforce plan. The gender balance in the United States has been tipped towards women for the last few years and is going to continue in this direction. In the year 2050, there will be 150 women for every 100 men in

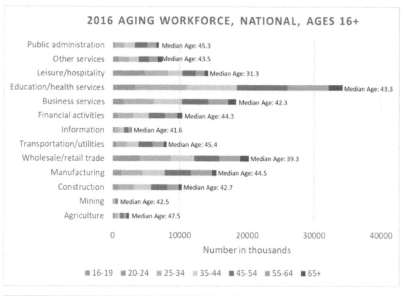

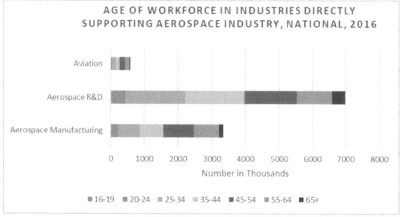

Figure 3.5 National workforce population age statistics. (From Bureau of Labor and Statistics from the Current Population Survey, 2017.)

the US In addition, there has been a continual increase in African American and Hispanic populations. A successful organization will figure out how to recruit, hire and retain a diverse workforce (Figure 3.6).

Current state of the industry

With the clusters defined, a deeper dive has provided some unexpected projections. In this chapter, the authors use a modeling tool which considers demand based on federal data sources that project supply and demand using algorithms applied to historical trends, the supply of workers completing higher education programs that meet those demands, and churn. Included in this data is the Labor Market Information data that Ohio Means Jobs uses to identify the most in-demand jobs. By layering on top of this data

Figure 3.6 Diversity in the workforce. (From Bureau of Labor and Statistics from the Current Population Survey, 2017.)

set the supply data and the churn data, a more detailed picture develops. The bottom line in many of the tables in this chapter is that Ohio is not necessarily underproducing STEM graduates in the context of the staffing patterns that support the defined industry clusters outlined.

STEM talent analysis

To validate this conclusion, National Science Foundation data was studied to understand if this was a projection unique to the staffing patterns evaluated or true of the broader STEM workforce (Figure 3.7).

The data indicate that the number of Science and Engineering degrees conferred in Ohio increased by 30 percent over ten years; however, the percentage of Science and Engineering degrees compared to all higher education degrees conferred has remained relatively flat,

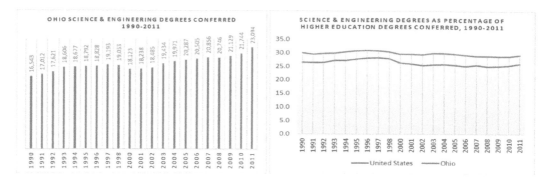

Figure 3.7 Changes in science and engineer degrees, 1990–2011. (From NSF, Science and Engineering Indicators State Data Tool, 2014.)

consistent and below the national trend. At the same time, occupations requiring those degrees have seen increases, but were also consistent and below the national trend (Figure 3.8).

This data suggests that the state is seeing an increase in STEM degrees which are meeting the increasing workforce demand and is likely generating a surplus of Science and Engineering degrees conferred. It is important to note that prior to the recession, Ohio employed more engineers than the national average, but the loss of large tech companies, like Delphi and NCR, greatly impacted the overall engineering occupations. Data show the state once again reached 2003 ratios in 2012, but as Ohio recovered from the losses, the national average continued to climb, resulting in a larger gap with the state lagging behind the national average. This loss of jobs led to a large shift in the available engineers seeking jobs which may have impacted some of the shifts in data which suggest degrees conferred are sufficient to meet business demand. The bottom line: STEM degrees may have increased in numbers but not as a percentage of total degrees. Ohio's growth in those areas has lagged the national trends; therefore, more work is needed to advance a STEM workforce that is not just meeting Ohio business demand, but could help meet national business demand or drive economic growth through the return of an innovative workforce focused on market trends and consumer demand. As the staffing patterns in the clusters identified in this report are compared to the supply, it will be important to keep in mind the overall STEM workforce supply.

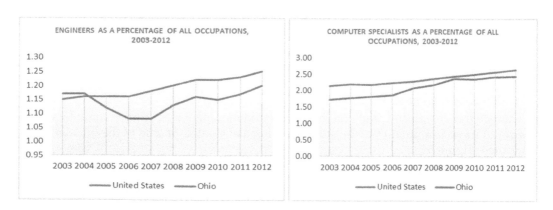

Figure 3.8 Changes in engineers and computer specialists, 2003–2012. (From NSF, Science and Engineering Indicators State Data Tool, 2014.)

Concerning the Aerospace Manufacturing, R&D, Federal/Military and Aviation clusters, the talent supply appears to be more than sufficient to meet the workforce demand. Many factors are coming into play: (1) concerted efforts by a robust network of STEM initiatives are making headway and addressing the STEM talent supply gap, (2) there are major gaps in skilled jobs that do not require higher education and are thus harder to measure supply statistics (e.g., Team Assemblers), (3) the only cluster with growth projected to meet or exceed national growth rates is the R&D cluster, the other clusters are flat or declining, (4) completion rates are up, but retention of students in the state of Ohio is declining and that brain drain is impacting the shortages that have been reported by businesses, and (5) this data is based on historical trends and is from secondary data sources only, it's possible there are STEM-related workforce shortages that could exist due to niche sectors not accurately portrayed by the available data. Regarding factors three and four listed earlier, a recommendation section has been produced that highlights the need for Network strategies.

Talent retention

One item that is often discussed when reviewing degrees conferred is the retention of the talent. The "Brain Drain" is a concern, especially when considering in-demand STEM degrees. A review of America Community Survey data from the US Census Bureau can provide some insight on degree holders in the State of Ohio and within each JobsOhio region. Ohio's percent of degree holders has experienced small, steady increases over the past six years according to the census five-year trend data. The rate of increase has been just above the US average, but Ohio still lags the nation in the percent of the population holding bachelor's degrees and graduate or professional degrees. The State's growth rate in these demographics was comparable with the overall population growth over the same period, a 1 percent growth. The national growth rate, however, was 6 percent while the growth in degree holders was less than 1 percent (Figure 3.9).

When considering the challenges reported by industry to find qualified workers, a closer look was warranted. When considering the JobsOhio regions, a positive trend was also evident.

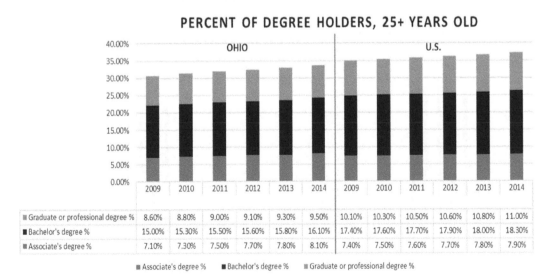

Figure 3.9 Percent of Degree Holders in Ohio and the US, 25+ years old, 2009–2014. (From American Community Survey, US Census Bureau, five-year estimates.)

PERCENT OF DEGREE HOLDERS, 25+ YEAR OLDS BY JOBSOHIO REGION

			Central						Northeast						Northwest						Southeast						Southwest						West			
	09	10	11	12	13	14	09	10	11	12	13	14	09	10	11	12	13	14	09	10	11	12	13	14	09	10	11	12	13	14	09	10	11	12	13	14
Graduate or professional degree %	10.4	10.6	10.9	11.0	11.3	11.6	8.41	8.66	8.79	8.85	9.08	9.32	7.04	7.22	7.35	7.41	7.56	7.90	4.91	5.07	5.20	5.27	5.41	10.2	10.7	11.0	11.0	11.5	11.4	9.53	9.91	10.1	10.2	9.00	10.5
Bachelor's degree %	20.1	20.2	20.5	20.7	20.8	20.9	14.7	15.1	15.3	15.4	15.6	15.8	11.9	12.4	12.4	12.5	12.6	12.8	7.88	7.98	8.02	8.42	8.60	8.76	17.9	18.2	18.5	18.6	19.6	19.3	15.9	16.2	16.3	16.4	13.1	17.1
Associate's degree %	6.47	6.67	6.89	7.09	7.09	7.17	6.87	7.10	7.21	7.39	7.54	7.69	8.40	8.57	8.73	8.96	9.15	9.47	7.07	7.26	7.67	8.03	8.41	8.65	6.92	7.17	7.15	7.43	7.65	7.84	7.41	7.63	7.73	7.89	8.64	8.43

■ Associate's degree %   ■ Bachelor's degree %   ■ Graduate or professional degree %

Figure 3.10 Percent of degree holders by JobsOhio regions, 25+ years old, 2009–2014. (From American Community Survey, US Census Bureau, five-year estimates.)

The Central region leads the state in the percentage of degree holders, followed closely by the Southwest region. Each region experienced a positive trend in degree holders (Figure 3.10).

This trend continues when analyzing just 24- to 35-year-olds in each JobsOhio region. This age group is often considered the most critical retention demographic. Connecting graduates to employment in Ohio is critical to retaining the needed talent. Except for a minor dip in the West region in 2013, each region experienced modest growth during each year analyzed. In addition, for each region except for the Northeast region, the overall population of 25- to 34-year-olds decreased. This suggests that the major migration out of each region other than the Northeast does not consist of predominantly bachelor's degree holders (Figure 3.11).

This data leads to questions that could be further explored in each region. More research must be done to determine if the reported talent gap by businesses experiencing

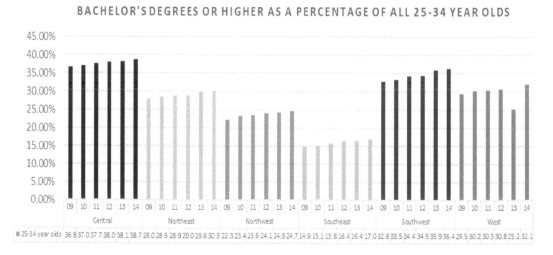

BACHELOR'S DEGREES OR HIGHER AS A PERCENTAGE OF ALL 25-34 YEAR OLDS

			Central						Northeast						Northwest						Southeast						Southwest						West			
	09	10	11	12	13	14	09	10	11	12	13	14	09	10	11	12	13	14	09	10	11	12	13	14	09	10	11	12	13	14	09	10	11	12	13	14
25-34 year olds	36.8	37.0	37.7	38.0	38.1	38.7	28.0	28.6	28.9	29.0	29.8	30.3	22.3	23.4	23.6	24.1	24.3	24.7	14.9	15.1	15.8	16.4	16.4	17.0	32.8	33.5	34.4	34.5	35.9	36.4	29.5	30.2	30.5	30.8	25.2	32.1

■ 25-34 year olds

Figure 3.11 Percent of bachelor's degree holders by JobsOhio regions, 25–34 years old, 2009–2014. (From American Community Survey, US Census Bureau, five-year estimates.)

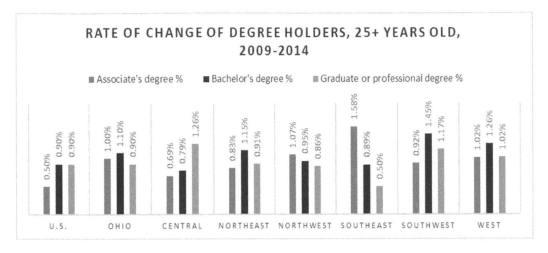

Figure 3.12 Rate of change of education attainment by region, 25+ year olds, 2009–2014. (From American Community Survey, US Census Bureau, five-year estimates.)

hiring difficulties is due to a mismatch in degrees conferred compared to industry need or if there is a disconnect between jobseekers and businesses seeking talent.

The rate of change in each region could serve as a metric to identify pockets of growth or areas that are improving because of initiatives to retain young talent (Figure 3.12).

Aerospace manufacturing

Aerospace Manufacturing directly supports more than 20,000 jobs in Ohio, ranking it eighth in the country. However, growth predictions for Aerospace Manufacturing in Ohio for the next decade are relative flat.

The location quotient (occupation percent of local employment/occupation percent of national employment) permits a comparison of the relative importance of the occupation in the regional employment mix relative to the country (Figure 3.13).

State Name	2015 Jobs	2023 Jobs	2015-2023 Change	2015 -2023 % Change	2023 Location Quotient
California	115,584	98,890	(16,694)	(14%)	1.19
Washington	96,669	100,328	3,659	4%	6.19
Texas	54,895	58,843	3,948	7%	0.96
Arizona	31,790	30,065	(1,725)	(5%)	2.25
Kansas	31,297	30,190	(1,107)	(4%)	4.37
Connecticut	30,978	28,854	(2,124)	(7%)	3.59
Florida	30,046	31,303	1,257	4%	0.79
Georgia	23,094	25,084	1,990	9%	1.23
Ohio	21,845	21,714	(131)	(1%)	0.87
New York	20,749	20,713	(36)	(0%)	0.48
Massachusetts	19,840	19,923	83	0%	1.18
Missouri	19,583	23,430	3,847	20%	1.78
Pennsylvania	18,266	18,882	616	3%	0.70
Maryland	13,659	13,605	(54)	(0%)	1.08
Alabama	13,458	13,955	497	4%	1.48

Figure 3.13 Top states for aerospace manufacturing jobs.

MSA Name	2015 Jobs	2023 Jobs	2015 - 2023 Change	2015 - 2023 % Change	2023 Location Quotient
Cincinnati, OH -KY-IN	11,731	12,061	330	3%	2.54
Cleveland -Elyria, OH	4,264	3,742	(522)	(12%)	0.80
Dayton, OH	2,971	3,072	101	3%	1.85
Akron, OH	1,205	1,075	(130)	(11%)	0.72
Mansfield, OH	555	688	133	24%	2.93
Columbus, OH	534	574	40	7%	0.12

Figure 3.14 Ohio aerospace manufacturing jobs by MSA. (From EMSI, 2015.2—QCEW Employees, Non-QCEW Employees, and Self-Employed.)

Although Ohio's growth is relatively flat in Aerospace Manufacturing, Ohio's metropolitan statistical areas (MSAs) vary. For the most part, Aerospace Manufacturing in Northeast Ohio is projected to decline, while Southwest Ohio is predicted to grow. The vibrancy and growth of the Aerospace Manufacturing industry earned the Southwest Ohio Aerospace Region (SOAR), a manufacturing designation from the Economic Development Administration's Investing in Manufacturing Communities Partnership (IMCP) initiative. Together, the Dayton and Cincinnati metro area support more than 14,000 jobs, representing two-thirds of Ohio's Aerospace Manufacturing industry (Figure 3.14).

The Aerospace Manufacturing workforce requires some of the highest-skilled manufacturing workers. Almost 20 percent of its workforce are engineers—aerospace, industrial, mechanical, electrical, and others. Also required are machinists and other highly-skilled production workers, technicians, software publishers, and managers at various levels.

The following table shows the top jobs in the industry, making up just over 50 percent of all jobs in Aerospace Manufacturing. Since the growth of the statewide Aerospace Manufacturing industry is relatively flat, the occupational demands are not dramatic. Although logisticians show up to 15 percent growth, the number remains relatively small—64 positions over 10 years (Figure 3.15).

However, looking at the overall occupational demand for those positions across all industries shows a much different picture. Ohio's aging workforce, replacements due to retirement and other churn will impact many occupations, particularly within the manufacturing industry. Regardless of whether manufacturing for the auto or aerospace industry, the higher skilled positions are required in all manufacturing, increasing the demand for those skills (Figure 3.16).

While many of the production occupations do not necessarily require postsecondary training, they do require on-the-job training. Manufacturers across Ohio have expressed difficulty attracting a skilled workforce. These projections suggest that demand will only increase and could potentially have negative effects on Ohio's Aerospace Manufacturing.

Research and development

Ohio's R&D has strong growth opportunity in the aerospace and defense industry. With many research universities across the State, the Air Force Research Laboratory (AFRL), NASA Glenn, and a large concentration of R&D companies doing Small Business

SOC	Description	Employed in Aerospace Industry (2015)	Employed in Aerospace Industry (2023)	Change (2015 - 2023)	% Change (2015 - 2023)	Median Hourly Earnings
17-2011	Aerospace Engineers	1,941	1,958	17	1%	$47.76
17-2112	Industrial Engineers	1,084	1,144	60	6%	$36.03
51-4041	Machinists	888	914	26	3%	$18.55
51-9061	Inspectors, Testers, Sorters, Samplers, and Weighers	844	927	83	10%	$16.93
51-2092	Team Assemblers	686	646	(40)	(6%)	$15.01
51-4011	Computer -Controlled Machine Tool Operators, Metal and Plastic	632	681	49	8%	$17.29
51-2022	Electrical and Electronic Equipment Assemblers	619	609	(10)	(2%)	$13.65
17-2141	Mechanical Engineers	582	564	(18)	(3%)	$34.50
15-1132	Software Developers, Applications	515	506	(9)	(2%)	$39.64
49-3011	Aircraft Mechanics and Service Technicians	486	478	(8)	(2%)	$29.05
13-1023	Purchasing Agents, Except Wholesale, Retail, and Farm Products	436	427	(9)	(2%)	$28.61
13-1081	Logisticians	418	482	64	15%	$33.49
51-1011	First-Line Supervisors of Production and Operating Workers	409	391	(18)	(4%)	$25.63
17-2199	Engineers, All Other	394	381	(13)	(3%)	$39.83
43-5061	Production, Planning, and Expediting Clerks	345	336	(9)	(3%)	$20.04
11-9041	Architectural and Engineering Managers	335	341	6	2%	$56.01
15-1121	Computer Systems Analysts	332	343	11	3%	$37.68

Figure 3.15 Top jobs in the aerospace manufacturing industry in Ohio. (From EMSI, 2015.2—QCEW employees, Non-QCEW employees, and self-employed.)

Innovation Research (SBIR) to support federal agencies, Ohio has a strong R&D foundation, which is projected to grow even stronger during the next decade. Although the R&D concentration is lower than the national average, Ohio ranks 15th in the number of R&D jobs, with a projected growth of 25 percent in the next decade.

Organizations such as Brookings have reported that more than 90 percent of a community's job growth is organic. Less than 10 percent will come from recruiting businesses from other communities. With this being the case, investing in the recruitment of major aerospace manufacturers and even an aircraft manufacturer could be viewed as a flawed strategy. This becomes clearer once one considers the possibility for organic growth to occur from a healthy, growing research and development sector. Investing in the spin out of new ventures from innovations in aerospace is the state's best chance at turning the aerospace manufacturing industry around and into strong growth numbers. This strategy is a long-term strategy as the job numbers would likely follow a logarithmic or exponential curve, meaning initial investments will typically not result in headline-worthy job numbers, but over time would produce greater results than incentive programs for business relocation (Figure 3.17).

Like Aerospace Manufacturing, the Aerospace R&D industry is also more heavily concentrated in Southwest Ohio, especially in the Dayton metropolitan area, which houses

Defense Innovation Handbook

SOC	Description	2015 Jobs	2023 Jobs	2015 - 2023 Change	2015 - 2023 % Change	Annual Openings	Ohio Completions (2013)	Gap
17-2011	Aerospace Engineers	4,479	4,670	191	4%	131	250	119
17-2112	Industrial Engineers	12,286	12,398	112	1%	403	291	(112)
51-4041	Machinists	28,283	29,086	803	3%	796	118	(678)
51-9061	Inspectors, Testers, Sorters, Samplers, and Weighers	27,662	27,943	281	1%	719	65	(654)
51-2092	Team Assemblers	62,688	62,340	(348)	(1%)	1,296	0	(1,296)
51-4011	Computer-Controlled Machine Tool Operators, Metal and Plastic	13,223	14,007	784	6%	484	26	(458)
51-2022	Electrical and Electronic Equipment Assemblers	8,354	7,822	(532)	(6%)	103	20	(83)
17-2141	Mechanical Engineers	12,487	12,725	238	2%	474	1,378	904
15-1132	Software Developers, Applications	27,816	32,126	4,310	15%	931	1,736	805
49-3011	Aircraft Mechanics and Service Technicians	3,231	3,534	303	9%	131	266	136
13-1023	Purchasing Agents, Except Wholesale, Retail, and Farm Products	12,619	12,751	132	1%	259	67	(192)
13-1081	Logisticians	6,031	6,811	780	13%	173	968	795
51-1011	First-Line Supervisors of Production and Operating Workers	28,724	27,866	(858)	(3%)	460	572	112
17-2199	Engineers, All Other	8,579	8,462	(117)	(1%)	163	621	458
43-5061	Production, Planning, and Expediting Clerks	13,592	13,814	222	2%	388	0	(388)
11-9041	Architectural and Engineering Managers	7,458	7,690	232	3%	217	6,470	6,253
15-1121	Computer Systems Analysts	27,453	31,496	4,043	15%	976	2,006	1,030

Figure 3.16 Gap analysis of top aerospace manufacturing jobs across all industries. (From EMSI, 2015.2—QCEW employees, Non-QCEW employees, and self-employed.)

State Name	2015 Jobs	2023 Jobs	2015 - 2023 Change	2015 - 2023 % Change	2023 Location Quotient
California	525,048	644,455	119,407	23%	1.40
Texas	259,765	309,109	49,344	19%	0.91
Virginia	178,311	232,274	53,963	30%	2.24
New York	145,959	171,515	25,556	18%	0.72
Florida	136,609	165,449	28,840	21%	0.76
Washington	130,740	161,961	31,221	24%	1.81
Michigan	129,196	149,260	20,064	16%	1.39
Massachusetts	119,603	142,261	22,658	19%	1.53
Pennsylvania	112,133	135,624	23,491	21%	0.91
Maryland	107,168	131,244	24,076	22%	1.90
Illinois	100,395	116,016	15,621	16%	0.77
Colorado	92,790	112,949	20,159	22%	1.56
New Jersey	92,664	108,499	15,835	17%	1.08
Georgia	86,492	108,246	21,754	25%	0.96
Ohio	83,863	104,944	21,081	25%	0.76
North Carolina	75,334	94,891	19,557	26%	0.82
Minnesota	48,625	56,750	8,125	17%	0.77
Missouri	44,106	51,799	7,693	17%	0.71
Alabama	43,162	54,681	11,519	27%	1.05

Figure 3.17 Top states for research and development jobs. (From EMSI, 2015.2—QCEW employees, Non-QCEW employees, and self-employed.)

MSA Name	2015 Jobs	2023 Jobs	2015 - 2023 Change	2015 - 2023 % Change	2023 Location Quotient
Columbus, OH	21,010	26,960	5,950	28%	1.02
Cincinnati, OH -KY-IN	18,721	23,565	4,844	26%	0.90
Cleveland -Elyria, OH	14,969	18,280	3,311	22%	0.71
Dayton, OH	11,111	13,586	2,475	22%	1.48
Akron, OH	4,172	5,320	1,148	28%	0.65
Toledo, OH	3,608	3,945	337	9%	0.55
Canton -Massillon, OH	1,220	1,498	278	23%	0.35
Huntington -Ashland, WV -KY-OH	1,159	1,242	83	7%	0.37

Figure 3.18 Ohio R&D jobs by MSA. (From EMSI, 2015.2—QCEW employees, Non-QCEW employees, and self-employed.)

AFRL and a strong concentration of small businesses. The Columbus region has the highest number of R&D jobs, followed by Cincinnati (Figure 3.18).

The workforce in Aerospace R&D is dominated by technical professionals. Other than the support occupations, most jobs require postsecondary education in a STEM field. The following jobs account for more than 50 percent of the industry, and almost all of them are expected to grow during the next decade (Figure 3.19).

SOC	Description	Employed in R&D Industry (2015)	Employed in R&D Industry (2023)	Change (2015 - 2023)	% Change (2015 - 2023)	Median Hourly Earnings
15-1132	Software Developers, Applications	6,531	9,075	2,544	39%	$39.64
15-1121	Computer Systems Analysts	4,143	5,537	1,394	34%	$37.68
17-2051	Civil Engineers	3,936	4,856	920	23%	$35.13
17-2141	Mechanical Engineers	2,313	2,670	357	15%	$34.50
15-1151	Computer User Support Specialists	2,138	3,209	1,071	50%	$20.97
13-1111	Management Analysts	1,881	2,355	474	25%	$33.71
11-1021	General and Operations Managers	1,706	2,199	493	29%	$43.21
15-1131	Computer Programmers	1,682	2,222	540	32%	$32.12
41-3099	Sales Representatives, Services, All Other	1,650	2,230	580	35%	$23.06
15-1133	Software Developers, Systems Software	1,643	2,504	861	52%	$41.78
11-3021	Computer and Information Systems Managers	1,559	2,106	547	35%	$54.31
43-6014	Secretaries and Administrative Assistants, Except Legal, Medical, and Executive	1,558	1,957	399	26%	$15.40
41-4011	Sales Representatives, Wholesale and Manufacturing, Technical and Scientific Products	1,546	1,974	428	28%	$32.63
11-9041	Architectural and Engineering Managers	1,530	1,787	257	17%	$56.01
43-9061	Office Clerks, General	1,507	1,824	317	21%	$13.47
15-1142	Network and Computer Systems Administrators	1,410	1,864	454	32%	$32.87
17-2199	Engineers, All Other	1,311	1,441	130	10%	$39.83
17-2071	Electrical Engineers	1,192	1,436	244	20%	$37.01
13-1161	Market Research Analysts and Marketing Specialists	1,184	1,666	482	41%	$27.68
17-3011	Architectural and Civil Drafters	1,162	1,155	(7)	(1%)	$22.58
17-2011	Aerospace Engineers	1,160	1,382	222	19%	$47.76

Figure 3.19 Top jobs in the R&D industry in Ohio. (From EMSI, 2015.2—QCEW employees, Non-QCEW employees, and self-employed.)

SOC	Description	2015 Jobs	2023 Jobs	2015-2023 Change	2015-2023 % Change	Annual Openings	Ohio Completions (2013)	Gap
15-1132	Software Developers, Applications	27,816	32,126	4,310	15%	931	1,736	805
15-1121	Computer Systems Analysts	27,453	31,496	4,043	15%	976	2,006	1,030
17-2051	Civil Engineers	7,943	9,081	1,138	14%	355	568	213
17-2141	Mechanical Engineers	12,487	12,725	238	2%	474	1,378	904
15-1151	Computer User Support Specialists	19,782	22,784	3,002	15%	716	2,178	1,462
13-1111	Management Analysts	24,314	27,238	2,924	12%	769	9,660	8,891
11-1021	General and Operations Managers	67,605	73,152	5,547	8%	2,016	10,605	8,589
15-1131	Computer Programmers	7,475	8,339	864	12%	324	1,381	1,057
41-3099	Sales Representatives, Services, All Other	33,828	36,695	2,867	8%	1,327	153	(1,174)
15-1133	Software Developers, Systems Software	6,769	8,511	1,742	26%	320	1,979	1,659
11-3021	Computer and Information Systems Managers	12,500	13,886	1,386	11%	358	2,348	1,990
43-6014	Secretaries and Administrative Assistants, Except Legal, Medical, and Executive	88,994	95,960	6,966	8%	2,003	388	(1,615)
41-4011	Sales Representatives, Wholesaleand Manufacturing, Technical and Scientific Products	22,278	23,061	783	4%	548	132	(416)
11-9041	Architectural and Engineering Managers	7,458	7,690	232	3%	217	6,470	6,253
43-9061	Office Clerks, General	113,291	117,772	4,481	4%	2,988	70	(2,918)
15-1142	Network and Computer Systems Administrators	14,053	15,096	1,043	7%	365	755	390
17-2199	Engineers, All Other	8,579	8,462	(117)	(1%)	163	621	458
17-2071	Electrical Engineers	5,159	5,375	216	4%	142	936	794
13-1161	Market Research Analysts and Marketing Specialists	20,775	24,664	3,889	19%	801	2,600	1,799
17-3011	Architectural and Civil Drafters	3,407	3,363	(44)	(1%)	53	256	203
17-2011	Aerospace Engineers	4,479	4,670	191	4%	131	250	119

Figure 3.20 Gap analysis of top R&D jobs across all industries. (From EMSI, 2015.2—QCEW Employees, Non-QCEW Employees, and Self-Employed.)

With the growth in the R&D industry across occupations, Ohio will have increasing workforce demands in these STEM fields. Currently, regional completions in STEM fields are keeping pace with the workforce demand. The following table shows the R&D occupations across all industries (Figure 3.20).

Research expenditures

Understanding the workforce demand in the R&D cluster is only one critical factor to determining the health and potential growth in the state. Commercialization and entrepreneurship success cannot be measured using traditional metrics, like jobs.

Ohio agencies have consistently led the state in R&D expenditures, and this trend is expected to continue. As some R&D powerhouse states like California have seen a decrease in expenditures, Ohio's have continued to increase (Figure 3.21).

If this trend continues, Ohio has the potential to close the gap with California. As this process occurs, however, strategies must be put into place to leverage the innovations

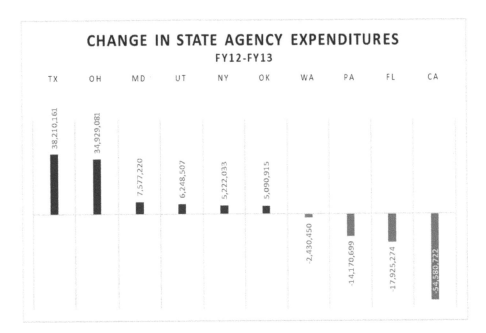

Figure 3.21 Change in R&D expenditures by state. (From National Science Foundation, National Center for Science and Engineering Statistics, Survey of State Government Research and Development, FYs 2012 and 2013.)

resulting from these expenditures. R&D must be converted to commercial application if the state wishes to convert these investments into real economic growth. This could be a bigger challenge for Ohio than any of the states listed due to the source of these expenditures. Ohio has a disproportional emphasis on companies and individuals contracted to conduct this R&D compared to the top 10 states (Figure 3.22).

In line with R&D expenditures, and with federal organizations like Air Force Research Laboratory and NASA Glenn investing heavily in R&D, an important metric to monitor is Federal obligations. Ohio performs well in this category (in the top 15 states) and could be positioned to improve its federal contract wins (Figure 3.23).

Federal/military sector

Ohio ranks 14th in the US for the number of civilian and military jobs. Like most of the other states employing large numbers of civilian and military workers, Ohio's federal jobs are projected to decrease. This aligns with the decrease of federal budgets and expenditures (Figure 3.24).

Understanding occupational demand for federal jobs in Ohio is challenging, since 40 percent of the staffing pattern is made up of military positions which are not disclosed in labor market data. However, of the remaining 60 percent (civilian positions), most of the in-demand positions are declining (Figure 3.25).

The federal civilian occupations are also in-demand in other industries, so a view of occupational demand across all industries is a better view of where workforce shortages may exist. The following table also shows gaps in degree completions related to the specific occupation (Figure 3.26).

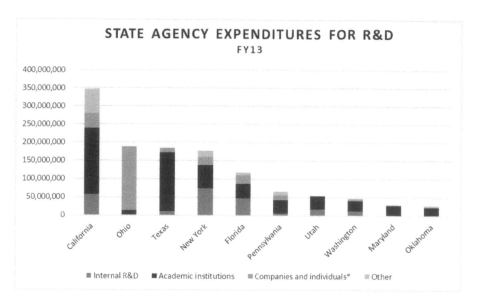

Figure 3.22 Top 10 states in R&D expenditures by source. (From National Science Foundation, National Center for Science and Engineering Statistics, Survey of State Government Research and Development, FYs 2012 and 2013. *Companies and individuals includes individuals under contract for research projects.)

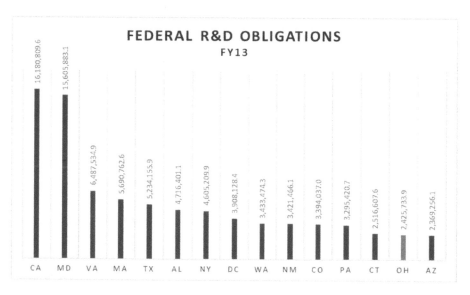

Figure 3.23 Federal R&D obligations by state. (From National Science Foundation, National Center for Science and Engineering Statistics, Survey of Federal Funds for Research and Development, FYs 2013–15.)

State Name	2015 Jobs	2023 Jobs	2015 - 2023 Change	2015 - 2023 % Change	2023 Location Quotient
California	397,567	373,641	(23,926)	(6%)	0.74
Texas	336,400	352,588	16,188	5%	0.95
Virginia	314,568	308,413	(6,155)	(2%)	2.70
District of Columbia	212,199	199,147	(13,052)	(6%)	9.34
Maryland	209,997	208,574	(1,423)	(1%)	2.74
Florida	201,604	201,496	(108)	(0%)	0.84
North Carolina	190,401	191,040	639	0%	1.49
Georgia	182,579	175,955	(6,624)	(4%)	1.42
Washington	142,833	138,811	(4,022)	(3%)	1.41
New York	133,781	129,409	(4,372)	(3%)	0.50
Pennsylvania	105,440	103,313	(2,127)	(2%)	0.63
Colorado	96,605	95,832	(773)	(1%)	1.21
Illinois	95,768	95,306	(462)	(0%)	0.57
Ohio	88,914	85,125	(3,789)	(4%)	0.56
Hawaii	87,342	85,827	(1,515)	(2%)	4.30

Figure 3.24 Top states for federal/military jobs. (From EMSI, 2015.2—QCEW employees, Non-QCEW employees, and self-employed.)

Aviation

As the aviation industry is considered a support industry that is driven by consumer demand, it's an industry that should be monitored. It is an industry that impacts the workforce of other industries, but it is not a driver industry for the state's aerospace and aviation networks (Figures 3.27 and 3.28).

Unmanned aircraft systems

The Unmanned Aircraft Systems (UAS) industry segment is still undefined. The primary organization used to benchmark data regarding this segment is the Association for Unmanned Vehicle Systems International (2013). AUVSI released an economic impact report in March 2013 entitled "The Economic Impact of Unmanned Aircraft Systems Integration in the United States." In this report, they applied an aircraft manufacturing staffing pattern to attempt to quantify the impact of UAS integration into the National Air Space (NAS) and used current aerospace activity and infrastructure to project job growth. Using this staffing pattern, they project job growth because of UAS usage for public safety and precision agriculture. One major concern with the use of this staffing pattern is that the UAS that will likely be produced will be significantly smaller than commercial aircraft and will be heavily reliant on sensors and other electronic components unique to the systems. This means greater attention on information technology and light weight polymers would be a good strategy verses expecting existing aircraft engine and parts manufacturers to expand their networks naturally. Ohio performs well using this staffing pattern because of the state's aircraft business; however, it's unlikely the scale and complexity of the engines produced in Ohio would be applicable.

SOC	Description	Employed in Fed/Mil Industry (2015)	Employed in Fed/Mil Industry (2023)	Change (2015 - 2023)	% Change (2015 - 2023)	Median Hourly Earnings
55-9999	Military occupations	35,482	35,353	(129)	(0%)	$18.34
13-1199	Business Operations Specialists, All Other	3,837	3,551	(286)	(7%)	$29.82
29-1141	Registered Nurses	2,307	2,120	(187)	(8%)	$29.24
15-1199	Computer Occupations, All Other	1,916	1,768	(148)	(8%)	$37.75
17-2199	Engineers, All Other	1,534	1,402	(132)	(9%)	$39.83
11-9199	Managers, All Other	1,438	1,338	(100)	(7%)	$31.74
13-1111	Management Analysts	1,437	1,326	(111)	(8%)	$33.71
29-1069	Physicians and Surgeons, All Other	1,281	1,169	(112)	(9%)	$88.50
43-4199	Information and Record Clerks, All Other	1,140	999	(141)	(12%)	$18.40
13-1041	Compliance Officers	1,134	1,057	(77)	(7%)	$27.70
13-1023	Purchasing Agents, Except Wholesale, Retail, and Farm Products	1,130	1,036	(94)	(8%)	$28.61
13-1031	Claims Adjusters, Examiners, and Investigators	994	921	(73)	(7%)	$29.42
13-1081	Logisticians	979	1,003	24	2%	$33.49
43-4061	Eligibility Interviewers, Government Programs	940	940	0	0%	$19.34
19-3099	Social Scientists and Related Workers, All Other	798	728	(70)	(9%)	$34.04
53-2021	Air Traffic Controllers	785	779	(6)	(1%)	$64.08
33-9093	Transportation Security Screeners	777	817	40	5%	$17.93
13-2099	Financial Specialists, All Other	776	708	(68)	(9%)	$33.67
13-2081	Tax Examiners and Collectors, and Revenue Agents	739	681	(58)	(8%)	$25.46
33-3021	Detectives and Criminal Investigators	672	678	6	1%	$29.95
11-1021	General and Operations Managers	653	609	(44)	(7%)	$43.21
13-1071	Human Resources Specialists	650	575	(75)	(12%)	$26.04
17-2011	Aerospace Engineers	645	582	(63)	(10%)	$47.76
23-1011	Lawyers	634	626	(8)	(1%)	$45.26
17-2072	Electronics Engineers, Except Computer	595	544	(51)	(9%)	$42.95
17-3029	Engineering Technicians, Except Drafters, All Other	593	546	(47)	(8%)	$27.60
29-2061	Licensed Practical and Licensed Vocational Nurses	586	539	(47)	(8%)	$19.46
21-1029	Social Workers, All Other	554	508	(46)	(8%)	$21.31
13-2011	Accountants and Auditors	540	503	(37)	(7%)	$29.24
43-9061	Office Clerks, General	512	460	(52)	(10%)	$13.47

Figure 3.25 Top jobs in the federal/military industry in Ohio. (From EMSI, 2015.2—QCEW Employees, Non-QCEW Employees, and Self-employed.)

One of the more insightful statements supplied in the report is that "states that create favorable regulatory and business environments for the industry and the technology will likely siphon jobs away from states that do not." While the data supplied in the report places Ohio among the top 15 states poised to grow because of UAS integration into the NAS, these are merely projections based on models and the state can

SOC	Description	2015 Jobs	2023 Jobs	2015 - 2023 Change	2015 - 2023 % Change	Annual Openings	Ohio Completions (2013)	Gap
13-1199	Business Operations Specialists, All Other	28,511	30,440	1,929	7%	647	130	(517)
29-1141	Registered Nurses	128,873	145,989	17,116	13%	4,840	13,662	8,822
15-1199	Computer Occupations, All Other	7,051	7,430	379	5%	169	1,254	1,085
17-2199	Engineers, All Other	8,579	8,462	(117)	(1%)	163	621	458
11-9199	Managers, All Other	21,275	21,717	442	2%	561	11,319	10,758
13-1111	Management Analysts	24,314	27,238	2,924	12%	769	9,660	8,891
29-1069	Physicians and Surgeons, All Other	16,638	18,186	1,548	9%	663	1,407	744
43-4199	Information and Record Clerks, All Other	4,664	4,647	(17)	(0%)	105	70	(35)
13-1041	Compliance Officers	6,784	7,219	435	6%	184	14	(170)
13-1023	Purchasing Agents, Except Wholesale, Retail, and Farm Products	12,619	12,751	132	1%	259	67	(192)
13-1031	Claims Adjusters, Examiners, and Investigators	10,517	11,390	873	8%	364	22	(342)
13-1081	Logisticians	6,031	6,811	780	13%	173	968	795
43-4061	Eligibility Interviewers, Government Programs	5,368	5,568	200	4%	135	66	(69)
19-3099	Social Scientists and Related Workers, All Other	1,452	1,435	(17)	(1%)	25	249	224
53-2021	Air Traffic Controllers	854	857	3	0%	41	7	(34)
33-9093	Transportation Security Screeners	800	842	42	5%	24	0	(24)
13-2099	Financial Specialists, All Other	4,835	4,921	86	2%	63	2,097	2,034
13-2081	Tax Examiners and Collectors, and Revenue Agents	1,758	1,704	(54)	(3%)	60	3,370	3,311
33-3021	Detectives and Criminal Investigators	2,159	2,176	17	1%	50	976	926
11-1021	General and Operations Managers	67,605	73,152	5,547	8%	2,016	10,605	8,589
13-1071	Human Resources Specialists	18,115	19,278	1,163	6%	488	1,128	641
17-2011	Aerospace Engineers	4,479	4,670	191	4%	131	250	119
23-1011	Lawyers	23,395	24,419	1,024	4%	510	1,699	1,189
17-2072	Electronics Engineers, Except Computer	4,142	4,198	56	1%	102	936	834
17-3029	Engineering Technicians, Except Drafters, All Other	4,087	4,084	(3)	(0%)	89	620	532
29-2061	Licensed Practical and Licensed Vocational Nurses	39,627	45,392	5,765	15%	1,773	3,506	1,733
21-1029	Social Workers, All Other	3,290	3,424	134	4%	88	2,114	2,026
13-2011	Accountants and Auditors	42,281	44,984	2,703	6%	1,698	3,530	1,832
43-9061	Office Clerks, General	113,291	117,772	4,481	4%	2,988	70	(2,918)

Figure 3.26 Gap analysis of top Ohio federal/military jobs across all industries. (From EMSI, 2015.2—QCEW employees, Non-QCEW employees, and self-employed.)

State Name	2015 Jobs	2023 Jobs	2015 - 2023 Change	2015 - 2023 % Change	2023 Location Quotient
Texas	82,344	89,687	7,343	9%	1.50
California	70,849	74,040	3,191	5%	0.91
Florida	57,549	59,644	2,095	4%	1.54
Georgia	46,823	47,761	938	2%	2.40
Illinois	40,804	43,150	2,346	6%	1.60
New York	39,761	39,771	10	0%	0.94
Virginia	20,512	20,593	81	0%	1.12
New Jersey	20,411	22,031	1,620	8%	1.23
Arizona	20,061	20,952	891	4%	1.60
North Carolina	19,729	20,361	632	3%	0.99
Colorado	18,567	21,081	2,514	14%	1.65
Washington	17,031	18,844	1,813	11%	1.19
Michigan	16,346	16,792	446	3%	0.88
Pennsylvania	15,774	14,665	(1,109)	(7%)	0.55
Minnesota	14,700	13,188	(1,512)	(10%)	1.01
Ohio	13,906	14,963	1,057	8%	0.61

Figure 3.27 Top states for aviation jobs. (From EMSI, 2015.2—QCEW Employees, Non-QCEW Employees, and Self-Employed.)

SOC	Description	Employed in Aviation Industry (2015)	Employed in Aviation Industry (2023)	Change (2015 - 2023)	% Change (2015 - 2023)	% of Total Jobs in Industry Group (2015)
53-2012	Commercial Pilots	2,498	3,148	650	26%	18.0%
49-3011	Aircraft Mechanics and Service Technicians	1,785	2,081	296	17%	12.8%
11-1021	General and Operations Managers	191	230	39	20%	1.4%
49-1011	First-Line Supervisors of Mechanics, Installers, and Repairers	226	258	32	14%	1.6%
39-6011	Baggage Porters and Bellhops	143	175	32	22%	1.0%
53-7061	Cleaners of Vehicles and Equipment	177	208	31	18%	1.3%
49-2091	Avionics Technicians	241	271	30	12%	1.7%

Figure 3.28 Top jobs in the aviation industry in Ohio. (From EMSI, 2015.2—QCEW Employees, Non-QCEW Employees, and Self-Employed.)

rise or fall based on their strategies to capture shares of the manufacturing market. Many of these jobs the UAS will perform may be new jobs that were impossible to accomplish without an unmanned system or these jobs could lead to the loss of jobs the UAS purpose would replace. Workforce strategies should be developed to train the workforce already in public safety and precision agriculture to adapt to this disruptive technology (Figures 3.29 and 3.30).

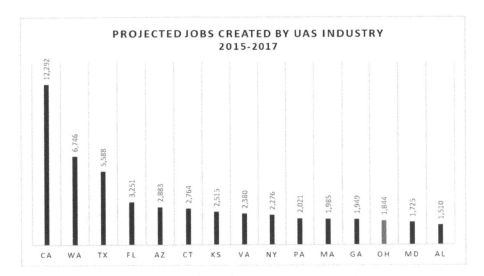

Figure 3.29 Projected jobs created by UAS industry 2015–2017. (From AUVSI, The Economic Impact of Unmanned Aircraft Systems Integration in the United States, March 2013.)

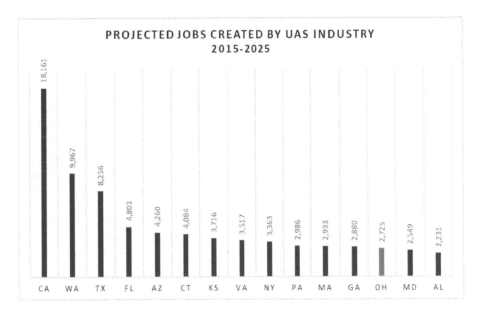

Figure 3.30 Projected jobs created by UAS industry 2015–2025. (From AUVSI, The Economic Impact of Unmanned Aircraft Systems Integration in the United States, March 2013.)

Building a workforce development ecosystem

The chapter thus far outlined an in-demand industry for a region of the country. In addition, the chapter specifically details how a region of the country could analyze an industry and study the demand and supply of its workforce. This type of analysis is the critical first step and ongoing need in building a robust ecosystem to support an industry. An ecosystem in a region includes many different entities. To build a robust ecosystem in support of an industry, it is important to have all stakeholders around the table. For example, the aerospace and defense industry ecosystem would include the following stakeholders:

1. Federal government
2. State government
3. Higher Education
4. Primary & Secondary Education
5. Industry
6. Non-profits
7. Professional Organizations

The Ecosystem should appoint a leader and develop a strategic focus and vision to be most effective. The goal is to work together in a collaborative way, for the betterment of all in the industry across the region. This initiative is more achievable with collaborative agencies sitting at the table and with incentives layered into the ecosystem through the Federal or State government. A healthy workforce ecosystem can have three states: meeting current industry demand (Steady State), anticipating future industry demand (Predictive State), or strategically driving industry growth (Driver State). Each state has common elements: investors, producers, consumers, and maintainers; however, the groups that comprise each element can be different for each state.

The lowest energy state is the Steady State focused on meeting the current demand, although this still requires healthy collaborations. In the Steady State, investments from government agencies and industry are directed by the source of investment to meet the greatest, current gap in workforce supply. Education and local unemployment providers serve as producers to direct the workforce into training programs that will meet the demand. Industry works as consumers, hiring and recruiting talent and ensuring on-the-job training is available to retain the talent. Non-profits serve as maintainers, ensuring the investors and producers are responding to changes in the ecosystem as changes naturally occur. This model ties overall economic development growth to the same growth rate as the existing industry. This can be effective if the existing industry is projected to grow.

The community should always maintain a healthy Steady State, but also develop a strategic vision. This can take two forms: anticipating future industry growth or driving growth. Both states require a robust network of implementers and funding sources with the flexibility to respond to opportunities. Both states represent risk to investors but with the participation of the correct collaborators and the availability of funding with some risk tolerance, these states can have the highest return on investment. The most important consumers are existing businesses, but the source of these consumers is different between the Predictive State and the Driver State.

In a Predictive State, the network is anticipating future growth. A strong economic development arm, producers, is needed with the ability to understand national and global industry trends, strong industry partnerships, and the ability to convene both producers and consumers for proactive activity. Critical to this system is engagement of the growing industries, the consumers. They may not be the industries with the strongest representation or the most capital to invest. In this state, the growing industries are not the investors; rather, they are those companies and entrepreneurs connected to the larger network (local, state, national of global networks) of future growth areas. If a critical mass of entrepreneurs can be organized, venture capital becomes the most important currency in this ecosystem, supplemented by the established industries which would benefit the most by the future growth. To ensure government investment, which is inherently risk adverse, is used with the largest impact, the workforce which would be the most impacted by a shift in industry focus should be identified and training programs should be developed by the providers in anticipation of the entrepreneurs' demand, either through direct jobs or indirect jobs. It's important this workforce investment does not occur too early in the process, training workers for jobs which are not yet available will lead to migration out of the community. The most important consumers in the Predictive State are start-ups and the maintainers are those non-profits providing entrepreneurial services.

The Driver State is the ecosystem state with the most energy required, longest timeframe, and highest risk. This model became popular amongst economic development groups employing industry cluster strategies. To be successful the following is required: flexible capital, information and data-driven strategies, a realistic vision of the time required to reshape the industry sectors, a high tolerance of risk, selective participation of stakeholders, responsive higher education, and industry expertise to aid in strategic planning. Huntsville, Alabama is a good example of this ecosystem being successful. The region committed to biotechnology growth and STEM education decades before they witnessed the benefits of those efforts. The Huntsville MSA was named the fastest growing tech hub of 2017 in a study published by ZipRecruiter with a tech job growth rate of 309 percent. In this ecosystem network, public-private partnerships were critical for long-term growth. This long-term growth has taken more than 60 years to realize. The Driver State is a marathon, not a sprint. The strategic plan starts with elementary school programming. In Hunstville, for example, second graders learn to code, and this focus on future skills continues throughout the education system, growing a workforce that will be capable of shaping the targeted industries. Producers in this state become the entire education continuum. Consumers include not just driver industries, but also infrastructure designed to shape the quality of life needed to retain talent. The maintainers are the most important members of this ecosystem. These are regional organizations which have cultivated the networks with the most influence in the community, but also have circles of influence outside the community. This could be the local government, military installations, collaboratives, or strong industry pillars.

For Southwest Ohio, the aerospace and defense industry may be an attractive target for the Driver State model. Like Huntsville, the federal installation R&D has the potential to produce the technology of the next century, but it must be harnessed today. Many attempts have been made to tap into the intellectual capital in AFRL represented by thousands of unlicensed patents. Continued efforts to break into this nearly pristine space are needed, and perhaps the best option for the region is to focus on a Predictive State, growing the entrepreneur space, in anticipation of a successful Driver State strategy.

To do this, leveraging the growth in the R&D sector will require a shift in business practices, using network models that enable exponential growth. For a network model to be successful, all nodes within the network must both give and receive value. From the organizations conducting research all the way to the customer who will buy a commercial product, all participants in the commercialization process must create value.

With respect to the potential loss of STEM graduates produced in Ohio, but not hired in Ohio, the best strategy to retain those workers is to connect them to nodes in a network for them that will be valuable before they graduate. In a recent report, a large IT company reported 80 percent of its US workforce was white. This stark data point speaks more about

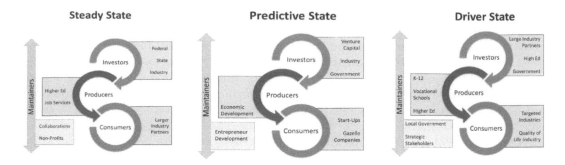

Figure 3.31 Graphics of steady state, predictive state, and driver state.

networks than it does to inclusion. Companies will hire from programs they're familiar with and students they know are up for the task. As a summarizing view, Figure 3.31 illustrates the elements of steady state, predictive state, and driver state that could serve as anchors for innovation and local business development.

References

Aerospace Industries Association, December 2016, *The State of the US Aerospace and Defense Industry*. Retrieved from http://www.aia-aerospace.org/wp-content/uploads/2016/12/AIA_StateOfIndusrtyReport_2016_V8.pdf.

Association of Unmanned Vehicle Systems International, 2013. *The Economic Impact of Unmanned Aircraft Systems Integration in the United States*. Retrieved from http://www.auvsi.org/our-impact/economic-report.

Boeing, 2016. *Aerospace Services Market Outlook*. Retrieved from https://www.boeing.com/resources/boeingdotcom/commercial/market/services-market-outlook/assets/downloads/2017-aerospace-smo.pdf.

Bureau of Labor Statistics, 2017. Multiple databases. Retrieved from https://www.bls.gov/.

Economic Modeling Specialists Incorporated, 2017. Multiple databases. Retrieved from http://www.economicmodeling.com/.

National Science Foundation Science and Engineering Indicators State Data Tool, 2017. Multiple databases. Retrieved from https://www.nsf.gov/statistics/2018/nsb20181/.

US Census Bureau, 2017, American Community Survey. Retrieved from https://www.census.gov/.

Ziprecruiter, 2017. *Fastest Growing Tech Towns in 2017*. Retrieved from https://www.ziprecruiter.com/blog/fastest-growing-tech-towns-in-2017/.

chapter four

Other transactions

Increasing importance in the Department of Defense

Sally J. F. Baron

Contents

> "It is in the DOD's interest to tap into the research and development being accomplished by non-traditional defense contractors, and to pursue commercial solutions to defense requirements."
>
> – Department of Defense, *Other Transactions Guide for Prototype Projects*, Version 1.2.0, January 2017, page 3

The other transaction

> "An OT is best defined by what it is not. An OT is a transaction (other than a contract, grant, or cooperative agreement) to which most of the laws and regulations governing federal contracts – including the FAR – do not apply."
>
> – DIUx Commercial Solutions Opening How-To Guide, November 2016, page 2

Background

Sometimes desperate times indeed call for desperate measures. In the late 1950s the United States watched in horror as the Soviet Union reached into space—and not likely for peaceful purposes. Just two years before that, Nikita Khrushchev was quoted with his infamous *"My vas pokhoronim!"* translated: ***"We will bury you!"*** (1956). He later clarified: *"… we* *will take a shovel, dig a deep grave and bury colonialism as deep as we can."* A short year later, the USSR had tested its first intercontinental ballistic missile (ICBM) on August 26, 1957, and shortly after, Sputnik was successfully launched into orbit on October 4, 1957. The *space race* was on. On April 20, 1961, the USSR once again scored a first against the United States: Cosmonaut Yuri Gagarin became the first human to orbit the Earth. Volley to the US: Astronaut John Glenn orbited not quite a year later on February 20, 1962 in *Friendship 7*. These *firsts*, achieved by the then Soviet Union did not go over well in the United States. Given the Soviet stated goal of global superiority Americans were nervous they would achieve it.

Khrushchev was still in office and his threats loomed large. He promised the USSR would bury colonialism, and he had the tools to do it—literally. The United States was still behind, scared and desperate. No time for government regulations or piles of paperwork to get in the way of progress. The US took action to ensure this new era of space exploration would not be interrupted. The action was to streamline the regulations that would slow down our entry to space. Representative Sam Rayburn and Vice President Richard Nixon signed the *National Aeronautics and Space Act of 1958* (NASA, 1958). Within this act, Congress clearly stated that space would be used for peaceful purposes, and perhaps more importantly, that newly created NASA now had within its power the authority to get to space with whatever means and using whatever types of contracting it deemed most efficient and effective. We could not afford to lose this one for whatever reason. Note the wording of this act as follows:

> *"to acquire (by purchase, lease, condemnation,* **or otherwise***), construct, improve, repair, operate, and maintain laboratories, research and testing sites and facilities, aeronautical and space vehicles, quarters and related accommodations for employees and dependents of employees of the Administration, and such other real and personal property (including patents), or* **any interest therein, as the Administration deems necessary within** *and outside the continental United States; to lease to others such real and personal property; to sell and otherwise dispose of real and personal property (including patents and rights thereunder) in accordance with the provisions of the Federal Property and Administrative Service Act of 1949, as amended (40 U.S.C. 471 et seq.); and to provide by contract or otherwise for cafeterias and other necessary facilities for the welfare of employees of the Administration at its installations and purchase and maintain equipment therefor;*
>
> *(4)* **to accept unconditional gifts or donations of services, money, or property, real, personal, or mixed, tangible or intangible;**
>
> *(5) without regard to section 3648 of the Revised Statutes, as amended (31 U.S.C. 529), to enter into and perform such contracts, leases, cooperative agreements,* **or other transactions as may be necessary in the conduct of its work and on such terms as it may deem appropriate,** *with any agency or instrumentality of the United States, or with any State, Territory, or possession, or with any political subdivision thereof, or with any person, firm, association, corporation, educational institution. To the maximum extent practicable and consistent with the accomplishment of the purpose of this Act, such contracts, leases, agreements, and other transactions shall be allocated by the Administrator in a manner which will enable small-business concerns to participate equitably and proportionately in the conduct of the work of the Administration;"* (NASA, 1958) (**Emphasis added.**)

This probably does not sound much like the government wording to which most of us are accustomed, but it illustrates that when the US government understands that time is of the essence it is willing to do away with its own laws and regulations. That's the sign of an attentive and flexible government. We are a nation of immigrants who value freedom and have proven historically we will fight hard to protect it. If that means circumventing our own regulations, then that's what we'll do. The need to move fast was at the forefront, and having excellent engineers, test pilots and acquisition professionals who could make sound judgements was key.

The US responded huge. Following the USSR Sputnik (1957), the US successfully orbited Explorer I in 1958; and Gagarin's orbit (1961) was followed by Glenn in 1962. President Kennedy took office in January 1961 and was not ready to let the USSR have the next *first*. On May 25, 1961 he spoke to Congress asking them to appropriate funds for a human trip to the Moon: "I believe that this Nation should commit itself to achieving the goal, before this decade is out, of landing a man on the Moon and returning him safely to Earth." Not coincidentally, this speech was a month after Cosmonaut Yuri Gagarin orbited the Earth. His speech was serious and he regretted asking the taxpayers for more money, but he was dedicated to this task. On September 12, 1962 he re-dedicated his administration to this task with a now famous speech at Rice University, Texas. It was about four months after Glenn orbited the Earth and Kennedy's speech took on a decidedly more uplifting tone as he denounced the hostile use of space, and asserted that space exploration would go ahead with or without the United States, and the presence of the US in the space domain would help ensure space was used for peace. "Whether it will become a force for good or ill depends on man, and only if the United States occupies a position of pre-eminence can we help decide whether this new ocean will be a sea of peace or a new terrifying theater of war." *Note*: In the 1960s using the term *man* as gender-neutral was acceptable, but today, it is not. *Human* (not *man* or *manned*) mission and *humankind* (not *mankind*) are accepted gender-neutral terms. Indeed, his demeanor was much more positive: "I believe we can do it." The US was making great strides technologically and there was good reason for optimism. We would not be buried by anyone.

Figure 4.1 President John F. Kennedy at Rice University. After Astronaut Glenn successfully orbited the Earth, Kennedy had reason to be optimistic. September 12, 1962. (Courtesy of Public domain.)

As we all know, the US lunar program had a remarkable and happy ending. Astronauts Neil Armstrong and Buzz Aldrin stepped foot on the Moon on July 20, 1969. Even after Kennedy's tragic assassination on November 22, 1963, the nation stayed committed to this goal. Ten missions were planned, but political pressure forced President Nixon to cut the last three. No human has walked on the Moon since Apollo 17 in December 1972. US actions to get there were decided, dedicated and swift. Using the OT as a tool to acquire technologies that were largely feasibility studies was critical to this success, and the government rightly and intentionally avoided unnecessary regulation. Without the use of the OT, it is not likely that this goal would have been achieved by the end of the decade, if at all. After the last Moon landing in 1972, and then the Skylab program, the space race was cooling off. During that time, however, the Cold War was heating up and an unprecedented defense build-up would mark the 1980s. With this fast ramp-up of defense spending, DOD was investigating improved acquisition methods, but it was not until 1989 that OTs were approved for use in the DOD. (See Figure 4.1.) Acquisition in the 1980s is discussed next.

Acquisition in the 1980s

The Cold War reached its apex during the 1980s and under President Reagan the defense build-up stood in stark contrast to the previous decade. As we have seen historically, where there is money, there is often mismanagement and sometimes malfeasance. In 1985, reports of defense contractors charging $400 for a hammer and $600 for a toilet seat began investigations on mismanagement. While there were clearly issues with contractors, what the government discovered was that there were also issues with the *process* required to bring a product to fruition. Marching a defense contractor, or any company, through the massive bureaucracy, extensive accounting and oversight was costly and slowed the process. In contracting, time is money, so costs skyrocketed. The president instructed David Packard, co-founder of Hewlett Packard and former secretary of defense, to head a blue-ribbon committee and examine the problems with defense acquisition and procurement. Perhaps one of the most important outcomes of that report was getting Dr. William Perry, a member of the panel, re-engaged with defense acquisition. Dr. Perry was appointed secretary of defense in 1994 and set out to streamline procurement. His first action, the now famous "Perry Memo," was fundamentally a directive for acquisition professionals to better leverage the commercial market. Perry asserted that a government system, by nature, could never achieve the efficiencies found in the commercial market, and with technologies such as commercial computing, both hardware and software going non-linear, there was every reason for defense to adopt superior technologies from the public sector (Carter and Perry, 1999).

By 1989, the DOD was authorized to use OTs for *prototyping*. It was a step in the right direction, but it was limited both in scope and budget. By 2002, the OSD (Office the Secretary of Defense) released *"Other Transactions" (OT) Guide for Prototype Projects*. (Ref: Undersecretary of Defense for Acquisition, Technology and Logistics, *"Other Transactions" (OT) Guide for Prototype Projects*, August 2002) This guide capped the use of OTs to $5M and again, only for prototype products. In 2009, the Undersecretary of Defense commissioned a report enunciating the rationale for the DOD to leverage the commercial market and additionally affirmed that the use of OTs would be essential to optimize the efficiency of commercial products. (Undersecretary of Defense, 2009) but the practice was still limited to prototyping. It is important for the acquisition professional to consider the meaning of prototype. Was the Saturn V booster a prototype? Indeed it was. It was the workhorse of the

Apollo era; there was nothing like it before that and has been nothing like it since. Was Space Transportation System (STS), better known as the Space Shuttle, a prototype? It was a feasibility study that never met its original specifications, nor did it meet operational status. It was a failed attempt at fast, cheap access to space, and also a prototype (Report of Columbia Accident Investigation Board, 2003).

Beyond the 1980s and into the next century, commercial technology was undergoing a revolution. Computing not only became common in every workplace, but moved into the home as well. Mr. John Neer eventually won the fight to get satellite imagery into the commercial sector as well, and the public grabbed on to these new technologies that once existed only behind locked doors. Companies popped up everywhere and by the twenty-first century, citizens were able to map locations on the Internet to see where they were going and what would be the fastest routes. The private sector was accelerating to velocities never before seen and the public sector took notice. How could the DOD leverage these amazing commercial technologies? The Perry Memo, and congressional action that followed, mandated that when feasible, the DOD must purchase products off the shelf rather than build them. The DOD needed a tool other than the FAR to buy commercial and the OT would fill that gap. The marriage of the OT and COTS (commercial-off-the-shelf) has been essential to leveraging the commercial market. A 2002 RAND report assessed the use of COTS products and OTs as ultimately beneficial to DOD: "Two consistent themes in the Department of Defense's acquisition reform over the last several decades have been: streamlining the process by reducing the burden caused by regulations and oversight procedures and adopting commercial practices and products. One reason to adopt commercial practices was to broaden the direct participation of commercial industry to DOD projects." (Ref: Smith, Giles, Jeffrey Drezner, Irving Lachow, *Assessing the Use of Other Transactions Authority for Prototype Projects,* RAND, National Defense Research Institute, 2002. https://www.rand.org/pubs/documented_briefings/DB375.readonline.html). More than that, the report concludes that this new system has brought a broader range of products to DOD *faster* and *cheaper.* It relies on the procurement officer to negotiate an appropriate agreement with the provider, rather than crawling through mounds of regulations that may not be relevant or helpful.

A recent article (Maucione, 2017) calls OTs *new* and *cool.* For those who are making good use of this "new" authority, they *are* new and cool and most importantly *fast.* According to Lt Gen Jay Raymond, the Air Force has lost its ability to go fast, and we need it back. The OT is a tangible step that will help get us there, if utilized appropriately. Former Assistant Secretary of the Air Force for Acquisition the Honorable Sue C. Payton said, "The process is intended to be free of the FAR, but the acquisition profession is still expected to apply best business practices," during our interview. She continued, "We are not training OT officers. We *don't want* to train OT officers. We would rather have officers with good judgement who have an excellent understanding of the market, the needs of the warfighter, and technical expertise. Then they need to go figure out what to buy and how to buy it. Moreover, we know that cranking any product through the *horse blanket* (a colloquialism for the FAR process) will increase the cost and time to field multifold. We cannot afford to do this; either from a cost or strategic perspective."

Moving the focus away from the process (how we produce) and into the product (what we deliver) is vital. Many Air Force acquisition professionals have been frustrated for years watching their efforts go through years of reviews and processes that add nothing to the final outcome. Some critics say that slugging through the FAR process can add 20% to the final cost (Dunn, 2009); others say it's higher. Perhaps more importantly, though, is the valuable time lost in getting new technologies to the field.

Case study: Boosters and the other transaction

This section examines the evolution of US Air Force launch vehicles from the 1990s to present. The Evolved Expendable Launch Vehicle (EELV) was an attempt by the Air Force to make boosters more efficient. The concept was borne in 1994 and was postulated based on traditional government-industry relationships with several companies who were, at that time, the only companies to manufacture boosters. It is instructive to review the political climate of that era. The Cold War ended in 1989, the Berlin Wall came down: it was the end of the Soviet Union, and defense spending went from a high of about 26.5% of the federal budget to 16.1% in 1999 (https://www.learner.org/workshops/primarysources/coldwar/docs/dspend.html). In ten years defense spending dropped by over 10% of federal outlay. This was a huge change and had to be well managed so that critical resources remained stable while Cold War defenses were strategically handled. In the Defense Department, the 1990s would be defined by the term *downsizing*.

The idea of merging industrial capability in boosters was originally devised in *The Space Launch Modernization Plan* (Moorman, 1994) in an attempt to respond to decreasing budgets and, as the report notes, the fact that on-orbit assets were more robust and capable than expected—that they stayed on orbit and functioned longer than originally planned. The future did not have a need, the committee thought, to have multiple industries competing for launches. EELV was a good-faith effort by the Air Force to maintain access to space, but to do so within looming budget constraints. The Air Force would depend on legacy systems from traditional defense contractors—the only option in those days. In the beginning, there were four bidders: Lockheed Martin, Boeing, McDonnell Douglas and Alliant Techsystems (ATK). By 2003, Boeing had acquired McDonnell Douglas, and ATK merged with Orbital Sciences. This left Lockheed Martin and Boeing as the national launch providers, but rather than competing, the two companies merged their booster capabilities by creating the United Launch Alliance (ULA), a joint venture with each company owning a 50% stake. ULA was a monopoly, but the Air Force and DOD assured Congress this would ultimately result in a 25% reduction in recurring launch costs between 2002 and 2020 (GAO, 1998).

Meanwhile, changes in EELV contracting were a sign of the times. Initially EELV was planned as a cost-plus-award-fee contract with four competitors; however, when only two companies were left in the era of downsizing, the Air Force decided this would be more efficient with the use of the OT instrument. In 1998, GAO responded with a report noting that the Air Force and DOD analysis of cost savings was incomplete. They noted that the defense agencies would need to perform a thorough net present value analysis (NPV) with the best forecast data possible, to include not only government military satellites, but the growing requirement for commercial satellites as well. What both the DOD and GAO failed to predict was the entry of commercial boosters. Under President Reagan, the Commercial Space Launch Act of 1984 was passed to facilitate private enterprise finding a way to space. Many commercial companies were manufacturing satellites which would ride on US government boosters, but none had, as of 1984, successfully launched commercially. US commercial companies seeking affordable access to space looked to Europe, Russia and China for cheap, reliable rockets.

The following two decades (1990s–2000s) saw remarkable efforts to change that. A few companies met with limited success: *Scaled Composites* was the first US company to succeed when in 2004 they launched human beings into space (defined as 100 km) twice within two weeks. They won the AnsariX prize of $10M on October 4, 2004, though they spent considerably more on their SpaceShipOne and White Knight that got them there.

Research on this continues under Virgin Galactic. *Andrews Space* was founded in 1999 and attempted commercial access to space with space planes. They have since been bought out, and efforts continue. In 2000, Amazon entrepreneur Jeff Bezos founded *Blue Origin*. The company has had considerable success on traditional booster launches and landings, and expect their first crewed commercial flights in 2018, with commercial services following. But as of this writing, the most successful 100% commercial access to space company is Space Exploration Technologies Corporation, best known as *SpaceX*, founded in 2002 by entrepreneur Elon Musk.

Over the past two decades, the EELV government/ULA booster program has been successful, and booster technology is mature as launch failures are at an all-time low (Figure 4.2). The OT purchasing philosophy has proven more efficient than other FAR-based contracting, but EELV is still the most expensive access to space, and commercial satellite companies have looked overseas, and now to *SpaceX*. *SpaceX* has built boosters completely in-house as a true all-American company, constructing its boosters, engines and capsules in the United States. So when EELV programs sought new engines, *SpaceX* cried foul when they looked to legacy technology in the Pratt & Whitney RD-180 engine made with Russian technology. On October 19, 2017, *SpaceX* was awarded a $40,766,512 OT for its Raptor rocket engine to be integrated to the EELV (Foust, 2017). As of this writing, this was one of the latest in an OT authority and a win for the Air Force, taxpayers and commercial industry. The current phase is expected to be completed by April 2018—a remarkably fast turnaround for any government project, least of all space.

The OT authority has proven instrumental in bringing the EELV to maturity, as well as engaging commercial technology to improve it. It will continue to be an important tool for known and unforeseen technologies of the future.

Figure 4.2 The Atlas V is a primary workhorse of EELV, a joint Air Force and ULA program. It has been one of the great success stories of the OT concept.

How to leverage the other transaction—Implications for the acquisition professional

The name is deliberately vague. It is not a contract, but simply an agreement where, in this case, the Air Force has permission to make a transaction with a supplier, vendor, company or whomever to furnish a product or service.

It is important for the acquisition professional to understand the utility and implications of the OT. Who might have guessed that the OT could have been employed for the EELV? Though specifically stated for a prototype, the EELV could indeed be considered a prototype. What else is a prototype? In areas of unknown technology, as space and air often are, nearly anything is a prototype. What about stealth technology? How much have stealth aircraft been used? As much as boosters? This is a matter of good management and excellent judgement that the acquisition officer must employ. In this section, we examine two avenues created to best leverage the use of OTs with rising technologies.

Defense Innovation Unit Experimental (DIUx)

> *"We're a fast-moving government entity that provides non-dilutive capital to companies to solve national defense problems."* – DIUx website

Then Secretary of Defense Ashton Carter began *The Defense Innovation Unit Experimental* (DIUx) in 2015 to provide a channel between DOD and bleeding edge technology in the commercial market. It is no surprise that DIUx offices are in the heart of important US technology centers: Silicon Valley, Boston, Austin, and the Pentagon. This proximity gives highest tech companies an easy path to the DOD as DIUx reports directly to the Secretary of Defense. Removing barriers to entry and allowing non-traditional defense contractors to play in the defense industry is a way to allow access for superior technologies. Typically these types of technologies require no long development time, and save the taxpayers the cost of research. In the past, a few such technologies "dropped out" of selling to the military as they would have been required to go through lengthy approval processes which offered these companies no profit or advantage.

In November 2016 DIUx released a 16-page "how-to" guide. *Sixteen pages.* This is unprecedented for a government document. The introduction rightly acknowledges that commercial companies with technology of significance to DOD often avoid the government and focus on commercial customers.

While still for prototypes, the DIUx definition is as follows: "A prototype project can generally be described as preliminary pilot, test, evaluation, demonstration or agile development activity used to assess the viability, technical feasibility, application or military utility of a technology, process, concept, end item, system, methodology or other discrete feature. The quantity or tenure should be limited to that needed to effectively assess the prototype" (DIUx, 2016). Once again, the DOD has used an intentionally broad definition to facilitate creative use of prototyping and the incorporation of technologies that do not fit traditional protocols and mind-sets.

The acquisition professional first needs to study requirements from the end user then be ridiculously thorough about market research. We are living in a time where technologies are everywhere, not just behind concrete walls. If you want it, it probably already exists. And even if it does not exist, countless entrepreneurs could probably figure it out. This is where the acquisitioner earns his or her keep. DIUx is an excellent place to begin if you are looking to fill a need in the Air Force that has not been filled.

System of Systems Consortium (SOSSEC)

The SOSSEC is another excellent place to begin (Nunziato, 2016). This is a non-governmental organization that has been successful working with defense agencies, including the Air Force, Army, Navy, Coast Guard, Marine Corps, and others, getting their products to the end user with the ten-step process. *Ten steps.* Once again they follow the philosophy of keeping the focus on the product, not the process.

SOSSEC walks the acquisition professional through a flexible process, ensuring cost and product are sound and superior.

Naysayers and supporters

> *"The hardest thing is not to get new thoughts in to people's minds, but to get old thoughts out." – Dr. William LaPlante (Assistant Secretary of the Air Force for Acquisition, 2013–2015)*

The OT process has both critics and fans. Critics are often people who have been accustomed to old ways of doing business, and the mindset that the DOD must control every step of acquisition; indeed focusing on process rather than product. For example, "In March 1998, the Inspector General testified about a continuing concern regarding the lack of controls over the other transaction process since normal rules and procedures generally do not apply" (Hill, 1998). Consider this: the DOD purchases commercial software so that we can all use word processing, email, and other essential office applications. Would it make sense for government workers to oversee and have control over every step of the process with commercial software companies, such as Microsoft or Apple? What about other DOD commercial providers? There are countless companies who build remarkable software with military applications. Should the services oversee every line of code that they write? It would be impossible and ridiculous. Yet, some have suggested that the services need source code for critical products. If that is true, then why doesn't the military require source code for word processing or spreadsheets applications? How much oversight do we need or want? With too much of it, we can count on delays in getting superior technology to the warfighters, and can count on potential adversaries, without long approval processes, to have commercial technology first.

Another critic wrote: "OTAs, from the outset, lack process, procedure and forms. Each OTA requires hundreds of hours of consideration and negotiation of terms. OTAs lack a comprehensive body of case law supplying interpretation of specific regulatory contract language. OTAs require expert personnel; critical and creative thinkers and astute negotiators" (McMartin, 2016). This writer understands it *precisely*. *Yes*—it *is* an instrument by which the DOD will require smart, creative acquisition expert professionals to be inquisitive, tough, and imaginative. Though his remarks are intended to criticize the process, to OT supporters, it looks like a directive. All military processes, including contracting through the FAR, *should* have experts running them—in what case should this not be true? It is tempting to believe that Defense should control every step of the process that creates products critical to success, but this is not practical, possible or even ideal. This is certainly not a new concept as it was introduced in 1956 by scholar Herbert Simon in his concept of *bounded rationality* (Simon, 1947). Humans are rational, he postulates, but we are bounded by what we can control and a single human can only control or know a finite amount. We succeed, therefore, by a process called *satisfycing*, Simon writes, where we must rely on others to operate what we cannot. For example: most of us travel by air. We rely on the airlines to operate each part of the flight, from management to mechanics to the pilots. We must count on that—and we stake our lives on it each time we step on a commercial aircraft. We do not go about inspecting every aircraft we board or interview every pilot. Why would using commercial products in the military be any different?

That any process does not use the FAR scares some people since it is a regulatory process that has evolved over decades. With it has evolved an immunity system with antibodies from the people who benefit from residing in a large bureaucracy. Many have argued that a bureaucracy is necessary for accountability, but others have argued that within a bureaucracy, accountability is diffused (Light, 1995). There are countless cases where the government has failed and no one has been held accountable, for if everyone is accountable, then really, no one is accountable. This works well for the people trying to preserve the bureaucracy and their role in it and benefit from it, but for the people fighting a war, or the taxpayers supporting it, it is abuse of power.

Many others in DOD are much more enthusiastic. Secretary of Defense James Mattis spoke of the DIUx as a channel to commercial companies and superior technologies to the warfighter: "I don't embrace it; I enthusiastically embrace it... there is no doubt in my mind that DIUx will continue to exist, it will grow in its influence and its impact on the Department of Defense" (Goldstein, 2017). Major General Sarah Zabel, Air Force Director of IT Acquisition Process Development, remarked: "This mechanism is just so much faster and so much more attuned to getting something quickly that we want today and not have to spend a couple years going through a protest, going through this huge process to get something we wanted two years ago" (Maucione, 2017).

General Zabel is speaking not only of the FAR, but of the much practiced process of contractor protests. Former Secretary Payton also spoke of this: "We cannot allow this process of protests to continue. It slows us down tremendously. The problem is that there is no incentive for a contractor to *not* file a protest; because we allow it. Look what has happened with the tanker. A protest has cost the Air Force a decade so far in getting a tanker on the runway and we are still using tankers that could be up to 80 years old. This is completely unfair to pilots and warfighters that depend on tanking support. We have failed them completely. A process that allows this is completely flawed and it is time to change. (Payton, 2017)"

Conclusion

Oddly enough, in 1963, when the United States space programs were well on their way, both civilian (NASA) and military (DOD), Khrushchev backed off of his initial promise: *"I once said 'we will bury you' and I got in trouble with it. Of course we will not bury you with a shovel. Your own working class will bury you."*

Any social scientist will say that bullies only pick on someone they are not afraid of—someone without a weapon—and that's exactly what the Soviet Union leaders did. Lots of big talk when they had the upper hand in space, but when we matched them point for point, and then moved beyond, Khrushchev backed down. His Soviet successors did the same when the US with its free society, strong economy and innovation, built a defense so strong that the USSR was finally out-spent and out-smarted, they gave up and the Berlin Wall fell in 1989. It was a curious strategy that many questioned at the time, but it is difficult to argue with success.

In the 1950s when the United States government realized we were about to be overcome by communist bullies it reacted with the *other transaction* authority. *Other* means any and everything we needed to get into space. The programs that followed—Corona, Mercury, Gemini, Apollo, and others—were highly successful and came about quickly, and they were all acquired using the *other transaction* tool. There were failures, but failures bring us merely a step closer to success. Our adversaries then as now have only made us stronger.

Public servants exist to be stewards of the taxpayer dollar and advocates for the warfighter, not to perpetuate and even grow huge empires in which they can spend an entire career working on a single project. That's what we've grown to and it's time to stop. It's time to go fast again. The OT is a tool by which this can be done. The Defense Department can now use it to the benefit of the taxpayer and warfighter.

As with any tool, the acquisition professional will require expertise, diligence, tenacity and creativity to make new technologies, often in the commercial market, work for the warfighter. To suggest, as some have, that the OT tool will require *increased* expertise is flawed. Any military project requires expertise to bring a superior technology to the battlefield quickly. Small groups of experts work better than large groups of bureaucrats and overseers. Famous aerospace engineer Clarence "Kelly" Johnson said it best in his seven rules of management. *"Strong but small project offices must be provided both by the military and industry,"* and *"No reports longer than 20 pages or meetings with more than 15 people."* He probably would have been proud of the DIUx 16-page instructions for OTs, and perhaps the SOSSEC Ten-step process. Kelly applauded the small, fast team, and worked during a time when the US was able to design and build aircraft, and even spacecraft, within a few years and not a few decades.

Technology has come far since then, and management processes are beginning to catch up by allowing superior commercial technology to the warfighters without unnecessary bureaucracy. It is time to go fast again.

List of acronyms for Chapter 4

ARPA-E   Advanced Research Projects Agency
CAS   Cost accounting standards
COTS   Commercial-off-the-shelf
DHS   Department of Homeland Security
DIUx   Defense Innovation Unit Experimental
DNDO   Domestic Nuclear Detonation Office
DOD   Department of Defense
DOE   Department of Energy
DOT   Department of Transportation
EELV   Evolved Expendable Launch Vehicle
FAA   Federal Aviation Administration
FAR   Federal Acquisition Regulation
GAO   Government Accounting Office
HHS   Department of Health and Human Services
IR&D   Independent research and development
NASA   National Aeronautics and Space Administration
NIH   National Institutes of Health
OSD   Office of the Secretary of Defense
OT   Other transaction
OTA   Other transaction authority, other transaction agreement
SOSSEC   The Systems of Systems Consortium
TSA   Transportation Security Administration
ULA   United Launch Alliance

References

Carter, Ashton B., and Perry, William J., *Preventive Defense*, Brookings Institute Press, Washington DC, 1999.

Department of Defense, *Other Transactions Guide for Prototype Projects*, Version 1.2.0, January 2017. Department of Defense, VA.

DIUx. *Commercial Solutions Opening How-to Guide*, November 2016, page 5. (https://www.diux.mil/) Last Accessed date: October 2017.

Dunn, Richard L., Injecting new ideas and new approaches in defense systems – Are "Other Transactions" the answer? Presented at the Naval Postgraduate School Annual Research Conference, May 2009.

Foust, Jeff, Air Force adds more than $40M to *SpaceX* engine contract,*Space News*, October 21, 2017.

GAO, Report to the Chairman, Subcommittee on Appropriations, House of Representatives, DOD Guidance Needed to Protect Government's Interest, (GAO/NSIAD-98-151), June 1998.

Goldstein, Phil, The future of the Pentagon's DIUx unit seems bright. *FedTech Magazine*, August 2017.

Hill, Eleanor, Statement of Eleanor Hill, DOD IG, before the Subcommittee on Acquisition and Technology, Committee of Armed Services, US Senate, March 18, 1998.

Khrushchev, Nikita, *We Will Bury You*. https://en.wikipedia.org/wiki/We_will_bury_you, 1956.

Light, Paul, *The Thickening Government: Federal Hierarchy and the Diffusion of Accountability*. Brookings, Washington, DC, 1995.

Maucione, Scott, *OTA Contracts are the new cool thing in DOD acquisition*. Federal News Radio, October 19, 2017.

McMartin, Benjamin, *Other Transaction (OT) Authority Mythology: Reflections of the Cure-all of Defense Procurement*, Linked-in, December 17, 2016.

Moorman, Thomas S., *Space Launch Modernization Plan*, Department of Defense, April, 1994.

National Aeronautics and Space Administration, *National Aeronautics and Space Act of 1958 Public Law #85-568, 72 Stat., 426.*, Signed by Sam Rayburn, Speaker of the House of Representatives, and Richard M. Nixon, Vice President of the United States and President of the Senate, July 29, 1958.

Nunziato, John, *SOSSEC 10-Step Process for Project Proposals under the AFRL—Rome NY Other Transaction Agreement*, The System of Systems Consortium, August 2016.

Payton, Sue C., *Former Assistant Secretary of the Air Force for Acquisition*, interviews, 28 September 2017.

Raptor engine family; https://en.wikipedia.org/wiki/Raptor_(rocket_engine_family)

Report of Columbia Accident Investigation Board, 2003.

Simon, Herbert, Chester Barnard, *Administrative Behavior*, Macmillan, New York, 1947.

Smith, Giles, Jeffrey Drezner, Irving Lachow, *Assessing the Use of "Other Transactions" Authority for Prototype Projects*, RAND, Santa Monica, CA, 2002.

Undersecretary of Defense, *Buying Commercial: Gaining the Cost/Schedule Benefits for Defense Systems*, Report of the Defense Science Task Force on Integrating Commercial Systems into the DOD Effectively and Efficiently, February 2009, page 18.

Undersecretary of Defense for Acquisition, Technology and Logistics, *"Other Transactions" (OT) Guide for Prototype Projects*, August 2002.

chapter five

Commercial technologies in the Department of Defense

Technology evolution and implications for acquisition professionals

Sally J. F. Baron

Contents

Introduction

The Department of Defense (DOD) has long used commercial products and the cooperation between DOD and industry is nothing new. From weapons and materials in the Revolution, to trucks and tractors in the World War I era, to fabric for uniforms throughout its history, commercial products from industry have provided superior warfighting ability to the US.

The fact that the founding of an entire military service—the Air Force—was based on an invention made completely within the private sector—yes; the Wright Flyer—is often downplayed.

Throughout US history, the acquisition process has been both successful while also being plagued with inefficiency, complexity and occasionally malfeasance. Over the past several decades nothing has changed the acquisition world more than the onslaught of high-tech products in the commercial market. Technology, including computers, telephones, workstations, laptops, satellite imagery, antenna networks, medicine and much more have grown exponentially in both technological advancement and availability that most could not have predicted while Defense Department acquisition practices have not kept pace.

Acquisition procedures, outlined in the Federal Acquisition Regulations (FAR), control the purchase of products from industry for delivery to the battlefield, but these procedures have become cumbersome and outdated. The difficulty in this is that many companies with potentially superior technologies may not have the infrastructure, nor might they desire it, to compete for DOD contracts. The process typically requires staffs of lawyers and contract experts. Critics have asserted that the FAR does not offer a streamlined process and does more to harm than help acquisition.

The military acquisition force, both civilian and uniformed, will have to develop new procedures to keep top commercial technologies in the hands of the warfighter when they offer better solutions that can get to battle faster. This chapter reviews the history of industry and commercial products' role and importance in defense. It examines the early history of flight and how it changed defense acquisition in this country, as well as selected histories of computing and space. The chapter concludes by suggesting methodologies to keep the US DOD on the cutting edge in both high technology and efficiency.

Department of Defense acquisition background

The *Department of Defense* (DOD) has evolved since the Revolutionary War to the most powerful military on Earth. Much of that success is a result of the ingenuity of people when they are given the latitude and incentive to work and achieve in a free, capitalistic society, coupled with a people who are determined to protect their freedom. The United States has provided the most optimal environment for innovation and creativity ever in recorded history. Our industrial base is unparalleled as our founders and first citizens, most of whom escaped oppressive governments where even certain ideas were a punishable crime, recognized the need for a society where basic freedom and the pursuit of happiness were self-evident rights assured by the *Constitution*. Ideas and innovation have flourished. Predictably, in the early days, innovators from oppressed countries immigrated to the US free society by the millions. As a result, innovation and creativity abounded and in a very short time the US went from a poor, fledgling country to one where more inventions flourished and quality of life improved.

From the signing of the *Declaration of Independence* and the rights provided by the *Constitution*, we have long recognized the need to defend personal rights and freedoms. History has illustrated that our earliest wars were won with superior industry, as well as the tenacity of America to remain independent and free. In World War II, our industrial base out-produced our adversaries, manufacturing nearly 300,000 aircraft for the allied forces from Niagara, New York to Los Angeles, California. Automobile factories were converted to armored vehicle production facilities in months, and tens of thousands were produced. We were fast and furious.

Industry was part of warfare then and as warfare evolved, the very best innovations in the world continually came from US private industry. Though the DOD supported research and development, Silicon Valley entrepreneurs provided the country, and the world, with superior computing and related technologies such as communication and digital imagery. As the world embraced these new technologies, they became more available and less expensive due to economies of scale.

The Gulf War (Operation Desert Storm, 1991) was handily won in less than two months with our on-orbit assets and superior technologies guiding smart bombs to precision hits with little or no collateral damage. Space, computing and communication technologies combined to allow warfighters instantaneous information previously unavailable. Economics plays heavily into all warfare: *if we cannot afford it we cannot have it.* We must also recognize that if we cannot defend it, we cannot have it, and we must begin with the assumption that resources are *finite.* These are axioms of existence of a well-defended nation.

Today, the DOD is a highly developed, complicated organization with wide and varied tasks and comprises numerous sub-organizations. Its total 2016 fiscal year proposed budget was $585B, and active duty members totaled approximately 2,118,000 (World Almanac, 2017). How is all this effectively organized and managed? With a lot—arguably too much—bureaucracy. The FAR has grown since inception and continues to grow with each change or new law that affects acquisition. It is rarer, however, that antiquated policies are deleted. What are we left with? An enormous document that no one completely understands with policies and regulations that frequently contradict one another. What is an acquisition officer to do? Deal with it, change it, or work around it. Dealing with it is a short-term solution and changing it is the only long-term, albeit cumbersome, solution. A discussion of how to work around it will come later. The more technology changes, the harder the FAR will be to manage. Inevitably the FAR will become larger, longer, less relevant and more unmanageable as has been the trend. This simply needs to stop. The current administration (2017) has asked that adding a single new regulation will require that two old ones be deleted. This is the first tangible step to reducing the sheer mass of the FAR in recent history (Lam, 2017).

Meanwhile, to comprehend the technological evolution and the changes implied we shall take a brief look at defense acquisition history next.

Acquiring the best products for the warfighter: from the beginning

The early US military had its hands full. The Revolutionary War, a fight for our independence from England, was a fight against one of the then most advanced militaries on the planet. The US needed to equip its forces rapidly, and it did; but not without pain and problems. The government was young, small and inexperienced. Once independence was secured on July 4, 1776, the Congress realized that the nation would need to build a strong military both for defense and also for western exploration.

Over a century later, perhaps the greatest test of US sovereignty was the Civil War. The North clearly had an industrial advantage which would ultimately secure the United States. Pennsylvania's well-established iron mills out-produced those in the South by about 10:1 (Sorenson, 2009). One of the greatest technological advances of the Civil War was the move from vulnerable wooden ships to iron and steel. Again, although the South had iron ships, this new shipbuilding technology put the industrial North at a great advantage as it had greater resources with which to build iron and steel products, including ships. Other weaponry such as canons, rifles and pistols were more advanced in the North. The North also took great advantage of its superior rail system.

Congress spent great monies to improve industry and keep the Union together, but where there is money, there typically is malfeasance. Sadly, this has been omnipresent throughout the history of government; it was as true then as it is now. As a result of fraudulence in the early Civil War, in 1861, the Congress developed the Committee on Government Contracts (1861) as the first of many Congressional oversight select committees to guard against loss in the government contracting business. Adding oversight has become a common practice for the US government in response to criminal activity and poor management and this has led to layer upon layer of bureaucracy and inefficiency.

Continuing with military acquisition cases, the nation's fortitude would once again be tested in the late nineteenth century in the Spanish-American War. The demise and sinking of the *USS Maine ACR-1,* as well as the deaths of 252 of her 350 crewmen, would preface the Spanish-American War by about six months. The ship's construction history provides insight into the acquisition process of the time. The *USS Maine* was commissioned in September 1895. Though it had a steam engine (rather than wind power) and iron cladding, it was considered out of date by the time it reached service (*USS Maine,* 1895). Considered an armored cruiser, her construction came at a time when naval technologies and needs were changing rapidly. The ship was designed by Theodore T. Wilson, who was likely preferred because he was American, and the ship was to be the largest US Naval vessel to date at 324′4″. Congress authorized funds for construction in 1886, and the keel was laid in the Brooklyn Navy Yard in 1888. The nine-year building time was considered slow, and though she was armored, by 1895, other naval ships were armored with lighter-weight, stronger steel. As such, the *Maine's* role became ambiguous because she lacked both the firepower and armor to serve as a cruiser. Acquisition and procurement delays have a huge cost that is not always obvious.

As industry became stronger and more important to the United States during this time, in the late 1800s many inventors and aviation enthusiasts worldwide were pursuing heavier than air flight. The most successful team was, of course, the Wright brothers of Ohio. The following section offers a case study in their attempts to bring their invention to the US defense officials of the War Department.

Aircraft and Department of Defense

First flight: An examination of a new technology

Here we examine one of the greatest and most important inventions of all time and the War Department's response to this revolutionary technology. Orville and Wilbur Wright were brothers, inventors, and metallurgists. They had a bicycle shop in Dayton, Ohio, and from their work with bicycles they concluded that if a human could manage the balance of a bicycle, then, quite possibly, a human could control weight and balance of a flying machine. The brothers had been interested in the problem of human flight from boyhood, and their interest manifested itself in work beginning in about 1900, which was funded largely by family, friends and other interested parties. After reading *Progress in Flying*

Machines, by Octave Chaunte, the Wrights contacted the author, which began a long and important relationship.

After years of building, studying, and trial and error, the Wrights met with success on December 17, 1903, in Kitty Hawk, North Carolina. They continued to modify and perfect their flying machine and in 1904 made over one-hundred flights with improved controls and safe landings. By 1905 the Wright brothers were satisfied they had a practical, salable product, and having borrowed money during their five years of experimentation, they were anxious to pay debts. They approached the US War department through Representative Robert M. Nevin. Their exchanges are instructive and are included as follows:

> *Note*: All the Wrights' correspondence are represented here as they appear in *Miracle at Kitty Hawk*, edited by Fred C. Kelly (Kelly, 2002). They are unchanged from original form, with the exception of the **bold type** for emphasis as noted.

> ### *January 18, 1905; Letter from Wilbur Wright to Congressman Robert M. Nevin*
>
> The series of aeronautical experiments upon which we have been engaged for the past five years has ended in the production of a flying-machine of a type fitted for **practical use.** It not only flies through the air at high speed, but it also lands without being wrecked. During the year 1904 one hundred and five flights were made at our experimenting station, on the Huffman prairie, east of the city; and though our experience in handling the machine has been too short to give any degree of skill, we nevertheless succeeded, toward the end of the season, in making two flights of five minutes each, in which we sailed round and round the field until a distance of about three miles had been covered, at a speed of thirty-five miles an hour. The first of these record flights was made November 9th, in celebration of the phenomenal political victory of the preceding day, and the second on December 1st, in honor of the one-hundredth flight of the season.
>
> The numerous flights in straight lines, in circles and over "S" shaped courses, in calms and in winds, have made it quite certain that flying has been brought to point where it can be made a great **practical use** in varying ways, one of which is that of **scouting and carrying messages** in time of war. If the latter features are of interest to our own government, we shall be pleased to take up the matter either on a basis of providing machines of agreed specification, at a contract price, or of furnishing all the scientific and practical information we have accumulated in these years of experimenting, together with a license to use our patents; thus putting the government in a position to operate its own account.
>
> If you can find it convenient to ascertain whether this is a subject of interest to our own government, it would oblige us greatly, as early information on this point will aid us in making our plans for the future.

Nevin had promised to take the letter to the Ordinance Board (Part of the War Department) and speak on behalf of the Wrights, but he was unable to deliver it personally due to illness. Nonetheless, he indeed got it into the right hands and received this reply shortly after:

> *Reply to Congressman Nevin from the Board of Ordinance and Fortification, signed by Major General G. L. Gillespie [exact date not available]*
>
> I have the honor to inform you that, as many requests have been made for financial assistance in the development of designs for flying-machines, the Board has found it necessary to **decline to make allotments** for the experimental developments of devices for mechanical flight, and has determined that, before suggestions with that object in view will be considered, the device must have been brought to the stage of practical operation without expense to the United States.
>
> It appears from the letter of Messrs. Wilbur and Orville Wright that their machine has **not yet been brought to the stage of practical operation,** but as soon as it shall have been perfected, this Board would be pleased to receive further representations to them in regard to it.

What was going on here? From the most casual read, it looks like the Board did not carefully read the Wrights' proposal. The Wrights clearly state that they spent many years developing a product that is currently operational. The word *practical* appears three times in their proposal, and they even suggest a function: reconnaissance, yet the Board seems to ignore this. They offer two types of contracting and what they consider the product ready. Strangely, the Board responds by emphasizing that they will not offer financial assistance, something the Wrights never requested, and asks them to come back when they have a machine ready for practical use. This short, simple letter, signed by Major General G.L. Gillespie appears to be one of the biggest acquisition blunders in military history. The Wrights offered exclusive rights to their invention and the government all but ignored them.

The Wright brothers were more than disappointed but not defeated. Their next actions are a harbinger of many commercial companies to come: they sought customers overseas. Note: I have interviewed countless commercial companies that told me they seek overseas companies as a result of their frustration with the US acquisition procedures.

We shall follow their continued dialogue in the following.

> *May 28, 1905; Letter from Wilbur Wright to Octave Chaunte (author, supporter, and friend)*
>
> We stand ready to furnish a practical machine for use in war at once, that is, a machine capable of carrying two men and fuel for a fifty-mile trip. We are only waiting to complete arrangements with some government. **The American government has apparently decided to permit foreign governments to take the lead in utilizing our invention for war purposes.** We greatly regret this attitude of our own country, but seeing no way to remedy it, **we have made a formal proposition to the British Government** and expect to have a conference with one of its representatives very soon.

May 30, 1905; Letter form Octave Chaunte to Wilbur Wright

As an American I greatly regret that our government has apparently decided to allow foreign governments to take the lead in utilizing your invention. Please advise me, 1st, Whether you have approached our war office? &c 2nd, Whether you would object to my putting a flea in its ear?

June 1, 1905; Letter from Wilbur Wright to Octave Chaunte

We would be ashamed of ourselves if we had offered our machine to a foreign government, without giving our own country a chance at it, but our consciences are clear. At the Christmas holidays we talked with Mr. Nevin, congressman from this district, and he proposed that we write him a letter containing a general statement of our business, and that he take it to Mr. Taft and secure an appointment for us to meet with the War Department officials, thus saving us delay when we should visit Washington. But owing on sickness, he was compelled to turn over our letter without personally seeing Mr. Taft and shortly afterward received a letter from the Ordinance Department which I enclose. As we had made no request for appropriation, but on the contrary had offered to furnish a machine of "agreed specifications at a contract price," (which offer was entirely ignored,) we were driven to the conclusion that the letter of the War Department was intended as a flat turn down. We still think so.

A note to Col. Clapper informing him that we were ready to talk business with the British government soon brought a response from the English war office requesting us to make a definite proposition. We submitted our proposition, and now have an answer stating that an officer will be sent to see us.

It is no pleasant thought to us that any foreign country should take from America any share of the glory of having conquered the flying problem, but we feel that we have done our full share toward making this an American invention, and if it is sent abroad for further development the responsibility does not rest upon us. We have taken pains to see that "Opportunity" gave a good clear knock on the War Department door. It has been for years been our business practice to sell to those who wished to buy, instead of trying to force goods upon people who did not want them. If the American government has decided to spend no more money on flying machines till their practical use has been demonstrated in actual service abroad, we are sorry, but we cannot reasonably object. They are the judges.

The correspondence clearly illustrates the shock and frustration of the Wrights and Mr. Chaunte. Wilbur emphasizes that they are not in the business of marketing products to organizations or people who do not want them, and is adamant about justifying that they did their due diligence by offering the invention to the US War Department first. In spite of this, for a second time, in October 1905, Wilbur wrote directly to the Board of Ordinance to ensure there was no misunderstanding that the Wright Flyers were indeed ready for practical use. The US government responded with a nearly identical letter as the earlier 1905 correspondence. At this point the Wrights were convinced that they needed to move on; stating their practice of not being marketeers; but rather inventors. The "form letter" type of a response indicates that the War Department had not done their research. The Wrights heavier-than-air flyer is something that the War Department long sought. In fact, the department invested in such a concept to the tune of $50,000 then-year dollars given to Samuel Langley to come up with such a machine. Langley failed miserably and as a result, the War Department did not believe that it could be done. This may have tainted their view of the Wright Flyer.

Next, Chaunte expresses his disgust in the lack of interest on the War Department, and encourages the Wrights to pursue interests overseas.

June 6, 1905; Octave Chaunte to Wilbur Wright

My feelings were of mortification and regret that the United States war department should have extended to you a "flat turn down" as you express it. Now that I have cooled down I see some advantages to your being forced to consider the overtures made by Col. Clapper for the British Government, because: First, your invention is worth far more to the British than the United States government. Second, the **British are less hampered than we are in appropriating secret service funds,** so that you can probably get a better price, and sooner. Third, your invention will make more for peace in the hands of the British than in our own for its existence will soon become known in a general way and the knowledge will deter embroilments.

One need only speculate why Chaunte believes the flyer would be worth more to the British government. First, Great Britain is a group of islands; not part of a larger landmass as the United States, and second, their proximity to early twentieth century Germany is perhaps a greater concern. Chaunte implies that Great Britain may have been more desperate. Perhaps most interestingly, Chaunte observes that Great Britain is "less hampered" than the US. Could it be that even in 1905 the US War Department had already become its worst enemy with an encumbered ability to acquire new technology? Next, Wilbur responds to his longtime friend, Octave, defending their position and illustrating the Wrights' intention to move on.

June 18, 1905; Wilbur Wright to Octave Chaunte

We have no intention of forgetting that we are Americans, and do not expect to make arrangements which would probably result in harm to our native country. The exact date of meeting the British representative is not fixed but will probably be within a month. Meanwhile we have decided to complete the machine and **take the risk of making a few private trials of the improvements we have added to the machine**. The machine will probably be complete in a couple of days and we will be testing it the latter part of the week if the weather is suitable. Of course we would be glad to have you visit us and see it go, if it should suit your convenience and pleasure.

The doubts of Capt. Ferber and other foreigners worry us not at all. In fact they are rather an advantage to us while we are wishing to secure privacy. We certainly shall not disarrange our own plans to satisfy either public or private curiosity at this time.

We find that we underestimated the weight of our last year's machine. We carried a total weight of about 915 lbs. This includes about 70 lbs. of steel bars which we used as ballast. The new machine with water and fuel will weigh almost exactly 850 lbs., with one man.

We quite approve your decision to make only brief reference to our power machine in the Standard Encyclopaedia article. Until we are really ready to make the machine public there are many reasons why it is not best to say too much for publication.

It is critical that the acquisition professional understand that with commercial products, the private company assumes all risk in research and development—a very good deal for the taxpayer. Mr. Wright mentions here that they are in a continual improvement phase with the *flyer*—something that he understands is an assumption of risk. As with most new technologies, risk is very high. As technologies mature, risk typically decreases. Professor James March explains this best in his theories of *exploration* and *exploitation*. He notes that *exploration* is a riskier venture, requiring ventures into unknown paradigms and characterized by experimentation with new alternatives. "its returns are uncertain, distant, and often negative." When technologies mature, he notes, they do so as a result of *exploitation*, characterized by refinement of a known technology (March, 1991). Aircraft are a classic example of this. Where the Wright brothers absorbed the risk with ventures into the unknown, for at the time no one knew if a heavier than air machine would actually be possible, others benefitted from refinement of their invention. Over the past century, Martin, Airbus, Boeing, Cessna, Cirrus and others have refined aircraft far beyond what many could have fathomed. Indeed, the Wrights went on to refine their own invention.

Over the next two years, the Wrights entertained interests from Great Britain, France, Austria, Germany, and a US marketer who wanted to be a third party selling the machine to Russia. The greatest trouble they had both abroad and at home was disbelief in their invention. All parties wanted demonstrations, and several parties believed that once they understood flight they could more easily re-create it, saving funding, rather than purchasing from the inventors. It was not until 1907—nearly four years after the first flight—that the War Department made a solid offer to the Wrights. Simultaneously, the Wrights set up a company in France. The Wrights' offer to the US War Department follows. Recall that their initial offer suggested that the machine be used for reconnaissance.

May 17, 1907; Wright brothers to the US War Department

We have some flyers in course of construction and would be pleased to sell one or more of them to the War Department, if an agreement to terms can be reached.

These machines will carry two men, an operator and observer, and a sufficient supply of fuel for a flight of two hundred kilometers. We are willing to make it a condition of a contract that the machine must make a trial trip of not less than fifty kilometers at a speed of not less than fifty kilometers an hour, before its acceptance by the Department, and before any part of the purchase price is paid to us.

If the War Department is in a position to purchase at this time, we will be pleased to have a conference for the purpose of discussing the matter in detail, or we are willing to submit a formal proposition if that is preferred.

June 15, 1907; Wright brothers (Orville) to the War Department

The price quoted in our letter of May 31 should be understood as the price of the first flyer delivered to the Government and the instruction necessary to enable a representative of the War Department to operate it. The price does not include any period of time during which the use of the invention would belong exclusively to the United States, since a recent contract precludes our offering such a right.

While great reverence has been given to the Wright brothers historically speaking, again, little attention has been paid to what was quite possibly the greatest military oversight of all time: rejecting an invention that was practical and ready for use which was long sought and would be critical to warfare forevermore. Perhaps most importantly, the US government missed its chance at exclusivity to the Flyer design. Mr. Wright specifically states a "recent contract" precludes it. With two world wars just around the corner, one can only speculate what might have happened is the US had exclusivity on the flyer. History will never know.

It is instructive for the acquisition professional to examine why this happened. The Board of Ordinance barely even read the Wrights' letter. They assumed it was a letter seeking financial compensation, not a legitimate offer of a functioning commercial product. Who was Major General G.L. Gillespie, the signor of the letter? Major General George Lewis Gillespie Junior was a Medal of Honor winner from his action in the Civil War, and his headquarters assignment, 1904–1905 was his last in the service. While his military service as an overseer of great harbor construction projects earned him a reputation as a competent engineer, could it be that his knowledge of new technologies was sparse? And what was his staff doing? A special projects staff—then as now—should have been on top of new technologies especially the widely publicized Wright brothers' endeavors. The question is still asked today: what is the government's incentive to be effective or efficient? Their salary and benefits would not change and it is a robust theory that bureaucracies are a good place for the mediocre to hide as the bureaucracy's size diffuses responsibility. These are important issues then as now (Figure 5.1).

Clearly the military was interested in a heavier than air machine as they were funding efforts to create one; with little success. In 1898 the US government had given Samuel Langley $50,000 (then year dollars) to construct a heavier than air machine. By mid-1903 Langley had not gotten his Aerodrome off the ground, but he continued and secured $10,000 from the Smithsonian (his employer) and $12,000 from the Hodgkins fund. By 1904 the Aerodrome had two stupendous public failures in which Charles Manly, Langley's assistant, nearly drowned in the Potomac River. It was a huge embarrassment to the War Department and others who supported it. These plunders may have been why the War Department was not anxious to become involved in heavier-than-air flight.

Meanwhile, The Wright brothers' success was well reported in numerous periodicals, including *The San Francisco Call*, San Francisco, California ("Airship Flight is a Success," December 18, 1903), *The Times Dispatch*, Richmond, Virginia ("A Machine That Flies," December 19, 2017), *The Washington Times*, Washington, DC ("High Gale No Bar to Flying Machine," December 19, 1903), *The Minneapolis Journal* ("Airship was a Great Success: The

Figure 5.1 The Wright 1908 Model A Military Flyer was the first heavier-than-air military aircraft in history. It was purchased by the United States military from the Wright brothers in 1909. In this photo, it is arriving at Fort Myer, Virginia on a wagon. [No copyright on this photo as it is public domain; taken in 1909.]

Wright brothers Give Out a Statement Regarding Their Recent Experiments," January 6, 1904), and dozens more. How did Gillespie and his staff not notice the Wrights' success when it had appeared in so many newspapers?

The Wright brothers' successes did not escape the notice of a young Army Signal Corps Officer, William "Billy" Mitchell. He became passionate about flight and its applications in the military, and in 1920, when the Air Service was founded, he was promoted to brigadier general and was then their most senior ranking officer. He asserted his theory that properly armed aircraft could sink battleships and successfully demonstrated it in 1921. But his criticism of the Army and Navy earned him a court martial. He resigned from military service in 1926, and spent the rest of his life preaching air power (Howard, 1998). He died in 1936. Though Mitchell is widely known as the father of the United States Air Force, and was promoted to the rank of major general in 1942 (posthumously), it was not until after his death that the services appreciated his application of munitions to air power. In hindsight, his ideas changed the way war would be fought forever. Billy Mitchell had the foresight to see new applications for a commercial product. The military had finally recognized that products created in the private sector had military applications, but were not ready to listen to a change agent who had further applications for aircraft. There is an important lesson for the acquisition professional: open your mind to new ideas that may not seem completely logical at the time. Yet another management lesson from Major General Mitchell's life: new ideas are not always appreciated and are typically a risky business. As James March indicates, experiments into untested realms will often meet with resistance and possibly negative rewards.

One might ask: *What other technologies is the DOD missing today?* The lesson for the military professional is clear: superior products exist in the commercial market. What exists today? Are acquisition personnel duly diligent? Where would one find new technologies? Symposia, conferences, trade journals, and other publications are full of inventions. Their application to the military is up to the military's ability to understand new applications. Discussing ideas with colleagues and other professionals in industry is a good place to start.

History has shown that the eventual adoption of flight as a military tool has been a crucial advantage. It bears repeating that an entire new service—the Air Force—was created to support a commercial product from the very early twentieth century; yet the government itself out and out missed its first creation. Ideas for flight application were forthcoming, but the government did not always embrace them.

Other innovations would come in the twentieth century—innovation that hardy anyone predicted. The next case study examines computing and its role in defense.

Computing and defense

Computing was recognized very early on as important to a strong defense and the DOD had early and critical roles in computer advancement. In the 1960s computing moved from semi-conductors to the integrated circuit and the now famous "Moore's Law" (Intel CEO Gordon Moore predicted that chip density would double every 18 months) has held in principle (Ceruzzi, 2000). Today, laptops and hand-held computers have replaced the large, cumbersome, expensive, and sometimes classified computers of the past. Some of the first computers occupied large rooms in covert facilities. Now, complex problems can be solved by simple spreadsheets and problems that were at one time unsolvable, requiring millions of iterations, are now easily solved with simple programs. For example, optimization problems, formerly done by hand, may have required thousands if not millions of iterations and could not have been done by humans. Today, they are commonplace and have made manufacturing and production enormously efficient, saving consumers and taxpayers billions.

Both private and public sectors pushed and pulled computing technology and during the 1950s (post WWII) the government and especially the DOD invested huge monies in their development. Private companies saw applications and invested heavily in research and development. [See Figure 5.2 for a schematic on how the computing and space industries worked symbiotically over the past half-century.] Personal computing began to evolve in the 1980s. Desktop workstations first came into being by Hewlett-Packard, Apple, and later IBM and others. While few in the traditional, older computer companies had little hope for home computers, other entrepreneurs and hobbyists in Silicon Valley were determined to get them to market. Now workstations and laptops are not only a part of our daily lives at work, but home as well. The marriage of the Internet, a child of Defense Advanced Research Projects Agency (DARPA), along with home computing gave rise to millions of applications for home and business. New applications pop up by the millions—nearly all from innovators in the private sector. As Figure 5.2 illustrates, this huge demand for computing technology over the past 50 years caused a tremendous pull for both hardware and software. With great competition comes great efficiencies, and computing technology has gotten remarkably inexpensive.

Without computing, space, aircraft, and other important technologies would not be nearly where they are today. For example, satellite control, booster trajectories, and digital imaging were all the result of awesome computing power. These technologies continue to advance to this day, and many which were exclusively part of defense have crossed into

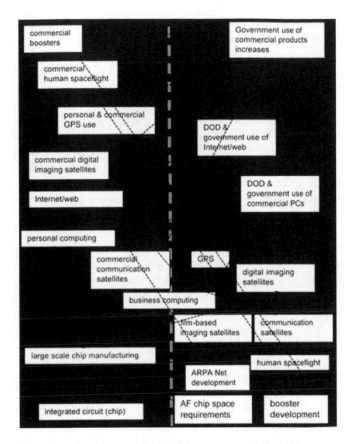

Figure 5.2 Symbiotic Relationship between Computing and Space in Public and Private Sectors.

the private sector as common, commercial technologies. In turn, efficiencies from competition have made them more affordable for the government, though the government is not always properly motivated to see these new efficiencies.

One of the most famous examples of military computing advances became evident to the public during the first Gulf War, Operation Desert Storm, 1991. Advances from the previous decade were incorporated into "smart" weaponry. The term "smart" is commonly used for weaponry that is able to direct itself even after released from the operator. For example, a "smart" bomb may be *heat seeking* and able to re-direct itself towards a target that emits heat. Other technologies allow guided weapons to avoid obstacles by pre-programming ground ephemera. Videos of guided weapons grabbed the attention of the world as they took out enemy targets with the precision of a surgeon's scalpel. This not only resulted in funding efficiency, but by targeting only military threats, collateral damage was reduced and innocent civilian lives saved. All this was the result of advanced computing and space technologies working together. The war was handily won in two months. This remarkable revolution in precision technology—combined with the fact that much of it could be seen by millions worldwide—did not go unnoticed by our adversaries.

When acquisition professionals consider computing, we must consider both hardware *and* software. Hundreds of COTS software companies have developed products that outpace what traditional defense contractors have produced; yet they still only fill the cracks of what is required in DOD. Under traditional FAR acquisition procedures COTS software could not be effectively provided to the government. The following quote is from an interview with the CEO of a COTS software company developing products for defense.

> "... There is no way that the use of COTS can exist under a rigid system. Part of the advantage to using COTS is the ability to refresh technology when new improvements come along. This requires constant contact with our government partner as we think it's more efficient to refresh the technology ourselves. This is complicated stuff—really rocket science. For us to build code and then have someone else patch it—I mean they have to go through the entire learning cycle of learning the code. Further, the government system of a maximum of 15% profit will also never work with COTS. Commercial products are already developed. With software products especially, the incremental cost of producing and selling one more copy is almost negligible. Depending on how much integration we need to do, our profit can be 90%–99%. We have absorbed all the development costs ourselves" (CEO Alpha software Vendor). Note: Based on Interview with software company CEO who preferred to be anonymous.

In the *design-and-build* acquisition philosophy, maximum profit is 15%. As the CEO points out, this would not work at all in the software industry. Most of what software companies are selling is *intellectual property*. Typically software programs evolve during years of development with new upgrades released periodically. The commercial software company absorbs all research and development costs. This is indeed not a small expense considering the hefty labor hours required for programming.

Yet, in the past, government agencies have not been anxious to embrace commercial software. In the 1990s a large civilian government agency was seeking new software for a major program upgrade. The software company mentioned earlier (Alpha) offered up a fully running commercial product for $5M. A traditional defense company (Beta) bid for the program on a design-and-build contract for $100M to be ready in five years. The government awarded the contract to Beta. Beta broke baseline and was approved for more funding and more time. In the end, they did not deliver, and the government returned to Alpha who sold them a functioning program. No one from the agency was punished for poor decision-making, and the loss was entirely the taxpayers. As with the Wright brothers, the government, with its layers of bureaucracy, was not held accountable, and the bad news is buried in history.

During a recent conversation with the Alpha CEO he remarked that he is taking his product overseas, and for the first time in Alpha's history, they are in litigation with the government. As with the Wright brothers, Alpha has done their due diligence in trying to sell to the US. Part of the problem with software is that it is not tangible; and for some, understanding its value is difficult. As with the Wright Flyer, people at the highest levels of government could not understand a heavier-than-air machine and were perhaps unable to envision applications. Today, aircraft are common. But perhaps defense professionals who did not grow up with computers have difficulty understanding all their potential and unique applications of hardware and software.

In modern military there are few who did not grow up with space. Though much of early space in the mid-twentieth century was classified, much has been de-classified and is available in the commercial world. This is discussed next.

Space and defense

Present day satellites and ground systems could not exist without remarkable computing technologies discussed earlier, and global communications could not exist without satellites. Space technology was once owned exclusively by the government. In the 1950s, 60s and 70s, the Air Force and other government agencies, such as the National Reconnaissance Office (NRO), and National Aeronautics and Space Administration (NASA) were among the first to work with government industrial partners in what were then feasibility studies. Today, launch technologies, space digital imaging, and human spaceflight are available commercially. With aerospace in its infancy, only the government could afford to play in such a risky venture, and all of the first space efforts in the mid-twentieth century were only in the public sector—the government and its industrial partners. Contracts were written so that companies would not go out of business if these high-risk, high-tech efforts failed. No one knew whether or not a human being could survive in a microgravity environment or the radiation astronauts would experience outside the Earth's protective atmosphere. No one knew if rocket boosters would actually get a satellite to orbit, and there was plenty of reason to believe that launch failure would be the rule and not the exception.

Space is a critical part of defense, and the first entrees to space were motivated by defense. The world's first steps into space came in the mid-twentieth century: boosters, satellites, robotic and human spaceflight. Getting a satellite into orbit is difficult to say the least. Boosters are the traditional method, and are large, expensive and risky. From the beginning and throughout their history, the trip has been highly risky. Next, we examine a brief history of one of the world's newest space companies: *SpaceX*.

Case study: Commercial boosters and innovation

It was 2002 and Silicon Valley entrepreneur Elon Musk took a trip to Russia to purchase rockets and returned empty handed. He had just failed to purchase two Russian boosters to refurbish them for his stated goal of making humankind a multi-planetary species— beginning with a Mars colony. The Russians offered $8M for one booster and Musk countered with a buy-one-get-one-free deal: $8M for two. The Russians rejected him outright, apparently not taking him seriously (Vance, 2015). Shortly after his disappointment, Musk, a talented engineer and known risk-taker, assembled a business plan for a rocket company which, in June 2002, became Space Exploration Technologies: *SpaceX*.

Up until then, the history of entrepreneurs succeeding in space booster companies was sparse. The enormous cost of one or more launch failures typically has put them out of business. In the early days of space (1950s–1960s), the government was able to absorb such losses, but with commercial rocketry, such failures typically cause investors to give up and the companies to completely fail. *SpaceX* did not want to be beholden to government restrictions so it began and remains a 100% commercial company. Many pundits were sure they would fail like the rest.

Within three short years from inception, in November 2005, *SpaceX* was ready to launch its first rocket, the Falcon 1, from Kwajalein, in the Republic of the Marshall Islands. The first attempt was a no-go, but at the second attempt, in March 2006, the rocket launched. However, about 30 seconds after lift-off, the rocket spun out of control. Its payload, the Air Force Academy cadet-built FalconSAT-2 was ejected and remarkably fell back to Earth, crashing through the storage shed and landing near its delivery crate. "I guess it just wanted to come home," jested Col. Martin France, head of the Academy's Astronautics department. He continued, "We are not sorry for taking a risk on *SpaceX*.

Space would not happen without new technologies and people willing to take risks. Our FalconSAT-6 is now waiting for a ride on the *SpaceX* Falcon 9. Falcon 9 has become a well-established launch vehicle in a short amount of time because people have been willing to take a chance on it. Because of new companies like *SpaceX*, the Air Force can leverage the commercial market and optimize budgets." The FalconSAT-2 flight model—that launched and returned to the Pacific, can be seen at the Air Force Academy's Space System Research Center (SSRC) Museum (Based on Interviews with Colonel Martin France, US Air Force, Permanent Professor and Department Head, Astronautical Engineering Department, United States Air Force Academy, July 13, 2017) (France, 2017).

After the first launch failure, Musk told his discouraged employees, "*SpaceX* is in this for the long haul, and come hell or high water we are going to make this work" (Vance, 2015). Others from the space community reminded Musk of the failures of every booster known to humankind. Their first successful launch came in September 2008; only six years from the company's founding. To date, the tenacity of *SpaceX* has paid off. It has a respectable success rate and has added the *Dragon Capsule* to its inventory. As of this writing, the Dragon Capsule has resupplied the International Space Station (ISS) nine times. This is a feat that many experts said could not be done by any commercial company. The *Dragon V2* was introduced in 2014 and will carry up to seven astronauts and its engines are made entirely from additive manufactured components (3D printed), another huge advance in technology.

The company has accomplished remarkable feats that the government has never even attempted. In December 2015, *SpaceX* became the first organization to successfully land a rocket stage after launch. In March 2017, it became the first to land a re-furbished booster. In each case the booster landed on a *SpaceX*-owned sea-faring barge named *Of Course I Still Love You*. In June 2017, the company successfully refurbished a *Dragon Capsule* to supply the ISS. Companies putting their payloads on refurbished boosters will receive a discount; bringing the cost of a Falcon 9 booster launch from about $62M to perhaps as low as $40M. This is lower than current defense contractor offerings and highly competitive with overseas launches. *SpaceX* proudly makes their boosters in California with nearly all parts made in the United States meeting environmental guidelines not found overseas and personnel who are working there by choice. It is truly an American company.

How can they compete? Follow the money. *SpaceX* has always been dollar-conscious. Rocket companies have already existed for over a half-century; booster companies can provide boosters for hundreds of millions of dollars, but there is an axiom the government sometimes forgets: *if we cannot afford it we cannot have it.* Commercial companies can never forget this. In its very early history *SpaceX* came very close to being unable to financially cover its payroll and another launch failure may have put them over this cliff. No taxpayer backed bailouts—it could have been the end. Now, the commercial space company is robust with an impressive launch manifest. They are low cost and reliable. *SpaceX* has been called "disruptive" (Fernholz, 2014) to the status quo. Thank goodness for disruption: this is why the US is the economically strongest country in the world: we innovate. This is why Elon Musk immigrated here: to pursue his innovation dreams. Twenty years ago he was homeless, now *SpaceX* has already outperformed its government predecessors and current competitors (Figure 5.3). What next? Mr. Musk says the next goal is turning around a booster in 24 hours, and in a more recent statement, announced that *SpaceX* will provide the technology for humans to settle Mars (Pasztor, 2017).

Meanwhile, Musk has become a viable and highly competitive government supplier, but with new so-called *disruptive* organizations comes skeptics, and companies with Pentagon ties that are decades-old tend to be favored. In March 2014, the Air Force awarded some

Figure 5.3 Falcon 9 first stage landing on drone ship *Of Course I Still Love You*. This historic landing was the first of a refurbished booster, bringing access to space to a new level of efficiency. March 30, 2017. (Courtesy of *SpaceX*.)

$11Billion in contracts to ULA on a sole source basis. The cost was considerably higher than *SpaceX* offered on an unsolicited basis. In April 2014, *SpaceX* filed a lawsuit against the Air Force. Elon Must explained: "This is not *SpaceX* protesting and saying these launches should be awarded to us. We're just protesting and saying that these launches should be competed." The government argued that they could not allow *SpaceX* to compete until they had passed Air Force certifications, but Musk retorted that officials intentionally procrastinated certification in hopes of later securing post-government employment at ULA's parent companies. Military and civilian government defense employees frequently seek post-government retirement employment by defense contractors. On January 2015, *SpaceX* announced they would drop their lawsuit and that an agreement had been reached with the Air Force to certify them by mid-2015. *SpaceX*'s Falcon 9 rocket passed certification "ending a monopoly held by ULA, a joint venture of Lockheed Martin Corp. and Boeing Co., since its creation in 2006" (Shalal, 2015). Air Force Secretary Deborah James applauded this effort noting that it should drive down launch costs.

For the acquisition professional, there are many lessons. Once again, we have seen a superior technology originate in the private sector. Though critics have argued that *SpaceX* has had its share of launch failures, the early history of other booster companies is similar. Moreover, *SpaceX* has not only attempted re-use, but has succeeded; something traditional booster companies have not. In the early days of rocketry, booster companies were also given government funding on what were then feasibility studies: no one knew whether or not they would actually work. The government did not want to see high-tech companies dissolve—adversely affecting the US capabilities in space. These types of contracts were sensible at the time, but over the decades have caused defense industrial partners to become non-competitive. Meanwhile, commercial companies such as *SpaceX* and *Blue Origin* have taken these opportunities to innovate.

Department of defense contracting

"Never tell people how to do things. Tell them what to do and they
will surprise you with their ingenuity."

George S. Patton Jr.

What is the point of describing the evolution of cutting-edge technologies? To clarify that
they have advanced, and to emphasize that government acquisition policies typically do
not advance with them; surely not nearly as fast. The larger and more hierarchical an
organization, the slower it changes (Brown, 1998). We've come a long way technologically,
but the DOD has had trouble keeping up. *Military specifications* (*milspecs*) are a method by
which the DOD can get precisely what it wants, at least in theory. But *milspecs* have a well-
earned reputation for getting in the way of progress. A commercial product is simply not
created to adhere to *milspecs*. *Milspecs* exist based on the incorrect assumption that the best
ideas come from the government. History has shown that the very best innovation comes
from the private sector.

United States polices have fallen behind the commercial market. Visionary and states-
man The Honorable William Perry has pushed DOD policy in an effort to match the
pace of the technologies themselves, but adaptation has been lagging. Dr. Perry, once the
Undersecretary of Defense for Research and Engineering (1977–1981) and later Secretary
of Defense (1993–1997) was passionate about the need for change in the DOD. Having
had a long and successful career as a businessperson and engineer, Dr. Perry noted that
there was no chance that the government could possibly keep pace with the free market
and its incentives for leading-edge, affordable technologies, and enacted policy that called
for DOD to use commercial products when available (The Perry Memo, 1994) (Carter &
Perry, 1999). His efforts to streamline acquisition processes are some of many. Each one has
had its share of successes, but it is difficult to change a system in which the rewards for
spending are plentiful. Some of those barriers were identified: *misaligned reward systems,
entrenched networks* and *historical precedent* (Baron, 2004). The article's findings include Air
Force personnel lacking knowledge of the commercial market as well. This must change.

Implications for the Department of Defense acquisition professional

In this chapter, we have established the superior efficiencies of the free market for defense
innovation and production. We have also established the existence of entrenched systems
that prevent the DOD from changing rapidly. The acquisition professional—indeed all
those in DOD—should consider accordingly. How can our team get the best products and
services to the warfighter and the best value to the taxpayer?

Clearly not all defense products are available in the commercial market. For example:
stealth fighter aircraft. Though someday they may be purchased off-the-shelf; not today.
From an optimization viewpoint, the most efficient way the services can perform acquisition is to leverage the commercial market for the items available there. The DOD does this
to a considerable extent already. For example, there was a time when computing was not
available off the shelf and the DOD was building workstations. The government now purchases workstations off-the-shelf from companies like Dell and Hewlett-Packard. Similarly,
software to operate such workstations is also purchased off-the-shelf. It would be mind-
boggling to think of the government trying to compete with these commercial products.
By extension, the acquisition professional should consider other software available off-the-
shelf and there is plenty. Many companies build off-the-shelf software and often a market

search is not that difficult. For example, countless private companies build and operate satellites. They have the same issues with privacy and hackers that the government has, and have excellent products. Try a Google search of satellite software and see what you find. The acquisition professional should purchase these items on a firm-fixed basis and "fly-before-buy;" that is; ensure the product is functioning prior to purchase. To be an excellent steward of the taxpayers' money, the acquisition professional should get the very best product for the very best price.

Commercial-off-the-shelf & market search

Let's say you are asked to purchase a refrigerator for the Air Force. How would you go about it? Would you give specifications to a company and have them build it to your exacting needs? Of course not! You would probably go down to Home Depot or Sam's Club and see what they have. Better yet, you may be more likely to hop online and see what is available that most closely meets the Air Force needs for the best price. This is the essence of a market search. The internet is quite possibly the best tool the acquisition professional has ever had, and it doesn't just work for refrigerators. Try Googling *COTS satellite ground control*. What did you find? Once you slug through all the government sites or people selling books, and maybe even a bedding company selling camping cots, you will likely find a several companies that perhaps you've never heard of who make products for use with actual satellites. Dig deeper and you will see animations and find people who you can actually talk with and will be more than delighted to show you their product. Be careful and thorough. You will also find what one of my interviewees called "Trojan Horse COTS." Large prime contractors who do not actually build commercial products are wise to the government's interest in becoming more efficient with the use of commercial products. As well, they are wise to the government mandates that require the use of commercial products when available and they want government business. You must do your research diligently and thoroughly to be a good steward for the taxpayer as well as get the best product to the warfighter.

Other transaction authority (OTA)

Though a child of the 1950s; when the US faced Soviet threats, the OTA has morphed over the past 60 years and DOD professionals are bringing it into its own. With the realization that commercial products need to be part of the warfighters' tools, the OTA is an excellent tool for bringing them from the private sector to the government. In essence the OTA is a superb way to speed up procurement by working outside the FAR. The OT or OTA is discussed in Chapter 4.

Summary

The Air Force came into being as a separate service based on the development of a commercial product: the aircraft. And we almost missed it. So what does the future hold? It is impossible to predict what technologies will be critical, but what is for sure is that as long as there is freedom of thought and a capitalistic market, innovators are motivated to invent as they always have, and top products will come from the commercial market.

At the time of the development of a heavier-than-air machine, it is not likely that anyone predicted the countless applications for which it would be the basis, nor the many other new technologies that could be combined in different ways to aid in the defense of this young country. Once human flight was established, the next logical step was space-flight which was undertaken by the Air Force and other government agencies working together with industrial partners; motivated by the threat of communism.

The Air Force is a remarkable service that has largely utilized cutting-edge technologies, but not always as efficiently or effectively as possible. The future acquisition professional needs to embrace such technologies and should consider the following.

1. *Research what is out there.* The commercial world has always been a source of innovation from the Wright brothers, to the Silicon Valley entrepreneurs, to *SpaceX*, and many, many more. This is truly the most important job an acquisition officer needs to do well, constantly and with a vengeance. The commercial market is constantly changing—not just year-to-year, but day-to-day. Companies that have had past failures, such as the three mentioned earlier, ultimately had huge successes. The acquisition professional must bring the best technologies to defense, or we will fall behind. Technologies exist in the open market that meet and *exceed* what the government owns. You just have to find them. Gadgets and gizmos with countless military applications are available to all: our friends and enemies. Stay on top of trade journals and other periodicals.

2. *Ignore sunk cost.* The sunk cost effect is a major human shortcoming. We tend to consider sunk cost when considering future investment in technologies. "But we've already put so much into this, we cannot change now!" Any economist will know that an expenditure already spent is gone and should not be considered in future investments. If the Air Force or any other organization has invested heavily in a technology development or a company that is not producing as promised with little hope of future success, it is essential to re-examine those technologies and consider other providers or a different path.

3. *Attend trade shows and symposia.* These are not a waste of time as some may think. The diligent professional will use these to leverage the best of breed technologies from the market and educate oneself with face-to-face learning often from actual inventors and developers. Ask questions, attend meetings and study literature. These can be extremely valuable.

4. *Watch what other countries are doing.* Private industry has no obligation to the DOD. Other countries without the enormous infrastructure of the United States' tremendous universities and grant procedures look toward commercial products where no development investment is required. Recall that the Wright brothers' first customer was not the US War Department as they had hoped and envisioned.

5. *Beware entrenchment.* Don't just think outside the box, live outside the box. We are creatures of habit and not all our habits are good. Examine the way "we've always done things" and think of how we could do them better. Beware of becoming entrenched in the old ways. If you are a supervisor, try to *reverse socialize*. That is, rather than teaching new people *the way we do things here*, try to learn from them and ask how they would do it. The youth of our country is the future.

Closing thoughts

As The Honorable Roberts Gates (Secretary of Defense; 2006–2011) told the Air Force Academy cadets in a lecture in 2010, as an officer you need to have the courage to tell blunt truths; though it will not likely be popular (Gates, 2010). History has shown that people often shy away from new ideas, and military history has its share of people willing to stick their necks out to share new ideas that were not always popular. Air Force legends Billy Mitchell and Hap Arnold were both criticized for pressing forth their idea of putting armament on aircraft in order to sink enemy ships. Bernard Schriever is known as the father of the ICBM, but he had to fight many in the Pentagon who believed that any nuclear weapon should be flown by a human pilot.

What are we missing today? What commercial innovations will you speak up for use in the battlefield?

List of acronyms for Chapter 5

DOD	Department of Defense
COTS	Commercial-off-the-shelf
FAR	Federal Acquisition Regulations
ISS	International Space Station
NRO	National Reconnaissance Office
OTA	Other transaction authority

References

Baron, Sally J.F., Keeping pace: Organizational barriers to commercial product use in DOD, *Journal of Public Procurement*, 4(2), 182, 2004.

Brown, Shona L., and Kathleen M. Eisenhardt, *Competing on the Edge: Strategy as Structured Chaos*, Harvard Business School Press, Boston, MA, 1998.

Carter, Ashton B., and Perry, William J., *Preventive Defense*, Brookings Institute Press, Washington DC, 1999.

Ceruzzi, Paul, *A History of Modern Computing*, MIT Press, Cambridge, MA, 2000.

Fernholz, Tim, *The Right Stuff: What it Took for Elon Musk's SpaceX to Disrupt Boeing, Leapfrog NASA, and Become a Serious Space Company.* Quartz, October 2014.

France, Martin E.B., (Colonel, US Air Force and Permanent Professor and Department Head, Astronautical Engineering Department, United States Air Force Academy) interviews— July 13, 2017.

Gates, Robert, *United States Air Force Academy Lecture*, Colorado Springs, Colorado, US, April 2, 2010.

Howard, Fred, *Wilbur and Orville, A Biography of the Wright brothers*, Dover, IL, 1998.

Kelly, Fred, C. (Ed.) *Miracle at Kitty Hawk: The Letters of Orville and Wilbur Wright*, Da Capo Press, New York, 2002.

Lam, Bouree, Trump's two-for-one regulation executive order, *The Atlantic*, January 30, 2017.

March, James G., Exploration and exploitation in organizational learning, *Organization Science*, 2(1), 71–87, 1991.

Pasztor, Andy, New space race to Mars pits NASA vs. *SpaceX*, *The Wall Street Journal*, October 4, 2017.

Shalal, Andrea, US Air Force certifies *SpaceX* for national security launches, *Science News*, May 26, 2015.

Sorenson, David S., *The Process and Politics of Defense Acquisition*, Praeger Security International, Westport, CT, 2009.

United States House Select Committee on Government Contracts, https://en.wikipedia.org/wiki/United_States_House_Select_Committee_on_Government_Contracts, 1861.

USS Maine, https://en.wikipedia.org/wiki/USS_Maine_(ACR-1), 1895.

Vance, Ashlee, *Elon Musk*, HarperCollins, New York, 2015.

World Almanac, *Military Affairs*, World Almanac Books, 2017.

chapter six

A system and statistical engineering enabled approach for process innovation

Darryl Ahner, Sarah Burke, and Aaron Ramert

Contents

Introduction

Engineering expertise alone used to be sufficient to inform whether a Department of Defense (DoD) weapon system would perform its tasks and meet requirements. More recently, these weapon systems have many more capabilities and consequently are extremely complex. With more capability, however, come more subsystems that must meet their own requirements. There is then a need to understand the performance, reliability, integration, and interactions of these subsystems early in the weapon system's development before large corrective costs manifest. The complexity arising from these integrated and interacting subsystems and larger systems can no longer be adequately informed by engineering expertise alone. To make informed decisions on these increasingly complex systems, a culture shift within DoD acquisitions must occur so that it moves from a culture reliant on engineering judgment to one that is information-based.

Typical DoD weapon system development transitions through four phases: capabilities identification, technology development, systems technology integration for engineering and manufacturing development, and production. The capabilities identification phase conducts analysis to identify what future capabilities are needed and determine specific measureable requirements of the potential system. The technology development phase identifies the technology that currently exists, its level of maturity, and facilitates rapid advancement of any technology that is needed to achieve the system requirements. The systems technology integration phase for engineering and manufacturing development focuses on the systems engineering (SE) task of integrating technology and re-engineering any shortfalls. Finally, the production phase begins manufacturing the system *en masse*. It is at this point that any system requirement not met or any technology integration issue not successfully addressed typically becomes prohibitively expensive to correct.

Moving from one phase to another requires decisions that are dependent on assessments. These assessments typically rely on subject matter expert judgement as well as a process known as test and evaluation. While subject matter assessments alone are not adequate, when coupled with efficient and effective testing and the corresponding quantitative analysis, a powerful means of knowledge development will result. This knowledge more effectively informs both SE and acquisition decisions. In this chapter, we discuss the inherent complexity of weapon systems, provide a brief history of knowledge development within the DoD, discuss innovation of the defense acquisition program through culture change, highlight the importance of scientific test and analysis techniques (STAT) to develop the foundation of the culture change, and finally provide the future direction of innovating defense acquisition.

Complexity of Department of Defense weapon systems

Before discussing methods to drive innovative practice into the defense acquisition process, we first explain what makes modern weapon systems so complex. These systems are inherently complex because of their systems of systems (SoS) nature, their reliance in development on modeling and simulation (M&S), their reliance on software, net-centricity, and, in the future, their ability to act more autonomously. All of these aspects make efficient and effective testing and assessment of these systems challenging. In the following subsections, we expand on each of these components, all of which are current challenges within defense acquisition.

Systems of systems architecture

DoD SoS engineering is the design of systems that satisfy specific requirements and is performed under uncertainty of advancing technology and integration of component systems. It focuses on choosing the right systems and their interactions to satisfy requirements in complex environments. In DoD and elsewhere, SoS can take different forms. Based on a recognized taxonomy of SoS, the four types of SoS which are found in the DoD today are virtual, collaborative, acknowledged, and directed (Maier, 1998; Dahmann, 2008). Virtual SoS lack a central management authority and a centrally agreed upon purpose for the SoS. Collaborative SoS have the component systems interact more or less voluntarily to fulfill agreed upon central purposes. Acknowledged SoS have recognized objectives, a designated manager, and resources for the SoS; however, the constituent systems retain their independent ownership, objectives, funding, and development and sustainment approaches. Directed SoS are those in which the integrated SoS is built and centrally managed to fulfill specific purposes. Having independent, concurrent management and funding authority at both the component system and SoS levels is a dominant feature of acknowledged SoS. Typically, attention is focused on the management issues that result from the overlapping authority over decisions rather than the technical implications for SE (ODUSD(A&T)SSE, 2008).

There are seven core elements that characterize SE in SoS and which contribute to the complexity of weapon systems. These interconnected elements include: (1) translating SoS capability objectives into SoS requirements, (2) assessing the extent to which these capability objectives are being addressed, and (3) monitoring and assessing the impact of external changes on the SoS. Central to SoS SE is: (4) understanding the systems that contribute to the SoS and their relationships and (5) developing an architecture for the SoS that acts as a persistent framework for (6) evaluating SoS requirements and solution options. Finally, the SoS systems engineer (7) orchestrates enhancements to the SoS, monitoring and integrating changes made in the systems to improve the performance of the SoS. It is this lack of focus on the technical state of the SoS, the metrics that provide knowledge on the developmental state of these elements, and other technical aspects of the SoS that can have significant performance and financial implications during the latter stages of the acquisition process.

Modeling and simulation

Another source of complexity of weapon systems development is the use of simulations in evolutionary acquisition. Evolutionary acquisition consists of a baseline system being developed and produced with upgrades added at a later date. These upgrades can be either improvements of current capabilities or the addition of new capabilities that were not a part of the initial design requirement. When evolutionary acquisition is pursued, it is often useful to have a validated simulation that possesses appropriate mathematical models to assess performance. "Modeling and simulation (M&S) provides a technical toolset which is regularly used to support systems acquisition and engineering" (ODUSD(A&T) SSE, 2008). M&S is applied throughout the system development life cycle supporting early concept analysis, design, developmental test and evaluation (DT), integration, and operational test and evaluation (OT).

Because of the characteristics of SoS, M&S can be a particularly valuable tool. Models, when implemented in an integrated analytical framework, can be an effective means of understanding the complex and emergent behavior of systems that interact with each other. Models can provide an environment to help create a new capability from existing systems and consider integration issues that can have a direct effect on the operational user. M&S can support analysis of architecture approaches and alternatives as well as analysis of hardware and software requirements and solution options.

Because it can be difficult or infeasible to completely test and evaluate all the capabilities of a SoS, M&S can be effectively applied to support T&E at different stages in the weapon systems development process. In particular, M&S can be used to understand the end-to-end performance of the overall SoS prior to implementation. In some cases, it is advisable to adopt a model-based process for gaining knowledge of a system. Because of the importance of M&S, it is essential to include planning for M&S early in weapon systems development planning. This planning includes "the resources needed to identify, develop, or evolve and validate M&S to support SE and test & evaluation" (ODUSD(A&T) SSE, 2008). Effectively and efficiently planning M&S into T&E to learn more about the system under development, while a valuable method, adds additional complexity to the current weapon systems.

Reliance on software

Another source of complexity of weapon systems is their reliance on software. Software dependencies within and between systems are also complex requiring knowledge concerning their performance. "Nearly all modern technology systems depend on software to perform their functions. From remotely piloted aircrafts and smart bombs to self-driving vehicles and advanced fighter jets, software is crucial to the success of today's weapons systems" (IG 2016).

"The quantity of software that enables weapons systems today drives complexity in engineering, test, and evaluation. Defense systems use hundreds of millions of lines of code generated by defense teams, reused from known government or commercial-off-the-shelf systems, and incorporated from open sources" (Baldwin and Lucero, 2016). With this reliance on software, testing systems requires more than engineering expertise to effectively characterize the performance of the system and/or identify shortfalls of the system.

Net-centricity

Net-centricity is itself an innovative functional approach that requires an innovative knowledge development process. Along with significant increases in software have come increases in networking and information exchanges across countless combinations of system interfaces. Net-centric systems are often characterized as having a service-oriented architecture (SOA), a paradigm for organizing and utilizing distributed capabilities that may be under the control of different owners. A SOA has software architecture where functionality is grouped around processes or capabilities and packaged as interoperable services. A SOA possesses an information technology infrastructure which allows different elements to exchange data or functionality with one another as they participate in the process. The aim is a loose coupling of services and separation of functions into distinct units. This allows accessibility of services over a network in order that they can be combined and reused in the furtherance of mission accomplishment. These services communicate with each other by passing data from one service to another, or by coordinating an activity between two or more services (Dahmann, Baldwin, and Rebovich, 2009). Incorporating net-centricity into T&E planning of weapon systems is not an easy task because of the interconnected nature of many systems. Henry and Stevens (2009) state that the systems used by all users throughout the DoD are interconnected. These systems include "unmanned aerial systems, handheld systems, ground vehicles, ships, etc." T&E must adapt to requirements of interconnected systems. "The entire enterprise becomes a single complex system comprised of numerous component systems" (Henry and Stevens, 2009).

Autonomous systems

Finally, incorporating autonomy into modern weapon systems is a challenge during T&E. Autonomous systems have gained great interest in recent years and most likely will need to operate in unstructured, dynamic environments. A recent Defense Science Board (2016) study notes that "autonomous systems can be cyber-physical or totally cyber-dominated. In any case, these systems will be dominated by a software architecture and integrated software modules. The DoD historically has had difficulty in specifying, developing, testing, and evaluating software-dominated systems." Particular knowledge development challenges for autonomous systems are noted in a 2016 Scientific Test and Analysis Techniques Center of Excellence (STAT COE) workshop report in the areas of requirements and measures, test infrastructure and personnel, design for test, test adequacy and integration, testing continuum, safety and cybersecurity, testing of human system teaming, and post acceptance testing (Ahner and Parson, 2016). The current Research & Engineering Autonomy Community of Interest (COI) Test and Evaluation, Verification and Validation (TEVV) Working Group Technology Investment Strategy 2015–2018, signed by the Assistant Secretary of Defense for Research and Engineering (ASD(R&E)), states the need for rigorous test methods of autonomous systems:

> "Cumulative evidence through RDT&E, DT, & OT—Progressive sequential modeling, simulation, test and evaluation M&S and T&E at each Technical Readiness Level (TRL) and product milestone currently provide an invaluable resource not only to verify and validate that a system satisfies the user requirements, but also to aid in technology development and maturation. However, the development of effective methods to record, aggregate, and reuse T&E results remains an elusive and technically challenging problem."

As just discussed, the complexity of weapon systems is illustrated by their SoS nature, their evolutionary acquisition relying on simulations, their reliance on software, net-centricity, and their ability to act more autonomously. All of these aspects of system complexity make the efficient and effective testing and assessment of these systems challenging. The current acquisition process is heavily reliant on engineering subject matter assessments which alone are not adequate to make informed decisions. When combined with efficient and effective testing and analysis that generates quality and insightful system performance information, a powerful means of knowledge development results. This knowledge development process that informs DoD complex weapon systems development must support the innovation, agility, and quality of those weapon systems while addressing the aforementioned aspects of system complexity.

Brief history of test and evaluation in the Department of Defense and current innovative efforts

Before presenting a description and method of implementation of this innovative knowledge development process, it is useful to understand the history of knowledge development, namely T&E, within the DoD. We provide a brief history of T&E and attempts at innovation in the DoD from which we learn and develop our approach to changing the culture to achieve a system and statistical engineering-enabled approach for process innovation.

In 1971, in order to oversee both DT and OT, the office of Director, Defense Test and Evaluation, was formed under the Office of the Director of Defense Research and Engineering, who was responsible for major acquisitions. The office was formed through a series of three memoranda by Deputy Secretary of Defense David Packard in response to recommendations by President Nixon's Blue Ribbon Defense Panel of 1970. In 1977, the need for independent OT saw it moved under the responsibility of the Assistant Secretary Defense for Program Analysis and Evaluation. However, this change lasted only a short time and in late 1978 it was moved back to the Director, Defense Test and Evaluation. In 1983, independent OT again arose as an issue resulting in Congress establishing the current office of Director, Operational Test and Evaluation (DOT&E), once again separating the DT and OT functions. In 1994, with the reassignment of live fire testing to DOT&E, the Director Test, Systems Engineering, and Evaluation office was formed. On June 7, 1999 (28 years after Packard created it) Secretary of Defense William Cohen disestablished the test office within what had become the Office of the Undersecretary of Defense for Acquisition and Technology and realigned DT responsibilities as a function under other offices. During those first nearly 3 decades, all emphasis in T&E in the department continued to be on (OT), and Cohen's decision was intended specifically to strengthen the Office of the DOT&E; however, it virtually eliminated oversight of DT. Congress would reverse this 10 years later with passage of the 2009 Weapon Systems Acquisition Reform Act that established the Director, Developmental Test and Evaluation (DDT&E) position (Fox, 2011).

This fluid history of overseeing T&E does not lend itself to supporting innovative acquisition nor does it lend itself to having a system and statistical engineering enabled approach for process innovation. However, with the establishment of the DDT&E, which later changed to Deputy Assistant Secretary of Defense for Developmental Test and Evaluation (DASD(DT&E)), DT was about to begin a path toward innovation through the Scientific Test and Analysis Techniques in Test and Evaluation Implementation Plan.

As systems became more complex, testing techniques remained the same. "New programs appear to be more complex than their immediate predecessors in terms of technology, functionality, and, perhaps to a lesser extent, their operational concept" (Drezner, 2009). The relative complexity of the weapon system itself is captured in technical complexity. Elements of technical complexity include the use of electronics, information technology, and software to provide critical functionality and capability beyond more traditional means. That these are increasing can be measured by the percent of acquisition program funds devoted to these technologies. These technologies reside in sensors, data processing, automation, communication, and data exchange. Many recent weapon systems are multifaceted, multifunction, and multimission systems that include many more specific functions and performance capabilities than predecessor programs (Drezner, 2009).

The proliferation of electronics in both performance and quantity is a major contributor to increasing weapon system complexity (Dietrick, 2006). In directing programs that have been problematic, managers for the government, the prime contractors, and the commercial subcontractors shared one common feature: they underestimated the complexity of requirements, integration of subsystems, and the interaction of changes in one subsystem with new demands on others (Berteau, 2009).

As the DoD continues to push innovation within its acquisition process, several elements within a framework are required for this innovation (Drezner, 2009):

- National factors, which include education level, strength in science and technology, and supporting infrastructure (e.g., communication and transportation)
- Research & development investment in a wide variety of projects, technologies, and sectors
- Status and attractiveness of the sector (e.g., excitement and dynamism) as indicated by the degree to which industry in that sector is admired by consumers and students, the degree to which it is pushing the state of the art, and its ability to attract and retain top people

- Competition in the sector, as determined by company strategies, industry structure, and rivalry
- Demand conditions—in other words, the customer demanding capabilities requiring innovative new technologies
- Related supporting industries including lower tiers and science and technology (S&T) base

Additional factors affecting innovation or the conditions that facilitate innovation not explicitly identified in the earlier model include the following (Drezner, 2009):

- An institutional and regulatory environment that encourages new concepts
- Early adopters who are willing to buy and use initial versions of the innovation
- A potential for significant demand for the product
- High potential payoff
- Minimal barriers to entry

Innovation of defense acquisition program knowledge development through culture change

Cultural change is difficult to achieve, especially in large organizations, and the DoD is among the largest in the world. Components of an organization typically consist of purpose and tasks, intellectual or mechanical processes, hierarchy of authority or structure, and people. Within these components are an interlocking set of goals, processes, roles, collaboration, coordination, cooperation, values, attitudes, and assumptions. Over time, these components settle and become a reinforcing system that is difficult to change.

A Forbes article entitled "How Do You Change an Organizational Culture?" by Steve Denning (2011) presents a strategy that all organizational tools need to be put into play to increase the likelihood of success, but argues that the order matters. These tools consist of leadership tools, management tools, and power tools. Leadership tools entail developing a vision and providing inspiration for change. Management tools include the activities of strategic planning, role definition, incentives, and training. Power tools consist of coercion and regulations. While implementing these tools methodically may lead to cultural changes in some organizations, the DoD is usually considered a bureaucracy characterized by adherence to fixed rules, specialization of functions, and a hierarchy of authority. These characteristics may require an approach to culture change differing from smaller organizations. The adherence to fixed rules requires new processes to be addressed by those rules in the form of requirements or regulations. To perform a new function, either a current group must be identified to be trained or educated, or a new specialized group must be resourced and formed. Finally, the new process must be requested or demanded by leadership in authority. To implement an innovative knowledge development process, these challenges need to be overcome to achieve the cultural change desired.

Culture change and innovation are tightly coupled in large organizations. Process innovation intervention is the act of incrementally setting the conditions to achieve enterprise wide acceptance of an improved process throughout a large organization. A firm understanding of methods and practices, documented best practices and case studies, and well-written policies and regulations are necessary conditions before culture change can occur within a large organization. This process innovation is depicted in

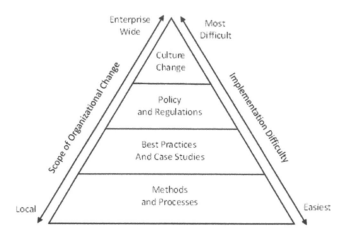

Figure 6.1 Process innovation intervention.

Figure 6.1 as a pyramid since each element is a foundation of the higher element. As seen in Figure 6.1, a culture change cannot occur without the structure of supportive policies and regulations, which in turn are built around established best practices that highlight the capabilities and strengths of the prescribed methods and processes.

In order to begin the cultural change to improve quantifiable knowledge for acquisition within the DoD, the DASD(DT&E) developed the Scientific Test and Analysis Techniques in Test and Evaluation Implementation Plan in coordination with the Army, Navy, and Air Force T&E executives and the office of the DOT&E. The plan calls for changes in policy, workforce development of the T&E workforce, and establishment of a STAT COE to achieve an innovative knowledge development process with more effective systems and statistical engineering. The STAT COE was established in 2012 to assist acquisition programs in developing rigorous, defensible test strategies as a source for the required highly qualified people with advanced degrees and knowledge of the acquisition process. These actions directly contribute to the lower two levels of the pyramid in Figure 6.1 by establishing and implementing STAT in DoD testing.

The DoD's efforts using STAT enable innovation in the form of a more information-based decision-making process in several ways. STAT requires a pool of highly qualified people with advanced technical degrees and knowledge of the acquisition process to be effectively implemented. Innovation is achieved incrementally through small continuous improvements in how T&E is conducted. STAT enables technology innovation by generating the knowledge to mature technologies that either fulfills a requirement gap or by generating the knowledge to inform the performance of a paradigm shift in new technologies, such as autonomous systems, hypersonics, and directed energy. The STAT COE innovates the DoD business model by the creation of a pool of highly qualified people that can be drawn upon by acquisition programs but still are considered an integral part of the program and not an outside entity, thus avoiding the problems of additional cost and hiring of scarce human capital. This pool of highly qualified people provides a significantly improved service that generates new customer value using (and not replacing) high quality T&E professionals in other areas. Finally, the STAT COE is innovative by adding this STAT capability without imposing a financial burden on the acquisition programs. The correct use of STAT lowers test costs by more efficiently making use of test resources and lowers acquisition costs through a more effective T&E process resulting in significantly higher return than the initial cost of the pool of highly qualified people.

Scientific test and analysis techniques

The STAT strategies allow programs to more effectively quantify and characterize system performance as well as provide information that reduces risk. STAT is defined as the scientific and statistical methods and processes used to enable the development of efficient, rigorous test strategies that will yield defensible results. STAT consists of methods and processes that encompass various techniques including design of experiments (DOE), observational studies, reliability growth, survey design and statistical analysis used within a larger decision support framework.

The primary challenge in applying STAT to DoD testing is the broad scale and complexity of the systems, missions, and conditions. While advanced STAT methods are used in industry, manufacturing, and healthcare environments more frequently, the DoD lags behind in adopting these techniques in part because of the complex and constrained nature of DoD testing as discussed in Section 2. In order to address this complex environment, the STAT COE has emphasized a SE approach to decompose the mission, system, or requirement into smaller pieces. One effective way these portions can then be readily translated into rigorous and quantifiable test designs and/or strategies is by using DOE. Figure 6.2 shows a flow diagram that summarizes the infusion of STAT into the DoD T&E process. The procedure begins with the requirement of interest and proceeds through the generation of test objectives, designs, and analysis plans, all of which can be traced directly back to the requirement.

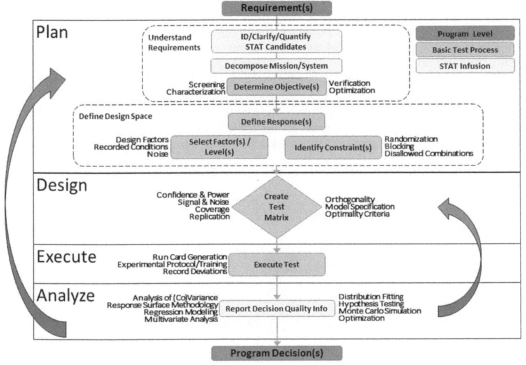

Figure 6.2 Schematic of STAT in the test & evaluation process.

DOE is the systematic integration of well-defined and structured strategies for gathering empirical knowledge about a process or system using statistical methods for planning, designing, executing, and analyzing a test. The end goal of a designed test is to produce clear results and meaningful analysis that leads to informed decisions and the best possible course of action. The DOE process can be explained by following the flow diagram in Figure 6.2.

Within the plan phase of Figure 6.2, there are two actions: understand the requirements and define the design space. The top portion, understand the requirements, is further subdivided into three actions: identify STAT candidates, understand the system and the mission, and determine the test objectives. At this point there is often a desire to rush to the computer and create a test design, but it cannot happen until T&E personnel fully understand the requirements of the system, the purpose of the test, and how to measure system performance. In other words, creating the test design cannot happen until the test is well-planned.

Planning is one of the most difficult phases of the process and cannot take place in a solitary environment. Planning requires input from operators, logistics personnel, analysts, subject matter experts, engineers, and the overall decision makers to understand the requirements and objectives of the test(s). There are often multiple documents which contain the necessary information and they can be incomplete and contradictory. Understanding these materials to develop an effective test plan requires a good test team to gather what they can, learn the rest, and (often) make educated assumptions and compromises.

The first step in successfully designing a test is to understand the requirements. Within the DoD, some common reference documents are the test and evaluation master plan and the capabilities development document. Requirements are the starting and end point for T&E. If the requirement is not understood clearly at the beginning of the process, the test team may plan a test that will not produce the data needed to address the requirement and adequately inform the decision maker. Understanding what is written, what is missing, and/or what needs to be clarified in the requirement is the first step in effective DOE implementation. Without a clear understanding of the requirement, what conditions it pertains to, how it factors into the mission, and how it can and should be tested, the T&E process as outlined in Figure 6.2 is unguided. This crucial first step drives the development of the test objectives, responses, factors, designs, and analysis plans. The amount of detail (or lack thereof) associated with a requirement will directly impact the amount, type, and quality of the data collected from any T&E event. Figure 6.3 depicts a translation of system requirements to performance measures. Key questions to ask when discussing what a requirement says and

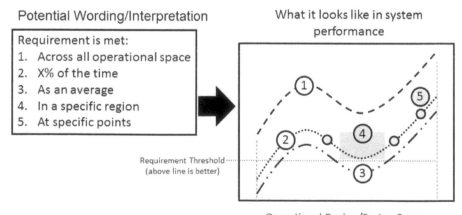

Figure 6.3 Translating requirements to performance.

how it should be evaluated include: (1) What remains to be clarified in the requirement? (2) What is my test objective to address the requirement? (3) Can I effectively characterize the system and if not, where are the un-testable regions? These questions will lead the test team into the design process and help with strategy development and resource planning. The test design must produce data that allows the analysis to address the requirement.

A clear understanding of all requirements by all stakeholders early in the process will inform the planning process, resulting in rigorous and defensible data. Many systems do not meet requirements at the outset and the program manager desires information on deficiencies so he can direct resources to correct them. In the SE process, testers help address performance issues and make improvements. Rigorous and defensible data focus this effort.

Once the requirements have been examined and distilled, the next step is to understand the system as it exists and the mission it is designed to perform. From these two steps, the team can define why the test is necessary and clearly define the test goal(s). The goals should be derived from the system requirements and be objective, unbiased, measurable, and of practical consequence (Coleman and Montgomery, 1993). The objectives should be referenced throughout the planning process to ensure that subsequent steps and decisions produce a relevant test plan. Objectives may be to characterize the performance of the system across several test conditions, identify factors that affect the response, validate performance in a simulation, optimize performance of the system in a specific region, or compare new versus legacy systems. If the test has some sort of pass/fail criteria, the consequences of a failure should be noted. They can include rejecting the system, requiring alterations, or simply purchasing an extra unit as a spare.

With the test objectives defined, the test team move to the second part of the plan phase in Figure 6.2, and the design space can be created. First it is important to determine the responses to record that will best address the objectives. The responses are the quantifiable dependent outputs of the system, which are influenced by the independent or controlled variables (factors). The responses must be observable, recordable, and should have a relationship to the test objectives. Defining the response(s) is not always an easy task. The response is ideally a continuous metric as opposed to a binary measure. Many response variables naturally tend to be binary such as whether a weapon hit or missed a target, a go/no go decision, or whether something is operationally effective or not. However, continuous metrics provide more information than binary metrics and result in more efficiently designed tests since they require fewer runs. Careful thought should be made to translate binary metrics into continuous metrics. For example, rather than measure hit or miss, measure distance from the aim point. Instead of pass/fail of a quality characteristic crossing a threshold, measure the change in that metric.

After the responses have been established, the next step is to determine the factors, the system inputs that potentially have an effect on the response (and therefore the performance of the system). There are three types of factors: control, hold constant, and noise. Control factors are purposefully varied during the test so that their specific impact on the response can be measured. Hold constant factors may also have an impact on the response, but are held constant and remain unchanged during the test. Hold constant factors may not be of primary interest or may be too difficult to control during the test. Noise factors likely influence the response, but cannot be controlled in real life and/or during the test. The noise factors can be further broken down into measurable and unmeasurable noise factors. The measurable factors, such as wind speed or component age, are recorded so their potential influence on the response can be accounted for. The unmeasurable (and often unknown) factors are best nullified by randomizing the test sequence. Randomization, one of the core principles of DOE, minimizes the effect of lurking variables—those factors that were not accounted for in the test.

A designed experiment focuses on control factors. Typical factors may include configurations, physical and ambient conditions, operator considerations, etc. Factors should not be excluded from consideration without careful thought and analysis. Brainstorming using a cause-and-effect diagram is one of the most effective methods to develop a comprehensive list of potential factors to include in the test. This process must be done collaboratively by the test team so that no factors are unintentionally excluded.

Levels are the values that each factor is purposefully set to during the test. For each factor included in the test, we need to determine the levels that it will be set to. Ideally, the factors are continuous (can take on an infinite number of possible values) since continuous measures provide the most information on system performance. Using continuous factors also allows you to make predictions at values of the factors that were not specifically observed in the test. While we gain the most information from a continuous factor, we typically set it to 2 or 3 settings initially in a designed experiment. Restricting the number of levels of a continuous factor is done in initial phases of testing because it is easier to identify factor effects. Additional levels can be included in later phases of testing to refine models if necessary, which we discuss in more detail later in this chapter. The next best option is an ordinal factor which can be set to a number of fixed ordered values between settings. The final option is a categorical factor, which can have any number of levels with no fixed relationship between them. However, more levels require more test runs to be able to model and determine the impact the factor has on the response.

When all of the factors and levels are agreed upon, there may still be some work to determine if there are constraints. A factor may be restricted from a level because of limitations to the system or test facility, because of safety or any other prudent reason. A disallowed combination occurs when a given set of levels for more than one factor cannot (or should not) be set at the same time. Possible reasons for declaring a set of levels to be a disallowed combination include the inability of the system to operate in that configuration, because the testing facility will not accommodate the configuration, or because there is no value in the information obtained when testing the combination. Examples of these situations are testing a car's cruise control in reverse, attempting a 500 yard shot at a 300 yard shooting range, and testing night vision devices during the day.

One of the primary advantages of using DOE in DoD testing is the ability to build an empirical model of the performance metric. This model allows you to identify the important factors affecting the response in addition to the magnitude and direction of that effect. Well-designed tests also allow you to efficiently identify any interaction effects. Two-factor interactions occur when the effect on the response of one factor depends on the level of another factor. Figure 6.4 shows an example of an interaction plot for a notional test with

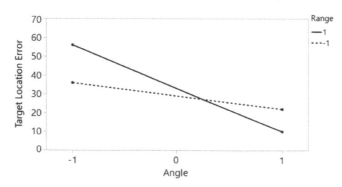

Figure 6.4 Notional example of a two-factor interaction.

factors range and angle on the response, target location error (TLE). In this example, the difference in TLE due to angle is stronger when range is at the high level (solid line). Because the factor levels are actively manipulated by experimenters during the test (rather than simply observed), DOE also allows you to establish causal relationships between the factors and the response.

Creating a test matrix

With the plan phase in Figure 6.2 complete, the next step is to create the test matrix. This is done with software which will take into account the difficulty in changing the factors and the purpose of the test. An ideal design is completely randomized to reduce noise, but if some factors are very difficult to change this may not be feasible. The software will take that into account and randomize where possible. The test matrix is designed in accordance with the test objective whether that is to screen for important factors, model the response in a specific operating envelope, or optimize the performance of the system.

The choice of the test matrix should be based on the objective of the test itself. A well-designed test will allow you to create a statistical, empirical model to quantify the effects of the factors on the response across the design space. The choice of design will determine the type of model that can be estimated using the results of the test. The model can then be used to predict values of the response throughout the test space, even at conditions not tested. Different objectives will lead to different design choices. For example, if the objective of the test is to identify the factors that have the most impact on the performance measure, a screening design such as a factorial or fractional factorial design is often a good choice. If the objective is to identify the factor levels that optimize a performance measure, a response surface design such as a central composite design may be appropriate. If the test is a computer experiment that has a deterministic response (i.e., you observe the same response value when the same factor levels are input into the experiment), then a space filling design such as a sphere packing or Latin hypercube design are common design choices. Computer-generated optimal designs are common design choices when there are constraints on the test space or when the expected model of the response has an unusual form (i.e., there are high order effects such as cubic terms in the model). Following the process outlined in Figure 6.2 by identifying the test objectives, responses, factors, and any constraints will generally lead to a clear design choice.

When planning a test, it is commonly recommended that at least 80% of the test process should be devoted to the planning phase (Montgomery, 2017). One method to do this is to evaluate a proposed test matrix prior to selecting a final design for the test. One of the biggest constraints in the DoD is a limited test budget, leading to a small number of runs available in the test matrix. The final choice of design must then balance the tradeoffs of the run size and the various properties of the design. One metric commonly advocated by T&E personnel is the power of the test, or the probability that a factor effect (main effects and/or two-factor interactions) will be detected given that the effect actually has an impact on the response. The power of the test can be estimated *prior* to testing using common assumptions associated with the empirical model (see Montgomery [2017] for complete details). Power is dependent on the signal-to-noise ratio (SNR), a ratio of the size of the effect that is practically important to be able to detect divided by the estimated variability due to noise in the system. Power greater than 80% for the desired SNR is a typical threshold when evaluating a design. One of the best (but most expensive) ways to increase the power of the test is to increase the run size.

While power is an important metric used to evaluate a proposed design, there are many other metrics that should be considered when comparing test matrices, including the confounding or alias properties of the design and the prediction variance. The STAT COE is working to build the foundation of the culture change pyramid (Figure 6.1) by advocating the use of these additional metrics when planning a test in DoD acquisitions. We highlight just a few of these methods in the following.

Confounding of effects occurs when a term in the proposed model cannot be distinguished from another term in the model. For example, a two-factor interaction between factors A and B may be confounded with the two-factor interaction between factors C and D. Once the test has been executed and a model is fit with the data, if the interaction term AB is statistically significant, we cannot resolve whether this effect is actually due to the interaction between AB or CD. An ideal test matrix will have little to no confounding so that conclusive decisions can be made after the test. One way to evaluate a design in terms of aliasing is a color map of correlations (Jones and Montgomery, 2010). This plot shows the correlation between each pair of terms in the proposed model. If the correlation between terms is 0, then there is no confounding. If the correlation is 1, there is perfect confounding between those two terms. Any number between 0 and 1 indicates partial confounding. Figure 6.5 shows two examples of a color map of correlations for a design with 8 factors. A black square represents a correlation of 1, the lightest gray represents a correlation of 0, and a medium shade of gray represents correlation between 0 and 1. The correlation plot for this example is a 36 × 36 matrix where each row and column represents the 8 main effects and 28 two-factor interactions possible for an eight factor test. The main effects are listed first and two-way interactions second. The plot can be thought of as having 3 sections of interest: the top left corner (8 × 8) shows the pattern of correlation between main effects. The bottom right corner (28 × 28) shows the correlations between all pairs of two-factor interactions. The off-diagonal section (top right (8 × 28)) shows the correlation structure for main effects by two-factor interactions. For the design in Figure 6.5a, there are several two-factor interactions that are perfectly confounded with each other, as seen in the lower right corner of the figure. The design represented in Figure 6.5b has fewer runs, resulting in more aliasing between model terms. Note that there are only a

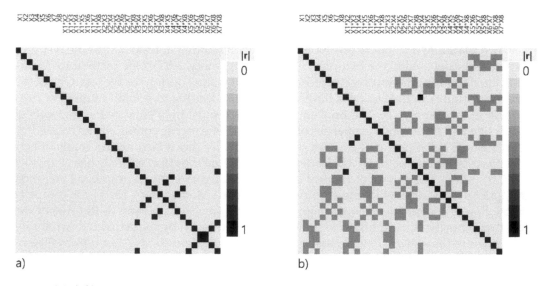

a)   b)

Figure 6.5 (a,b) Example color maps of correlations.

few terms perfectly confounded with each other; however, there are many effects with partial confounding. An ideal color map of correlations plot will have all light gray in the off-diagonals as this indicates that no model term is confounded with another. However, this often comes at the expense of additional runs in the experiment.

An additional design metric that is frequently considered is the prediction variance. A common objective of a test is to make predictions of future values with a given degree of confidence. To make these prediction intervals meaningful, the test points should cover the design space such that the prediction variance throughout the design region is low. One method to analyze the prediction variance is to use a fraction of design space (FDS) plot (Zahran et al., 2003). FDS plots display the prediction variance of the response for a given model across regions of the design space. These plots provide a simple way to compare designs in their potential ability for prediction. Figure 6.6 shows two examples of an FDS plot from the same designs represented in Figure 6.5. The ideal plot is flat across most of the design region with low values in prediction variance. Note that because of the smaller run size, the prediction variance is much higher for the design in Figure 6.6b.

Another consideration to make when choosing a test matrix is the type of model that can be fit. Screening designs typically allow you to estimate main effects and some two-factor interactions, but not higher order terms such as quadratic terms. This is because the goal of a screening experiment is to identify the important effects; follow-on testing can be used to refine the model of the response as necessary. If there is previous testing or subject matter expertise that suggests that a higher order model will be necessary to adequately model the response, the test strategy and choice of design should reflect that knowledge. However, one of the greatest capabilities of using DOE in testing is the ability to test sequentially.

Knowledge discovery using sequential experimentation

As discussed previously, the results of the test can be used to build an empirical model of the response. Many tests initially have a long list of potential factors that may have a statistically significant effect on the response in some way. A common assumption in DOE is the sparsity of effects principle (Montgomery, 2017). Sparsity of effects means that the variability in the response can typically be explained by only a subset of the potential factors in terms of main effects and two-factor interactions. The principle states that higher order terms are frequently negligible. One large experiment that tests whether every factor

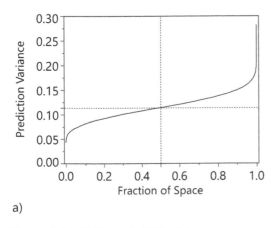

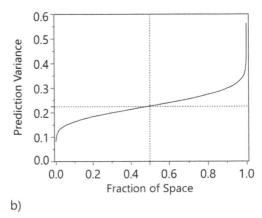

a)   b)

Figure 6.6 (a,b) Example FDS plots.

Block	Design	Model
1: Screen – linear model	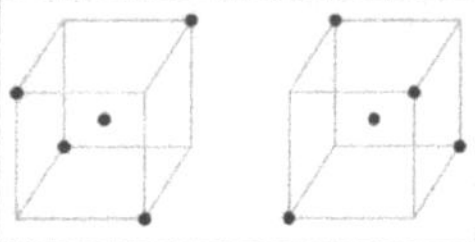	$Y = \beta_0 + \sum_{i=1}^{k} \beta_i x_i + \text{some} \sum_{i<j} \beta_{ij} x_i x_j + \varepsilon$
2: Augment – Interaction model	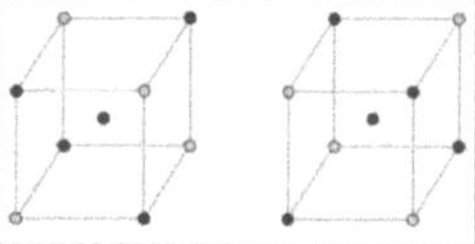	$Y = \beta_0 + \sum_{i=1}^{k} \beta_i x_i + \sum_{i<j} \beta_{ij} x_i x_j + \varepsilon$
3: RSM – 2nd order model	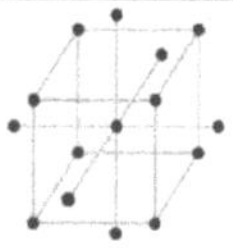	$Y = \beta_0 + \sum_{i=1}^{k} \beta_i x_i + \sum_{i<j} \beta_{ij} x_i x_j + \sum_{i=1}^{k} \beta_0 x_i^2 + \varepsilon$
4: Validation runs	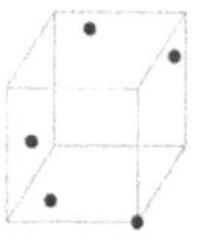	**Actual** 0.315 | **Predicted** (0.30, 0.33) | **Valid** ✓

Figure 6.7 Potential sequential test strategy for a test with four factors. (From Simpson, J., *Testing via Sequential Experiments*, Scientific Test and Analysis Techniques Center of Excellence, Dayton, OH, 2014.)

has a high order effect on the response is, therefore, an inefficient test. A better approach is to build the empirical model in stages by testing in phases so that you start with an initial test that allows you to screen for significant factors and then disregard factors that are not statistically significant in future stages of testing. This process allows you to isolate the key factors that influence the response and refine the empirical model by fitting a potentially higher-order model with the remaining factors.

Figure 6.7 shows a potential progression of testing for a test with four factors. The initial stage allows you to determine the significant factors by focusing on the main effects and some (but not all) two-factor interactions. This initial stage of testing may take 10 runs or more if replicates are done. The second phase of testing allows you to resolve any confounded interaction terms among the factors to determine which two-factor interactions are actually causing a change in the response. After the initial two stages of testing, suppose only three factors were determined to be significant. Including center runs (the points in the center of the design space) allows you to determine whether quadratic terms should be included in the model. Stage 3 progresses testing to estimate and determine the statistical significance of the quadratic terms of only the three remaining factors. The final stage typically consists of 5 to 10 test points to validate that the model performs well for conditions in the interior of the design space, typically at locations not previously tested.

In total, the number of runs for this sequential test is between 40 and 45. This efficiency in runs is possible because you can leverage information learned in previous testing to inform the next stage of test. For example, the initial ranges of factors may be quite wide to identify active main effects of a factor. As testing progresses, the range of the levels may decrease in order to focus on a particular region of the design. Alternatively, because there may not be a clear understanding of the factors of a system, the initial ranges of a factor may not be wide enough and follow-on testing is adjusted to move outside the original test space.

A sequential test strategy allows better knowledge discovery of the system under test. Questions that can be answered more easily using a sequential approach rather than a "one-shot" test include: (1) how do the factors affect the response and by how much; (2) how is the system expected to perform at conditions not tested; (3) are there areas of the design space that perform better or worse than the specified requirements; (4) which conditions provide optimal system performance; and (5) what are the tradeoffs in the system if there are multiple, competing objectives. One large test will not be able to answer all of these questions.

Example of sequential testing

Consider for example a notional test to characterize the performance of a missile warning system. It is unknown how the warning system will behave under a variety of conditions. Responses of interest include the time it takes to signal the presence of a missile and the missile detection rate. There are many factors that may affect the performance of the system as determined by several subject matter experts. These include: the environment (urban vs desert), time of day (day vs night), target intensity (low watts vs high), approach angle (0 vs 5 degrees), angular motion (0 vs 0.1 rads/sec), sensor resolution (0.1 vs 6 millirads), sensor sensitivity (0.1 vs 5 picowatts/cm^2), and frame rate (30 vs 60 Hz). Not all of these factors will likely affect the performance of the warning system due to the sparsity of effects principle. A sequential test strategy can first identify the critical few that impact performance. Follow-on testing may then be done to refine the empirical model and perform validation runs.

The subject matter experts initially identified the time of day to be a potential factor and specified two levels: night vs day. As discussed previously, the information obtained from categorical factors is much more limited than that obtained from continuous factors. An alternative measurement of time of day that can be used is illuminance, a measure of the intensity of illumination on a surface. Illuminance, measured in lux, can be as high as 100,000 in direct sunlight and as low as 0.0001 on a moonless, cloudy night. For this test, the illuminance was chosen to range from 5, which represents a dark night, and 10,000, which represents full daylight, but not in direct sun. The target intensity was also initially classified as a categorical factor (low vs high watts). To make this a continuous factor, these levels can be translated into a numeric low and high level using subject matter expertise.

A potential initial test is shown in Table 6.1, with all factors coded to be between –1 and 1. The design is a fractional factorial design that will allow you to determine the main effects and some two factor interactions that have an effect on the performance of the warning system. The power of the main effects is high (>0.95 for all factors for an SNR of 2) and four center runs are included in the design to determine if the response is characterized by any quadratic effects. With eight potential factors, this initial screening design can be followed by a second phase of testing once the primary factors driving changes in the response are identified in the preliminary phase 1 analysis. Figure 6.8 shows the color map of correlations and FDS plot for this design.

This example provides the initial phase of a potential sequential experiment. Because there are eight potential factors that may have an effect on the response (in addition to interactions and quadratic terms), designing one large to test to investigate all these potential effects is prohibitively large and therefore extremely inefficient. The current culture in the DoD, however, favors one large test plan. STAT emphasizes sequential learning so that future testing can incorporate the knowledge gained in previous testing. If the initial test in this example indicates that only 3 factors have a significant effect on the response, the next phases of testing are greatly reduced and focus on the vital few to best refine the empirical model of the performance metrics.

Table 6.1 Phase 1 screening design for missile warning system

Run	Environment	Illuminance	Target intensity	Approach angle	Angular motion	Sensor resolution	Sensor sensitivity	Frame rate
1	Urban	1	1	1	1	−1	−1	−1
2	Desert	−1	1	−1	1	−1	1	−1
3	Urban	−1	1	−1	1	1	−1	1
4	Desert	0	0	0	0	0	0	0
5	Desert	1	1	1	1	1	1	1
6	Desert	−1	−1	−1	−1	1	1	1
7	Urban	−1	−1	−1	−1	−1	−1	−1
8	Desert	0	0	0	0	0	0	0
9	Urban	1	1	−1	−1	1	1	−1
10	Urban	−1	1	1	−1	−1	1	1
11	Desert	1	1	−1	−1	−1	−1	1
12	Desert	1	−1	1	−1	−1	1	−1
13	Urban	0	0	0	0	0	0	0
14	Desert	−1	−1	1	1	−1	−1	1
15	Desert	1	−1	−1	1	1	−1	−1
16	Desert	−1	1	1	−1	1	−1	−1
17	Urban	1	−1	1	−1	1	−1	1
18	Urban	−1	−1	1	1	1	1	−1
19	Urban	1	−1	−1	1	−1	1	1
20	Urban	0	0	0	0	0	0	0

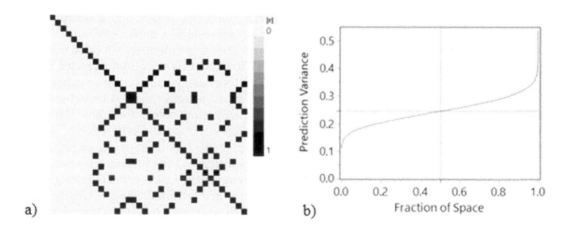

Figure 6.8 (a) color map of correlations and (b) FDS plot for the phase 1 screening design.

Innovation success indicators

Looking at success through the Process Innovation Intervention pyramid depicted in Figure 6.1, several key elements have been achieved without achieving the cultural change. Some methods and processes were already well established such as DOE (Coleman and Montgomery, 1993) while others have been developed, such as that depicted in Figure 6.2, which form a solid foundation of methods and processes. In its first five years, the STAT COE has partnered with over 41 major acquisition programs and has developed and disseminated best practices and case studies to illustrate how to implement these methods and processes within the schedule and budgetary constraints of real life program management. Acquisition policies and regulations have been changed to require the use of STAT for both DT and operational acceptance testing. The elements in Figure 6.1 appear to have been met without the resulting culture change to a more information based decision process being fully realized. So one might wonder, "What is missing?"

In the 2016 DOT&E Annual Report to Congress, the Director stated:

> Since 2012 when the STAT COE was formed, I have noted that programs who engage with the STAT COE early have better structured test programs that will provide valuable information. The STAT COE has provided these programs with direct access to experts in test science methods, which would otherwise have been unavailable.

The Director insightfully noted that the improved outcomes of an information-based decision process that used both engineering judgement and STAT over a mainly engineering judgement based process occurred for the "programs who engage[d] with the STAT COE early." The initial successes of the STAT COE are encouraging but many programs still do not have access to personnel that provide the capability of STAT. The STAT COE currently is only able to interact with a subset of major defense acquisition programs but without additional support the full potential of the COE cannot be realized. With adequate support from DASD(DT&E) and the program managers the STAT COE can provide expertise to all levels of acquisition programs and build the base levels of the pyramid in Figure 6.1. This would result in more efficient testing and analysis for the increasingly complex systems throughout the DoD.

The DOT&E recognized this when he stated:

> However, the COE's success has been hampered by unclear funding commitments. The COE must have the ability to provide independent assessments to programs (independent of the program office). Furthermore, the COE needs additional funding to aid program managers in smaller acquisition programs. Smaller programs with limited budgets do not have access to strong statistical help in their test programs and cannot afford to hire a full-time PhD-level statistician to aid their developmental test program; having access to these capabilities in the STAT COE on an as-needed basis is one means to enable these programs to plan and execute more statistically robust developmental tests. Finally, the STAT COE has also developed excellent best practices and case studies for the T&E community (DOT&E, 2016).

The DOT&E statement suggests that although the methods and processes are developed, they are only accessible to programs through organizations like the STAT COE; and although best practices and case studies are developed, their accessibility is a challenge and is currently not adequate to achieve the information-based outcomes desired.

In addition to the accessibility of using STAT capability, policy and regulations requiring STAT are necessary, but not wholly sufficient. Without thought leaders who require from their acquisition organizations a transition from an industrial-era engineering judgement decision making culture to an information-based decision making culture that uses both STAT and engineering judgement, the gains of using STAT cannot be fully realized. Culture change to a STAT-infused T&E process cannot propagate throughout the DoD without leadership championing the cause. Similar to the success of Six Sigma in industry (Snee and Hoerl, 2003), unless leadership understands and advocates the use of information-based decision making through DOE and other statistical techniques, change will not occur.

Conclusion

Innovative knowledge development processes are essential to support DoD complex weapon systems development, a process characterized by rapidly maturing technologies. The complexity of these systems arises from their SoS nature, their evolutionary acquisition relying on simulations, their reliance on software, net-centricity, and, in the future, their ability to act more autonomously. All of these aspects of system complexity make the efficient and effective testing and assessment of these systems challenging. To meet this challenge requires a transition from an industrial-era engineering judgement focused decision making culture to an information-based decision making culture.

The history of developmental test and evaluation within the DoD is fluid with a dedicated emphasis on DT within the Office of the Secretary of Defense that is inconsistent. With more complex systems becoming the norm, the need for a more information-based decision process that uses both engineering judgement and STAT over the most recent mainly engineering judgement focused based process is required.

The Scientific Test and Analysis Techniques in Test and Evaluation Implementation Plan addresses these changes by incrementally setting the conditions to achieve enterprise wide acceptance of an improved, innovative knowledge development process throughout the DoD in the form of STAT. This process accomplishes its development objectives through close integration with the SE process. Several components inform SE design decisions, including contractors, program engineers, system subject matter experts, and test results. Through collaboration with the program engineers and subject matter experts, the STAT COE is at the forefront of changing the DoD acquisition culture by injecting rigorous, defensible, and innovative sequential test methodologies and processes into this knowledge accumulation.

STAT are methods and processes that encompass various techniques including DOE, observational studies, reliability growth, survey design, and statistical analysis used with a larger decision support framework. DOE is one STAT technique in particular that has a wide, powerful application. We use this technique to illustrate the STAT innovative knowledge development process.

The initial implementation of STAT has yielded significant results for those programs having STAT expertise available. Engineering expertise alone used to be sufficient to inform whether a DoD weapon system would perform its tasks and meet requirements. More recently, these weapon systems now have many more capabilities and consequently are extremely complex. There is ample evidence suggesting that the current system lacks such innovative approaches. The implementation of an innovative knowledge development process requires a culture change beyond only relying on engineering expertise judgement to one that is information driven.

To achieve the cultural change desired, both greater accessibility to using STAT capability and thought leaders requiring STAT developed information are needed. Through collaboration with the program engineers and subject matter experts, the STAT COE is at the forefront of changing the DoD acquisition culture by injecting rigorous, defensible, and innovative test methodologies and processes into this knowledge accumulation.

References

Advancing Open Standards for the Information Society (OASIS), *Reference Architecture Foundation for Service Oriented Architecture Version 1.0*, OASIS, Burlington, MA, 2012.

Ahner D. and Parson C., *Test and Evaluation of Autonomous Systems*, Scientific Test and Analysis Techniques Center of Excellence, Dayton, OH, 2016.

Baldwin K. J. and Lucero S. D., Defense systems complexity: Engineering challenges and opportunities, *The ITEA Journal*, 37(1), 10–16, March 2016.

Berteau D., Forward, in *Organizing for a Complex World: Developing Tomorrow's Defense and Net-Centric Systems*, G. Ben-Ari and P. A. Chao, (Eds.), Center for Strategic & International Studies, Washington, DC, 2009, p. ix.

Coleman D. E. and Montgomery D. C., A systematic approach to planning for a designed industrial experiment, *Technometrics*, 35, 1–27, 1993.

Dahmann J., Systems engineering for department of defense systems of systems, in *System of Systems Engineering*, M. Jamshidi, (Ed.), Wiley, Hoboken, NJ, 2008, pp. 218–231.

Dahmann J. S. and Baldwin K. J., Understanding the current state of US defense systems of systems and the implications for systems engineering, in *2nd Annual IEEE International Systems Conference*, Montreal, Canada, 2008.

Dahmann J., Bladwin K. J., and Rebovich G. Jr., System of systems and net centric enterprise systems, in *7th Annual Conference on Systems Engineering Research*, Loughborough, UK, 2009.

Defense Science Board, Autonomy, Office of the Under Secretary of Defense for Acquisition, Technology and Logistics, Washington, DC, 2016.

Denning S., How do you change an organizational culture? Forbes, *Forbes Magazine*, July 23, 2011. Retrieved from www.forbes.com/sites/stevedenning/2011/07/23/how-do-you-change-an-organizational-culture/# 6339756639dc Last accessed: June 5, 2018.

Dietrick R. A., Impact of weapon system complexity on systems acquisition, in *Streamlining DOD Acquisition: Balancing Schedule with Complexity*, Center for Strategy and Technology, Air War College, Air University, Montgomery, AL, 2006, p. Chapter 2.

Director, Operational Test and Evaluation, FY 2016 Annual Report, Department of Defense, Washington, DC, 2016.

Drezner J. A., *Competition and Innovation under Complexity*, RAND Corporation, Santa Monica, CA, 2009.

Fox J. R., Defense Acquisition Reform, 1960–2009, *An Elusive Goal*, Center of Military History, United States Army, Washington, DC, 2011.

Henry W. and Stevens J., Net-centric system development, in *3rd Annual IEEE International Systems Conference*, Vancouver, Canada, 2009.

Inspector General of the DoD, *DoD Needs to Require Performance of Software Assurance Countermeasures During Major Weapon System Acquisitions*, Department of Defense, Washington, DC, 2016.

Maier M. W., Architecting principles for systems-of-systems, *Systems Engineering*, 1(4), 267–284, 1998.

Montgomery D. C., *Design and Analysis of Experiments*, 9th ed., Wiley, Hoboken, NJ, 2017.

Montgomery D. C. and Jones B., Alternatives to resolution IV screening designs in 16 runs, *International Journal of Experimental Design and Process Optimization*, 1(4), 285–295, 2010.

Office of the Assistant Secretary of Defense for Research & Engineering, *Technology Investment Strategy 2015–2018*, Department of Defense, Washington, DC, 2015.

Office of the Deputy Under Secretary of Defense for Acquisition and Technology, *Systems and Software Engineering, Systems Engineering Guide for Systems of Systems, Version 1.0*, Department of Defense, Washington, DC, 2008.

Simpson J., *Testing via Sequential Experiments*, Scientific Test and Analysis Techniques Center of Excellence, Dayton, OH, 2014.

Snee R. D. and Hoerl R. W., *Leading Six Sigma: A Step-by-Step Guide Based on Experience with GE and Other Six Sigma Companies*, FT Prentice Hall, New York, 2003, p. 41.

Zahran A., Anderson-Cook C., and Myers R. H., Fraction of design space to assess the prediction compatibility of response surface designs, *Journal of Quality Technology*, 35, 377–386, 2003.

chapter seven

Building resilient systems via innovative human systems integration*

Mica R. Endsley

Contents

Introduction

Imagine a land called Nonods in which the people built a great many bridges. These bridges had a tendency to collapse frequently, however, killing or injuring a number of Nonods in the process. The bridges were also fairly rickety requiring lengthy training as well as many procedures to avoid falling off of them, significantly slowing traffic across the land. Now within Nonods there were many civil engineers who had amassed significant knowledge about how to build strong bridges that would not fall and that would support much more rapid traffic. However, the Nonod bridge builders generally ignored these engineering principles. "Why, we cross bridges all the time," they said, "so we know perfectly well how to build bridges." As a result, the Nonods continued spending a great deal of their treasure on building bridges that worked poorly, and periodically a number of Nonods were killed trying to use them. "Oh, well," they would say. "Bridges fall down. Not much one can do about that." Or they would say, "The people walking on them must have done something wrong to make them fall." And thus the Nonods were quite unprepared to move their people across the land quickly when they needed to repel an invasion from the north and they were summarily defeated in battle. The Nonods were no more.

The story of our imagined Nonods illustrates a reality in our acquisition system. But the problem is not that of building bridges but systems that allow for effective human

* Reprinted with Permission from Endsley, M.R., Building resilient systems via strong human systems integration, *Defense AT&L*, January-February, 2016, pp. 6–12.

performance. Like the Nonods, many program managers believe that "people just make errors, and that is not something that can be remedied." However, there is a strong base of scientific research and engineering foundation in the field of human factors, developed over the last 60 years, that provides a rich basis for developing robust systems that can significantly reduce human error.

Human factors engineering is based on the scientific understanding of how people perceive and process information, their physical characteristics, and how people make decisions and carry out tasks with the use of technology.

One can substantially improve human performance and reduce the likelihood of errors, simply by designing a system that is compatible with the characteristics of the people who must operate and maintain it. For example, research shows that simply making text a combination of capital and small letters (rather than all capitals) can improve reading time for lines of text by between 10 percent and 15 percent and reduce errors by about 12 percent, according to Sanders and McCormick in "Human Factors in Engineering and Design" (1993). If displays use colors consistent with human expectations (e.g., red for stop and green for start), performance will be significantly faster and people will make far fewer errors than when the colors are the opposite of expectations. These are two very simple examples, but they demonstrate the significant improvements in human performance that can be made with design features that cost almost nothing to implement. And I have found systems in the military that violate both principles, leading to unnecessary problems and poor performance. Figure 7.1 illustrates a quadrant of good, bad, start, and stop design choices.

By applying human factors principles during the design and development of our military systems, we can significantly reduce instances of catastrophic failures that lead to crashed aircraft or fratricide. And we can significantly reduce the ongoing operations and maintenance costs that eat into our limited budgets.

For example, today's manned aircraft have benefited significantly from the application of good human factors principles during system design. Early flight experience during World War II led aviation experts to realize that perfectly good aircraft were crashing because pilots had difficulty integrating and understanding displays that worked in nonintuitive and inconsistent ways and that were prone to spatial disorientation and other hazards.

The field of human factors developed to address these problems and the incidence of "human error" decreased rapidly. Military Standards such as MIL-STD-1472 and MIL-STD-1295 were developed to codify this work. However, acquisition changes in the 1990s led many programs to stop requiring attention to these human factors design standards and we saw a resurgence of problems. For example, the grounding of the F-22 fleet of tactical fighter aircraft amid concerns about pilots' hypoxia-like symptoms was found to be due to the lack of a critical backup for the Onboard Oxygen Generation System (OBOGS). That backup system was eliminated to reduce weight, even though there had been insufficient

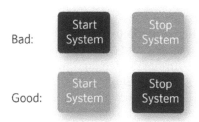

Figure 7.1 Poor versus proper interface design.

modeling and testing of the life-support system to support the decision or detect problems with the pressure vests used by the pilots. The Air Force's failure to incorporate Human Systems Integration (HSI), including human factors, in its requirements and acquisition process was a major contributing factor to this problem, according to the Air Force Scientific Advisory Board that investigated the incidents.

Today, we see similar problems with many remotely piloted aircraft. Basic human factors design principles were not applied during the initial development of the Predator ground stations. Recent analysis by the Air Force Safety Center shows that our unmanned aircraft have 6 times more Class A mishaps than our manned aircraft, and 73 percent of these were associated with human-factors problems. While the loss of an unmanned aircraft generally does not involve loss of life, it does involve loss of an expensive asset and of mission capability.

The costs of ignoring human factors during system design are too great. How people perform with technology is a critical component of total system performance. While our systems development processes often focus only on the mechanical performance of the technology, it is important to remember that our job is not only about the technology; it's also about how well the technology will support the people who need to use it to accomplish their missions.

Human systems integration

The military has worked to improve the incorporation of human-factors design principles into the development of its programs through HSI, which is a disciplined, unified and interactive systems engineering approach for integrating human considerations into system development, design and life-cycle management. This works to both improve total system performance and reduce costs of ownership across the system's life cycle. It incorporates nine key areas: manpower, personnel, training, human factors engineering, environment, safety, occupational health and survivability.

HSI takes into consideration human factors engineering principles, along with plans for the numbers and qualifications of the people assigned to use the system, and the amount and type of training needed to operate the system. This helps achieve effective system designs by simplifying the actions required for use, providing compatibility with human capabilities, and significantly easing training and manpower requirements in many cases. The environment in which the system must operate, along with various important safety factors, also is addressed in developing systems to support robust human performance. Table 7.1 presents the major domains of HSI.

HSI provides a detailed process for determining and incorporating requirements for effective human performance and safe operations, for applying sound engineering principles, and the metrics and analysis for enhancing overall system performance in a wide variety of demanding situations. The Department of Defense (DoD) has mandated inclusion of HSI in the development of our military systems. DoD Instruction (DoDI) 5000.02, Enclosure 7 addresses HSI, stating that the program manager should plan for and effect HSI, beginning early in the acquisition process and throughout the product life cycle, charging the program manager with responsibility for ensuring that HSI is considered at each program milestone.

The US Army addresses HSI with its longstanding HSI (formerly MANPRINT) program through Army Regulation 602-2. The Navy has developed an HSI Management Plan for carrying out DoDI 5000.2. And the Air Force has incorporated HSI into its Air Force Instruction (AFI) on Life Cycle Management and has developed an *HSI Guidebook,*

Table 7.1 Human systems integration (HSI) domains

Manpower	The determination of total personnel required to operate, maintain and sustain a system in order to achieve full operational capabilities.
Personnel	The determination of total human characteristics and skill requirements for a system to support capabilities necessary to fully operate, maintain and support a system.
Training	The use of analyses, methods and tools to ensure systems training requirements are fully addressed and documented by systems designers and developers. This is necessary to achieve the level of individual and team proficiency required to successfully accomplish tasks and missions.
Human Factors Engineering	The consideration and application of human capabilities and limitations throughout system definition, design and development to ensure effective human and machine integration for optimal total system performance.
Environment	The considerations of environmental factors, such as water, air and land, and the interrelationships between a system and these factors.
Safety	The consideration and application of system design characteristics that serve to minimize the potential for mishaps that could cause death or injury of operators and maintainers or threaten the system's survival and/or operation.
Occupational Health	The factors in system-design features that minimize the risk of injury, acute or chronic illness, or disability and/or that reduce job performance of personnel who operate, maintain or support the system.
Habitability	The consideration of system-related working conditions and accommodations necessary to sustain the morale, safety, health and comfort of all personnel.
Survivability	The consideration and application of system-design features that reduce the risk of fratricide (the death of one's own forces), the probability of detection, the risk of attack if detected and damage if attacked.

HSI Requirements Guide, and *Air Force Pamphlet 63-128* with mandatory requirements for conducting HSI as a part of systems development.

Nevertheless, in my travels across the Air Force, I have found that many programs still lack adequate consideration of HSI. Experience within the Army and Navy has been similar. While some programs manage to include HSI, in many cases HSI requirements take a back seat to other engineering considerations or are missing completely. It turns out that, like the Nonods, some program managers do not fully appreciate the ways in which HSI can improve system performance, or they remain confused about how to effectively incorporate HSI into their programs. This is due to a number of fundamental gaps in understanding about HSI. Figure 7.2 summarizes the requirements of rapid prototyping of user interfaces.

Myth No. 1: Human systems integration means asking what users want

Often when I have asked program managers what sort of HSI considerations they have included in their programs, they proudly tell me, "We showed it to some users." While a good step, this unfortunately is quite insufficient. Human preference does not equal human performance. User input is very important to development of good systems. Users know a lot about what their jobs entail and where the difficulties are, and they can provide useful feedback when looking at new system designs or when trying them out during Developmental Test and Evaluation (DT&E) or Operational Test and Evaluation (OT&E). However, they generally are not experts at understanding the detailed physical, physiological, perceptual and cognitive processes, capabilities and

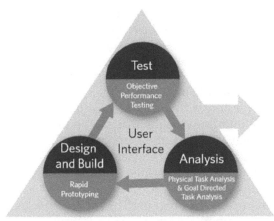

Rapid Prototyping of User Interface
- Creates a tested set of detailed visual requirements for system
- Supports architecture decisions
- Understanding of functionality
- Effective information and process flows
- Does all these things across the System of Systems

Figure 7.2 Use HSI tools and processes to define requirements and interfaces early.

limitations of humans, and they often will miss the many subtle features of technology that can negatively impact human performance.

Good HSI means applying known human engineering design principles and performing objective evaluations of the functioning of the system when in use by a representative sample of its intended users. Time to perform tasks, error rates, workload and situation awareness can all be objectively measured to find problems and make design trade-offs with the goal of creating effective total system performance. Just as we would not test an engine simply by having pilots look at it, we will not get a good assessment of the human interface just by having the user look at it.

Myth No. 2: Human systems integration means including the newest display techniques and hardware

At the opposite end of the spectrum from neglecting HSI, some programs go looking for HSI in all the wrong places. That is, they want to make really cool user interfaces by incorporating the latest ideas from science fiction movies or computer scientists. I have seen displays built into three-dimensional rotating cubes, displays that project information into holograms and virtual reality headsets, or those that involve large arm movements for extended periods to interact with displays. While well intended, many of these so-called advancements can be fatiguing, can reduce situation awareness in critical situations, and actually can lead to much slower performance and higher error rates on critical tasks. Cool does not equal effective. Good user interfaces may not always require the latest hardware and software concepts. Instead designers must pay attention to the requirements associated with users' tasks and match the most effective hardware and software approaches to those tasks.

Myth No. 3: Human systems integration should be done at the end of a program

Among program managers, one of the most pervasive misunderstandings is the belief that the user interface should be considered at the end of the program after the technology issues are sorted out. This is the worst time to do HSI. At that point, generally only small fixes can be applied to a system that has placed controls in the wrong places or that has software logic and layouts that fundamentally confuse users and do not provide the needed information in ways that will help users achieve good situational awareness or rapid performance. Just as one cannot really fix a poorly designed Nonod bridge with a few Band-Aids, one cannot fix a poor user interface with a few tweaks at the end of the program. And making the extensive changes needed is generally very costly at that point and causes program timelines to be exceeded. HSI should be started at the very beginning of a program. By conducting an early analysis of user requirements, tasks and information needs, an HSI team can create early prototype interface designs that can be tested with users early in the program. These prototypes then can create the foundation for software and hardware development. They provide a clear indication of what is needed before a penny is spent on bending metal or on expensive software coding of interfaces that will need to be changed repeatedly as users try them out.

This creates significant time and money savings for the program. The Air Force recently was forced to cancel its Expeditionary Combat Support System (ECSS) program, costing more than $1.1 billion and 8 years of effort. A major reason was the program's inability to understand the system requirements, leading to extensive churn in requirements and solutions and failed reprogramming efforts. Had this HSI process been employed early, there would have been a prototype system available for testing with the many users of the system. This would have established a means to ensure that the needed functionality and information flow was well understood before software development even started.

Myth No. 4: Anyone can do human systems integration

Just as the Nonods believed that they could design bridges because they were bridge users, many people believe anyone can do HSI because they are people and so they know what people need. However, even well-meaning people will not do an adequate job of HSI if they have not received the appropriate training—combining knowledge of human capabilities (physical, cognitive and perceptual) with knowledge on how to design systems, develop training or conduct the needed HSI domain analyses. As in other areas of engineering, there is a significant body of knowledge that needs to be acquired. Most HSI practitioners have advanced degrees in industrial engineering, psychology or physiology. However, HSI is a multidisciplinary profession, so practitioners may have a wide variety of degree titles that can leave some people confused as to how to find the right expertise. Just as you can hire a Certified Public Accountant (CPA) to do your taxes, you also can find an HSI expert for your team who is a Certified Professional Ergonomist (CPE)—after having passed the required exams and demonstrated proficiency in the field.

Myth No. 5: We can just train around human systems integration problems

There is a long history of trying to use training to compensate for poorly designed user interfaces. Unfortunately, training alone cannot overcome interfaces inconsistent with human expectations (e.g., requiring the user to push down on a lever to go up), that create known physiological problems (e.g., a lever that requires the pilot to move her head down and to the side during landing, resulting in the pilot's disorientation), or that require extensive time-consuming procedures for simple tasks. Even with extensive training, people will continue to make errors when the technology is incompatible with how they think and operate, particularly when under stress. And trainers will tell you that good HSI can significantly reduce the training time required for any system. Good training is important, but it is no substitute for good system design.

Myth No. 6: With automation, we don't need to worry about human systems integration

Many people believe that as systems become more automated, worrying about HSI or the human operators of the systems will become less important. However, exactly the opposite is true because almost all this automation still requires human interaction. Extensive experience with automated systems over the last 30 years has shown that automation actually can make the user's job more complicated. For example, pilots and system operators find that their cognitive workload can increase substantially as they work to understand how to properly program the automation during operations. And they can suffer from lower situational awareness when working with automation because it often leaves them out of the loop and struggling to understand what it is doing so they can supervise the automation and intervene in time-critical situations. The move toward more automation or autonomy in many systems requires that we pay even more attention to the user interface than ever to make the behavior of the system more transparent and understandable, creating effective human-automation teams.

Myth No. 7: Human systems integration costs too much

Actually, good HSI saves programs money, both during system development and later in operations. Attention to HSI early in a program can provide clear directions for system development, saving extensive rework later, when it is much more expensive to redo software or hardware. Attention to HSI also can save a great deal of money in the military's limited operations and maintenance budgets. Life-cycle costs account for between 35 percent and 70 percent of a system's overall costs. These costs can be significantly reduced if HSI is emphasized during system development. For example, attending to the design of the interface for a satellite control ground station or a command-and-control system can significantly reduce the number of operators required. Attending to the design of the aircraft for supporting maintainer tasks can significantly reduce the hours required for routine maintenance and increase its availability for flight. The truth is our development programs cannot afford a failure to apply good HSI.

The acquisition community is the linchpin for human systems integration

Acquisition professionals have a critical role in developing technology for their users. All of our airmen, soldiers and seamen have demanding and critical jobs to do that depend on well-designed systems that will work the way that they do—supporting the accomplishment of their tasks rapidly and effectively. It is critical that we avoid system designs that are obstacle courses of hidden hazards and latent failures.

Acquisition programs can accomplish these goals by first paying attention to HSI requirements when establishing program requirements. If these requirements are not spelled out in clear measurable ways, experience has shown that contractors will not, and often feel they cannot, spend any effort in ensuring that systems are easy to use or consistent with human capabilities and limitations. And if HSI requirements are not included in program documents, there is little that can be done to make contractors fix even egregious interface problems without making expensive program modifications.

Second, make sure not only to require that system developers create an HSI plan but that it is implemented early in the program, and include it as a critical part of design reviews.

In some cases, we have found programs that required an HSI plan but failed to require the contractor to actually implement it, which did no good at all. Design reviews should include not only a review of the contractor's progress on HSI tasks, but also a review of objective test metrics showing whether their work has been successful and identifying areas for further improvements.

Third, make sure you have the needed HSI professionals as a part of your program team. You won't be able to tell if contractors have done a good or a poor job if you don't have people with the required knowledge and experience to evaluate the system design, the methods used or the test results. In the Air Force, the 711th Human Performance Wing has a body of HSI professionals who can provide the expertise needed. The Army has the Army Research Laboratory Human Research and Engineering Directorate (ARL HRED), and the Navy has HSI professionals imbedded at the Naval Sea Systems Command (NAVSEA) and the Space and Naval Warfare Command (SPAWAR).

To learn more about HSI a number of resources are available. The Defense Acquisition University offers a 2-hour introductory course in Human Systems Integration (CLE 062). The Air Force Institute of Technology offers courses in Basic Human-Systems Integration (SYS 169), Intermediate Human Systems Integration (SYS 269), and a certificate in Human Systems Engineering, as well as advanced degrees. The Naval Postgraduate School offers an online Human Systems Integration Certificate, in addition to master's and doctoral degrees with emphasis in HSI.

The good news is that there is an extensive body of knowledge and expertise that can help all of our acquisition programs develop safe and resilient systems that promote effective human performance as a part of total system performance. Like the Nonods, we just need to apply that knowledge to our programs to be successful.

chapter eight

Innovative model for situation awareness in dynamic defense systems

Mica R. Endsley

Contents

Adapted and reprinted with permission from Endsley, Mica R.,
"Toward a Theory of Situation Awareness in Dynamic Systems,"
Human Factors, Vol. 37, No. 1, 1995, pp. 32–64.

Introduction

The range of problems confronting human factors practitioners has continued to grow over the past fifty years. Practitioners must deal with human performance in tasks that are primarily physical or perceptual, as well as consider human behavior involving highly complex cognitive tasks with increasing frequency. As technology has evolved, many complex, dynamic systems have been created that tax the abilities of humans to act as effective, timely decision makers when operating these systems. The operator's situation awareness (SA) will be presented as a crucial construct on which decision making and performance in such systems hinge. In this paper I strive to show (a) the importance of SA in decision making in dynamic environments and the utility of using a model of decision making that takes SA into account, and (b) a theory of SA that expands on prior work in this area (Endsley, 1988a, 1990c, 1993b). True SA, it will be shown, involves far more than merely being aware of numerous pieces of data. It also requires a much more advanced level of situation understanding and a projection of future system states in light of the operator's pertinent goals. As such, SA presents a level of focus that goes beyond traditional information-processing approaches in attempting to explain human behavior in operating complex systems. SA can be shown to be important in a variety of contexts that confront human factors practitioners.

Aircraft

In the area with perhaps the longest history, SA was recognized as a crucial commodity for crews of military aircraft as far back as World War I. SA has grown in importance as a major design goal for civil, commercial, and military aircraft, receiving particular emphasis in recent years (Federal Aviation Administration, 1990; US Air Force 57th Fighter Wing, 1986). In the flight environment, the safe operation of the aircraft in a manner consistent with the pilot's goals is highly dependent on a current assessment of the changing situation, including details of the aircraft's operational parameters, external conditions, navigational information, other aircraft, and hostile factors. Without this awareness (which needs to be both accurate and complete), the aircrew will be unable to effectively perform their functions. Indeed, as will be discussed further, even small lapses in SA can have catastrophic repercussions.

Air traffic control

In a related environment, air traffic controllers are called on to sort out and project the paths of ever-increasing numbers of aircraft in order to ensure goals of minimum separation and safe, efficient landing and takeoff operations. This taxing job relies on the SA of controllers who must maintain up-to-date assessments of the rapidly changing locations of aircraft (in three-dimensional space) and their projected locations relative to each other, along with other pertinent aircraft parameters (destination, speed, communications, etc.).

Large-systems operations

The operators of large, complex systems such as flexible manufacturing systems, refineries, and nuclear power plants must also rely on up-to-date knowledge of situation parameters to manage effectively. In their tasks, operators must observe the state of numerous system parameters and any patterns among them that might reveal clues as to the functioning of the system and future process state changes (Wirstad, 1988). Without this understanding and prediction, human control could not be effective.

Tactical and strategic systems

Similarly, firefighters, certain police units, and military command personnel rely on SA to make their decisions. They must ascertain the critical features in widely varying situations to determine the best course of action. Inaccurate or incomplete SA in these environments can lead to devastating loss of life, such as in the case of the U.S.S. *Vincennes*. Incorrect SA concerning an incoming aircraft (from confusing identification signals and a lack of direct information on changes in altitude) led to the downing of a commercial airliner and subsequent loss of all aboard. From reports of the accident (Klein, 1989a), it appears that the decision makers' SA was in error (perceived hostility of the incoming aircraft), not the decision as to what to do (if hostile, warn off and then shoot down if not heeded). This is an important distinction that highlights the criticality of SA in dynamic decision making.

Other systems

Many other everyday activities call for a dynamic update of the situation to function effectively. Walking, driving in heavy traffic, or operating heavy machinery surely call for SA. Roschelle and Greeno (1987) reported that experts in solving physics problems rely on the development of a situational classification. Gaba et al. (1995) describe the role of SA in medical decision making. As humans typically operate in a closed-loop manner, input from the environment is almost always necessary.

The need for SA applies in a wide variety of environments. Acquiring and maintaining SA becomes increasingly difficult, however, as the complexity and dynamics of the environment increase. In dynamic environments, many decisions are required across a fairly narrow space of time, and tasks are dependent on an ongoing, up-to-date analysis of the environment. Because the state of the environment is constantly changing, often in complex ways, a major portion of the operator's job becomes that of obtaining and maintaining good SA. This task ranges from trivial to one of the major factors determining operator performance. In analyzing the decision making of tactical commanders, Kaempf et al. (1993, p. 1110) reported that "recognizing the situation provided the challenge to the decision maker," confirming SA's criticality.

In each of the domains discussed, operators must do more than simply perceive the state of their environment. They must understand the integrated meaning of what they are perceiving in light of their goals. Situation awareness, as such, incorporates an operator's understanding of the situation as a whole, forming a basis for decision making. Researchers in many areas have found that expert decision makers will act first to classify and understand a situation, immediately proceeding to action selection (Klein, 1989b; Klein et al., 1986; Lipshitz, 1987; Noble et al., 1987; Sweller, 1988).

There is evidence that an integrated picture of the current situation may be matched to prototypical situations in memory, each prototypical situation corresponding to a "correct" action or decision. Dreyfus (1981) presented a treatise that emphasized the role of situational understanding in real-world, expert decision making, building on the extensive works of deGroot (1965) in chess, Mintzburg (1973) in managerial decision making, and Kuhn (1970) in science. In each of these areas the experts studied used pattern-matching mechanisms to draw on longterm memory structures that allowed them to quickly understand a given situation. They then adopted the course of action corresponding to that type of situation. Hinsley et al. (1977) have found that this situation classification can occur almost immediately, or, as Klein (1989b) has pointed out, it can involve some effort to achieve.

In his studies of fire ground commanders, Klein (1989b) found that a conscious deliberation of solution alternatives was rare. Rather, the majority of the time experts focused on classifying the situation in order to immediately yield the appropriate solution from memory. Kaempf et al. (1993) reported that of 183 decisions by tactical commanders, 95% used this type of recognition decision strategy, involving either feature matching to situation prototypes (87%) or story building (13%). Although much of this work emphasizes the decision processes of experts, novices must also focus a considerable amount of their effort on assessing the state of the environment in order to make decisions. Cohen (1993) pointed out that metacognitive strategies may become more important in these cases as forming an assessment of the situation becomes more challenging.

Given that SA plays such a critical role in decision making, particularly in complex and dynamic environments, there is a need to more explicitly incorporate the concept into human factors design efforts. A theory of SA that clearly defines the construct and its relation to human decision making and performance is needed to fulfill this mission.

A model of situation awareness

Because direct research on SA itself is limited and has been conducted only in recent years, a thorough and rigorously defined theory may not yet be possible. The present objective is to define a common ground for discussion using the information that is available in order to provide a starting point for future work on SA.

This information will be presented in a framework model—a model that is descriptive of the SA phenomenon and that synthesizes information from a variety of areas. It will explicitly address certain attributes of the construct. Specifically, Klein (1989b) stated that a desired theory of situation awareness should explain dynamic goal selection, attention to appropriate critical cues, expectancies regarding future states of the situation, and the tie between situation awareness and typical actions. Within this context, it is the goal of this effort to delineate what SA is and what it is not, to provide an understanding of the

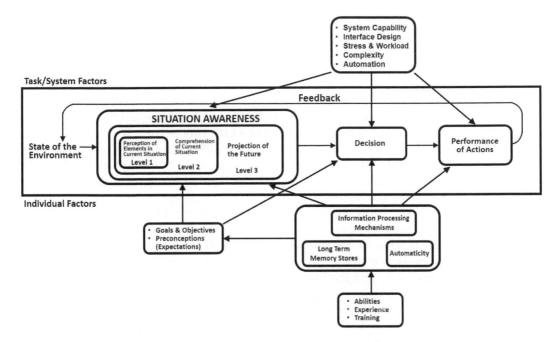

Figure 8.1 Model of situation awareness in dynamic decision making.

mechanisms that underlie the construct, and to discuss the factors that may influence it. The implications of the model for design, error investigation, and future research will be discussed. (This discussion will be illustrated by examples of SA from the aircraft domain; however, it applies equally to other contexts presented earlier.)

Figure 8.1 provides a basis for discussing SA in terms of its role in the overall decision-making process. According to this model, a person's perception of the relevant elements in the environment, as determined from system displays or directly by the senses, forms the basis for his or her SA. Action selection and performance are shown as separate stages that will proceed directly from SA.

Several major factors are shown to influence this process. First, individuals vary in their ability to acquire SA, given the same data input. This is hypothesized to be a function of an individual's information-processing mechanisms, influenced by innate abilities, experience, and training. In addition, the individual may possess certain preconceptions and objectives that can act to filter and interpret the environment in forming SA.

SA will also be a function of the system design in terms of the degree to which the system provides the needed information and the form in which it provides it. All system designs are not equal in their ability to convey needed information or in the degree to which they are compatible with basic human information-processing abilities. Other features of the task environment, including workload, stress, and complexity, may also affect SA. The role of each of these individual and system factors in relation to SA will be addressed.

Definitions and terminology

Contrary to Sarter and Woods (1991), who believe that developing a definition of SA is futile and not constructive, I believe it is first necessary to clearly define SA. The term has lately become the victim of rather loose usage, with different individuals redefining it at whim, leading to the recent criticism that SA is the "buzzword of the '90s" (Wiener, 1993, p. 4). Unless researchers stick to a clear, consistent meaning for the term, the problem will present a significant handicap to progress.

In conjunction with the model, therefore, a few issues will be stated explicitly to clarify the present formulation of SA. As a matter of consistent terminology, it is first necessary to distinguish the term situation awareness, as a state of knowledge, from the processes used to achieve that state. These processes, which may vary widely among individuals and contexts, will be referred to as situation assessment or as the process of achieving, acquiring, or maintaining SA. (This differs from recent efforts by Sarter and Woods (1991), who view SA as "a variety of cognitive processing activities," in contrast to most past definitions of SA, which have focused on SA as a state of knowledge. I am in full agreement with Adams et al. (1995) that there is great benefit in examining the interdependence of the processes and the resultant state of knowledge; however, in order to clarify discourse on SA, it is important to keep the terminology straight.)

Furthermore, SA as defined here does not encompass all of a person's knowledge. It refers to only that portion pertaining to the state of a dynamic environment. Established doctrine, rules, procedures, checklists, and the like—though important and relevant to the decision-making process-are fairly static knowledge sources that fall outside the boundaries of the term.

In addition, SA is explicitly recognized as a construct separate from decision making and performance. Even the best-trained decision makers will make the wrong decisions if they have inaccurate or incomplete SA. Conversely, a person who has perfect SA may still make the wrong decision (from a lack of training on proper procedures, poor tactics, etc.) or show poor performance (from an inability to carry out the necessary actions). SA, decision making, and performance are different stages with different factors influencing them and with wholly different approaches for dealing with each of them: thus it is important to treat these constructs separately. (This stance also differs from that taken by the US Air Force [Judge, 1992], which has adopted a definition of SA that includes action and decision making, in contrast to most prior research on SA.)

Similarly, SA is presented as a construct separate from others that may influence it. Attention, working memory, workload, and stress are all related constructs that can affect SA but that can also be seen as separate from it. Subsuming any of these constructs within the term situation awareness loses sight of the independent and interactive nature of these factors. SA and workload, for instance, have been shown to vary independently across a wide range of these variables (Endsley, 1993a), although workload may have a negative effect on SA in certain situations. These factors will be addressed more explicitly in a later section.

Although numerous definitions of SA have been proposed (Endsley, 1988a: Fracker, 1988), most are not applicable across different task domains. For the most part, however, they all point to "knowing what is going on." Referring to Figure 8.1, we use the following general definition of SA (Endsley, 1987b, 1988b):

Situation awareness is the perception of the elements in the environment within a volume of time and space, the comprehension of their meaning, and the projection of their status in the near future.

Each of the three hierarchical phases and primary components of this definition will be described in more detail.

Level 1 SA: Perception of the elements in the environment

The first step in achieving SA is to perceive the status, attributes, and dynamics of relevant elements in the environment. A pilot would perceive elements such as aircraft, mountains, or warning lights along with their relevant characteristics (e.g., color, size, speed, location). A tactical commander needs accurate data on the location, type, number, capabilities, and dynamics of all enemy and friendly forces in a given area and their relationship to other points of reference. A flexible manufacturing system operator needs data on the status of machines, parts, flows, and backlogs. An automobile driver needs to know where other vehicles and obstacles are, their dynamics, and the status and dynamics of one's own vehicle.

Level 2 SA: Comprehension of the current situation

Comprehension of the situation is based on a synthesis of disjointed Level 1 elements. Level 2 SA goes beyond simply being aware of the elements that are present to include an understanding of the significance of those elements in light of pertinent operator goals. Based on knowledge of Level 1 elements, particularly when put together to form patterns with the other elements (gestalt), the decision maker forms a holistic picture of the environment, comprehending the significance of objects and events. For example, a military pilot or tactical commander must comprehend that the appearance of three enemy aircraft within a certain proximity of one another and in a certain geographical location indicates certain things about their objectives. The operator of a power plant needs to put together disparate bits of data on individual system variables to determine how well different system components are functioning, deviations from expected values, and the specific locus of any deviant readings. In these environments a novice operator might be capable of achieving the same Level 1 SA as more experienced decision makers but may fall far short of also being able to integrate various data elements along with pertinent goals in order to comprehend the situation.

Level 3 SA: Projection of future status

The ability to project the future actions of the elements in the environment—at least in the very near term—forms the third and highest level of SA. This is achieved through knowledge of the status and dynamics of the elements and comprehension of the situation (both Level 1 and Level 2 SA). For example, knowing that a threat aircraft is currently offensive and is in a certain location allows a fighter pilot or military commander to project that the aircraft is likely to attack in a given manner. This provides the knowledge (and time) necessary to decide on the most favorable course of action to meet one's objectives. Similarly, an air traffic controller needs to put together information on various traffic patterns to determine which runways will be free and where there is a potential for collisions. An automobile driver also needs to detect possible future collisions in order to act effectively, and a flexible manufacturing system operator needs to predict future bottlenecks and unused machines for effective scheduling.

SA, therefore, is based on far more than simply perceiving information about the environment. It includes comprehending the meaning of that information in an integrated form, comparing it with operator goals, and providing projected future states of the environment that are valuable for decision making. In this aspect, SA is a broad construct that is applicable across a wide variety of application areas, with many underlying cognitive processes in common.

Elements

From a design standpoint, a clear understanding of SA in a given environment rests on a clear elucidation of the elements in the definition—that is, identifying which things the operator needs to perceive and understand. These are specific to individual systems and contexts, and as such are the one part of SA that cannot be described in any valid way across arenas. Although the pilot and power plant operator each relies on SA, it simply is not realistic or appropriate to expect the same elements to be relevant to both. Nonetheless, these elements can be, and should be, specifically determined for various classes of systems.

Endsley (1993c) presented a methodology for accomplishing this and described such a delineation for air-to-air fighter aircraft. Examples of elements in this arena include:

1. Level 1: Location, altitude, and heading of own ship and other aircraft; current target; detections: system status: location of ground threats and obstacles
2. Level 2: Mission timing and status: impact of system degrades: time and distance available on fuel: tactical status of threat aircraft (offensive/defensive/neutral)
3. Level 3: Projected aircraft tactics and maneuvers, firing position and timing.

One may also talk about awareness of certain subcategories of SA (usually system specific), which include requirements across all three levels of SA. For instance, spatial awareness or geographical awareness is frequently of concern in aircraft. Mode awareness, as discussed by Sarter and Woods (1991), is another example of a subset of SA that may be of concern in certain systems, across all three levels (e.g., "What is it doing, why is it doing that, what will it do next?").

Time

Several other aspects of SA should be mentioned at this point. First, although SA has been discussed as a person's knowledge of the environment at a given point in time, it is highly temporal in nature. That is, SA is not necessarily acquired instantaneously but is built up over time. Thus it takes into account the dynamics of the situation that are acquirable only over time and that are used to project the state of the environment in the near future. So although SA consists of an operator's knowledge of the state of the environment at any point in time, this knowledge includes temporal aspects of that environment, relating to both the past and the future.

Space

It has been observed that SA is highly spatial in many contexts. Pilots and air traffic controllers, for instance, are concerned with the spatial relationships among multiple aircraft, and this information also yields important temporal cues. Many other fields may also be concerned with the spatial as well as functional relationships among system components. In addition to its aspect as a frequent "element" of SA, spatial information is highly useful for determining exactly which aspects of the environment are important for SA.

An operator's SA needs to incorporate information on that subset of the environment that is relevant to tasks and goals. Within this boundary, the elements may be further subdivided into levels of importance for SA or may assume a relevance continuum, depending on the problem context. In a piloting context, for example, the relevance of different aircraft will depend on their location and speed relative to own ship and the pilot's goals (e.g., response to an immediate threat, tactics determination, or long-term mission re-planning): a different amount of relevance may be indicated for different goals. In other contexts, such as manufacturing or power plant environments, relevance of elements may be determined by the spatial, temporal, or functional relationships of elements to goals.

In this way, elements may vary in their relevance across time, although they do not generally fall out of consideration completely. At least some SA on all elements has been found to be needed, even if this conveys merely that the element is not very important at the moment. For instance, while in close combat, many pilots report that they are interested only in where their opponent is. Too frequently, however, though they are successful in avoiding enemy missiles, they end up flying into the ground with lethal results (Kuipers et al., 1989; McCarthy, 1988). In order to know that they can afford to pay less attention to altitude than to enemy aircraft, pilots need to know that they are at least above a certain level at all times. A certain amount of SA on other elements is required at all times in a similar manner.

Team SA

It is possible to talk about SA in terms of teams as well as individuals. In many situations several individuals may work together as a team to make decisions and carry out actions. In this case one can conceive of overall team SA, whereby each team member has a specific set of SA elements about which he or she is concerned, as determined by each member's responsibilities within the team.

SA for a team can be represented as shown in Figure 8.2. Some overlap between each team member's SA requirements will be present. It is this subset of information that constitutes much of team coordination. That coordination may occur as a verbal exchange, as a duplication of displayed information, or by some other means. As such, the quality of team members' SA of shared elements (as a state of knowledge) may serve as an index of team coordination or human-machine interface effectiveness.

Overall team SA can be conceived as the degree to which every team member possesses the SA required for his or her responsibilities. This is independent of any overlaps in SA requirements that may be present. If each of two team members needs to know a piece of information, it is not sufficient that one knows perfectly but the other not at all. Every team member must have SA for all of his or her own requirements or become the proverbial chain's weakest link.

For instance, in an aircraft cockpit, both the pilot and copilot may need to know certain pieces of information. If the copilot has this information but the pilot in charge does not,

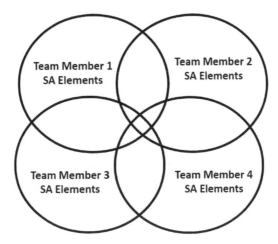

Figure 8.2 Team situation awareness.

the SA of the team has suffered and performance may suffer as well unless the discrepancy is corrected. How that information transmission occurs—the process of achieving SA—can vary. It may constitute a verbal exchange or separate, direct viewing of displays, with each individual independently acquiring information on the status of the aircraft. Higher levels of SA that may not be directly presented on displays may be communicated verbally, or, if the team members possess a shared mental model (Salas, Prince, Baker, and Shrestha, 1995), each team member may achieve the same higher-level SA without necessitating extra verbal communication. Mosier and Chidester (1991), for example, found that better-performing teams actually communicated less than did poorer-performing teams. In this case, the degree to which each team member has accurate SA on shared items could serve as an index of the quality of team communications (i.e., each member's ability to achieve the goal of communication as efficiently as possible).

Link to decision making

In addition to forming the basis for decision making as a major input, SA may also impact the process of decision making itself: There is considerable evidence that a person's manner of characterizing a situation will determine the decision process chosen to solve a problem. Manktelow and Jones (1987) reviewed the literature concerning deductive problem solving and showed, through numerous studies, that the situation parameters or context of a problem largely determines the ability of individuals to adopt an effective problem-solving strategy. It is the situation specifics that determine the adoption of an appropriate mental model, leading to the selection of problem-solving strategies. In the absence of an appropriate model, people will often fail to solve a new problem, even though they would have to apply the same logic as that used for a familiar problem.

Other evidence suggests that even the way a given problem is presented (or framed) can determine how the problem is solved (Bettman and Kakkar, 1977; Herstein, 1981; Sundstrom, 1987; Tversky and Kahneman, 1981). The simplest explanation for this is that different problem framings can induce different information integration (situation comprehension), and this determines the selection of a mental model to use for solving the problem. Thus it is not only the detailed situational information (Level 1 SA) but also the way the pieces are put together (Level 2 SA) that direct decision strategy selection.

Link to performance

The relationship between SA and performance, though not always direct, can also be predicted. In general, it is expected that poor performance will occur when SA is incomplete or inaccurate, when the correct action for the identified situation is not known or calculated, or when time or some other factor limits a person's ability to carry out the correct action. For instance, in an air-to-air combat mission, Endsley (1990b) found that SA was significantly related to performance only for those subjects who had the technical and operational capabilities to take advantage of such knowledge. The same study found that poor SA would not necessarily lead to poor performance if subjects realized their lack of SA and were able to modify their behavior to reduce the possibility of poor performance. Venturino, Hamilton, and Dvorchak (1989) also found that performance was predicted by a combination of SA and decision making (fire-point selection) in combat pilots. Good SA can therefore be viewed as a factor that will increase the probability of good performance but cannot necessarily guarantee it.

Human properties affecting and underlying SA

Within this basic model of SA, I will discuss the factors underlying and influencing the SA process. This discussion will first focus on characteristics of the individual, including relevant information-processing mechanisms and constructs that play a role in achieving SA. It will proceed to factors related to the system and task environment as they affect the operator's ability to achieve SA.

Although some researchers have continued to argue that relatively little is known about SA (Sarter and Woods, 1991), this belies the vast amount of highly pertinent work that has been done—specifically, research devoted to more general aspects of human cognition. Although members of the psychology community continue to debate the exact structure and nature of information-processing mechanisms, a detailed discussion of various theories regarding each lies beyond the scope of this paper. Thus, the relationship between SA and these mechanisms, as generally understood, will be explored.

In combination, the mechanisms of short-term sensory memory, perception, working memory, and long-term memory form the basic structures on which SA is based. Figure 8.3 shows a schematic description of the role of each of these structures in the SA process.

Pre-attentive processing

According to most research on information processing (for a review see Norman, 1976, or Wickens, 1992a), environmental features are initially processed in parallel through preattentive sensory stores in which certain properties are detected, such as spatial proximity, color, simple properties of shapes, or movement (Neisser, 1967; Treisman and Paterson, 1984), providing cues for further focalized attention. Those objects that are most salient, based on preattentively registered characteristics, will be further processed using

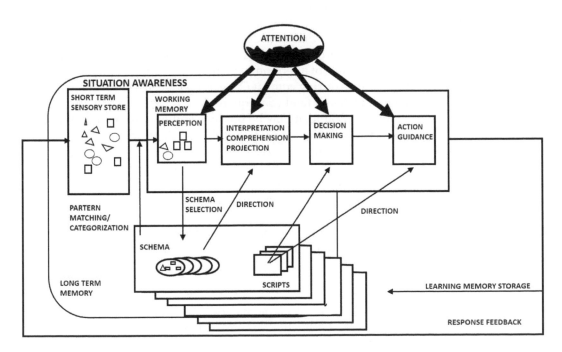

Figure 8.3 Mechanisms of situation awareness.

focalized attention to achieve perception. Cue salience, therefore, will have a large impact on which portions of the environment are initially attended to, and these elements will form the basis for the first level of SA.

Attention

The deployment of attention in the perception process acts to present certain constraints on a person's ability to accurately perceive multiple items in parallel and, as such, is a major limit on SA. Direct attention is needed for not only perceiving and processing the cues attended to but also the later stages of decision making and response execution. In complex and dynamic environments, attention demands resulting from information overload, complex decision making, and multiple tasks can quickly exceed a person's limited attention capacity.

Operators of complex systems frequently employ a process of information sampling to circumvent this limit. They attend to information in rapid sequence following a pattern dictated by the portion of long-term memory concerning relative priorities and the frequency with which information changes (Wickens, 1992a). Working memory also plays an important role, allowing one to modify attention deployment on the basis of other information perceived or active goals (Braune and Trollip, 1982). For example, perception of a strange noise may prompt a pilot to look at the engine status indicator. When involved in the goal of shooting at an enemy aircraft, attention may be directed primarily at that target. In addition to highly salient cues catching one's attention, therefore, people are active participants in determining which elements of the environment will become a part of their (Level 1) SA by directing their attention based on goals and objectives and on the basis of long-term and working memory (each of which will be discussed in more detail).

In a study of pilot SA, Fracker (1989) showed that a limited supply of attention was allocated to environmental elements based on their ability to contribute to task success. Because the supply of attention is limited, more attention to some elements (resulting in improved SA on these elements), however, may mean a loss of SA on other elements once the limit is reached, which can occur rather quickly in complex environments. In an investigation of factors leading to fighter aircraft accidents involving controlled descent into the terrain, Kuipers et al. (1989) cited lack of attention to primary flight instruments (56%) and too much attention to target planes during combat (28%) as major causes. Focusing on only certain elements led to a lack of SA and fatal consequences.

In addition to information sampling, it may be possible to work around attention limits in other ways to some degree. Kahneman (1973) stated that attentional resources can be increased somewhat by physiological arousal mechanisms. Further relief to attention limitations can be provided through people's ability to divide their attention under certain circumstances. Wickens's multiple resource theory (1992a) provides a model for determining which types of information can be most easily attended to in parallel. Damos and Wickens (1980) also found that attention sharing is a skill that can be learned and that some people excel at it over others. In addition, limitations of attention may be circumvented to some degree through the development of automaticity.

Perception

In addition to affecting the selection of elements for perception, the way in which information is perceived is directed by the contents of both working memory and long-term memory. Advanced knowledge of the characteristics, form, and location of information, for instance, can significantly facilitate the perception of information (Barber and Folkard, 1972; Biederman et al., 1981; Davis et al., 1983; Humphreys, 1981; Palmer, 1975; Posner et al., 1978). That is, one's preconceptions or expectations about information will affect the speed and accuracy of the perception of that information (Jones, 1977, pp. 38–39).

Repeated experience in an environment allows one to develop expectations about future events. In the aircraft environment, premission briefings typically build up preconceptions about what will be encountered during the mission. An air traffic controller's report of traffic at a particular altitude or a bill of lading that accompanies a shipment in a manufacturing environment each develops in recipients a certain expectation about what they will encounter that predisposes them to perceive the information accordingly. They will process the information faster if it is in agreement with those expectations and will be more likely to make an error if it is not (Jones, 1977).

Long-term memory stores also play a significant role in classifying perceived information into known categories or mental representations as an almost immediate act in the perception process (Hinsley et al., 1977). Categorization is based on integrated information and typically occurs in a deterministic, nearly optimal manner (Ashby and Gott, 1988). The classification of information into understood representations forms Level 1 SA and provides the basic building blocks for the higher levels of SA.

With well-developed memory stores, very fine categorizations may be possible. For instance, an experienced pilot will be able to classify observed aircraft into exact models (e.g., F-18c vs. F-18d). This highly detailed classification provides the pilot with access to detailed knowledge about the capabilities of the aircraft (from long-term memory). A novice may not be able to make this level of classification and would consequently have less information from the same data input.

The cues used to achieve these classifications are important to SA. With higher levels of expertise, people appear to develop knowledge of critical cues in the environment that allow them to make very fine classifications. The development of memory structures for this process will be discussed more fully subsequently. At this juncture it is important to note that the classification made in the perception stage (right or wrong, detailed or gross) is a function of the knowledge available for making such classifications and will produce the elements of Level 1 SA.

Working memory

Once perceived, information is stored in working memory. In the absence of other mechanisms (such as relevant long-term memory stores), most of a person's active processing of information must occur in working memory. New information must be combined with existing knowledge and a composite picture of the situation developed (Level 2 SA). Projections of future status (Level 3 SA) and subsequent decisions as to appropriate courses of action must occur in working memory as well. In this circumstance, a heavy load is imposed on working memory, as it is taxed with simultaneously achieving the higher levels of SA (Levels 2 and 3), formulating and selecting responses, and carrying out subsequent actions.

Wickens (1984, p. 201) has stated that prediction of future states (the culmination of good SA) imposes a heavy load on working memory by requiring the maintenance of present conditions, future conditions, rules used to generate the latter from the former, and actions that are appropriate to the future conditions. Fracker (1987) hypothesized that working memory constitutes the main bottleneck for SA. This is most likely the case for novices or those dealing with novel situations.

Long-term memory

In practice long-term memory structures can be used to circumvent the limitations of working memory. The exact organization of knowledge in long-term memory has received diversified characterization, including episodic memory, semantic networks, schemata, and mental models. This discussion will focus on schemata and mental models that have been discussed as important for effective decision making in a number of environments (Braune and Trollip, 1982; Rasmussen and Rouse, 1981) and that are hypothesized to play an important role in SA.

Schemata provide coherent frameworks for understanding information, encompassing highly complex system components, states, and functioning (Bartlett, 1932; Mayer, 1983). Much of the details of situations are lost when information is coded in this manner, but the information becomes more coherent and organized for storage, retrieval, and further processing. A single schema may serve to organize several sets of information and as such will have variables that can be filled in with the particulars for the case being considered. A script—a special type of schema—provides sequences of appropriate actions for different types of task performance (Schank and Abelson, 1977). Ties between schemata and scripts can greatly facilitate the cognitive process because an individual does not have to actively decide on appropriate actions at every turn but will automatically know the actions to take for a given situation based on its associated script.

A related concept is the mental model. Rouse and Morris (1985) defined mental models as "mechanisms whereby humans are able to generate descriptions of system purpose and form, explanations of system functioning and observed system states, and predictions of future states" (p. 7). They stated that experts will develop mental models in a shift from representational to abstract codes. From this definition, mental models can be described as complex schemata that are used to model the behavior of systems. Therefore, a mental model can be viewed as a schema for a certain system.

Related to this is the situational model (or situation model), a term used by VanDijk and Kintsch (1983) and by Roschelle and Greeno (1987), which will be defined as a schema depicting the current state of the system model (and often developed in light of the system model). Rasmussen (1986) also used the term internal dynamic world model with the same general meaning. The terms situation model and situation awareness will be defined here as equivalent.

A situation model (i.e., SA) can be matched to schemata in memory that depict prototypical situations or states of the system model. These prototypical classifications may be linked to associated goals or scripts that dictate decision making and action performance. This provides a mechanism for the single-step, "recognition-primed" decision making described earlier. This process is hypothesized to be a key mechanism whereby people are able to efficiently process a large amount of environmental information to achieve SA. A well-developed mental model provides (a) knowledge of the relevant elements of the system that can be used in directing attention and classifying information in the perception process, (b) a means of integrating the elements to form an understanding of their meaning (Level 2 SA), and (c) a mechanism for projecting future states of the system based on its current state and an understanding of its dynamics (Level 3 SA).

For example, a pilot may perceive several aircraft (considered to be important elements per the mental model) recognized as enemy fighter jets (based on critical cues) that are approaching in a particular spatial arrangement (forming Level 1 SA). By pattern-matching to prototypes in memory, these separate pieces of information may be classified as a particular recognized aircraft formation (Level 2 SA). According to an internally held mental model, the pilot is able to generate probable attack scenarios for this type of formation when in relation to an aircraft with the location and flight vector of his or her own ship (Level 3 SA). Based on this high-level SA, the pilot is then able to select prescribed tactics (a script) that dictate exactly what evasive maneuvers should be taken.

The key to using these models to achieve SA rests on the ability of the individual to recognize key features in the environment—critical cues—that will map to key features in the model. The model can then provide for much of the higher levels of SA (comprehension and projection) without loading working memory. In cases in which scripts have been developed for given prototypical situation conditions, the load on working memory for generating alternative behaviors and selecting among them is even further diminished.

A major advantage of this mechanism is that the current situation need not be exactly like one encountered before. This is a result of categorization mapping (a best fit between the characteristics of the situation and the characteristics of known categories or prototypes). Of prime importance is that this process can be almost instantaneous because of the superior abilities of human pattern-matching mechanisms. When an individual has a well-developed mental model for the behavior of particular systems or domains, the model will provide (a) for the dynamic direction of attention to critical cues, (b) expectations regarding future states of the environment (including what to expect as well as what not to expect) based on the projection mechanisms of the model, and (c) a direct, single-step link between recognized situation classifications and typical actions.

Development

Schemata and mental models are developed as a function of training and experience in a given environment. A novice in an area may have only a vague idea of important system components and sketchy rules or heuristics for determining the behavior he or she should employ with the system. With experience, recurrent situational components will be noticed along with recurrent associations and causal relationships. This forms the basis for early schema or model development.

Holland et al. (1986) provided a thorough description of the development of mental models. According to their description, an individual will learn (a) categorization functions that allow people to map from objects in the real world to a representative category in their mental model, and (b) model transition functions that describe how objects in the model will change over time. By repeatedly comparing the predictions of their internal model with the actual states of the system, individuals will progressively refine their models to develop more specific and numerous categorization functions which allow for more accurate predictions based on detailed object characteristics and better transition functions for these specialized categorizations. This process enables people to progressively refine their classification of a perceived object from an aircraft to fighter aircraft to F-18 to F-18c and gives them a more refined idea of the behavior and capabilities of the aircraft (in order to provide predictions). Their explanation also provides for two more features that are important to recognized attributes of situation awareness: default information and confidence levels.

Default information

Holland et al. (1986) explanation includes a "Q-morphism" in which default information for the system is provided in a higher layer of the model (i.e., a more general level of classification). These default values may be used by individuals to predict system performance unless some specific exception is triggered, in which case the appropriate transition function for that more detailed classification will be used. For example, a pilot will make decisions based on general knowledge of how fighter aircraft maneuver if the specific model of aircraft is not known. This feature allows people to operate effectively on the basis of often limited information.

In addition, default values for certain features of a system can be used if exact current values are not known. Fighter pilots, for example, usually get only limited information about other aircraft. They therefore must operate on default information (e.g., it is probably a MIG-29 and therefore likely traveling at certain approximate speed). When more details become available, their SA becomes more accurate (e.g., knowledge of the exact airspeed), possibly leading to better decisions, but they are still able to make reasonable decisions without perfect information. This provision of mental models allows experts to have access to reasonable defaults that provide more effective decisions than those of novices who simply have missing information (or poorer defaults). In many cases, experts may incorporate this type of default information in forming SA.

Confidence level

A second important aspect of situation awareness concerns a person's confidence level regarding that SA. People may have a certain confidence level regarding the accuracy of information they have received based on its reliability or source. The confidence level associated with information can influence the decisions that are made using that information (Norman, 1983). An important aspect of SA, therefore, is the person's confidence concerning that SA, a feature that has been cited by both pilots and air traffic controllers (Endsley, 1993c; Endsley and Rodgers, 1994).

Holland et al. (1986) hypothesized that there is a degree of uncertainty associated with the mental model's transition function that will provide confidence levels associated with predictions from the model. Similarly, one could hypothesize a degree of uncertainty associated with the validity of features used to make the mapping from the real world to categories in the model. For example, if three sources of information indicate a certain object is an apple but one source indicates it is an orange, the object may be characterized in the internal model as an apple but with an uncertainty factor attached to it.

VanDijk and Kintsch (1983), in work on speech understanding, have conceptualized a context model that allows uncertainties to be linked to information from various sources and taken into account in the decision process as well as the stated facts. Borrowing this concept, any given situation model may include a context feature representing the degree of uncertainty regarding the mapping of world information to the internal model and the projections based on the model. This feature allows people to make decisions effectively, despite numerous uncertainties, yet small shifts in factors underlying the uncertainties can dramatically change resultant conclusions (Norman, 1983).

Automaticity

In addition to developing mental models with experience, a form of automaticity can be acquired. Automatic processing tends to be fast, autonomous, effortless, and unavailable to conscious awareness in that it can occur without attention (Logan, 1988).

Thus automaticity of certain tasks can significantly benefit SA by providing a mechanism for overcoming limited attention capacity.

In relation to SA, automaticity poses an important question, however. To what degree do people who are functioning automatically have SA? SA, by definition, involves one's level of awareness, which implies consciousness of that information. With automaticity, however, certain features of cognitive processing occur below conscious awareness.

Logan (1988) provided a detailed discussion of automaticity in cognitive processing that he maintained occurs through a direct-access, single-step retrieval of actions to be performed from memory. This description of automaticity is consistent with the previous discussion on the use of schemata and mental models for matching recognized classes of situations to scripts for actions. In this process, "attention to an object is sufficient to cause retrieval of whatever information has been associated with it in the past" (Logan, 1988, p. S87)-that is, to activate the schema or mental model.

When processing in this way, an individual appears to be conscious of the situational elements that triggered the automatic retrieval of information from memory (SA), but he or she probably will not be conscious of the mechanisms used in arriving at the resultant action selection. That is, a person will know the Level 1 elements (e.g., there is an engine problem), even though he or she may not be aware of or be able to articulate the critical cues that led to that knowledge (e.g., a slight change in engine pitch: Nisbett and Wilson, 1977) and may not be able to identify the process used to arrive at a decision because it was directly retrieved from memory as the appropriate script for that situation (Bowers, 1991: Manktelow and Jones, 1987). As expressed by Dreyfus (1981), the individual knows the what but not the how. If asked to explain why a particular decision was made, an individual will usually have to construct some rationale using logical processes to provide an explanation of the action he or she actually chose in an automatic, nonanalytic manner (through the direct link of prototypical situations to scripts). The state of the situation itself (SA), however, can still be verbalized as it is in awareness. (This process has direct implications for the measurement of SA, which is addressed in the following article in this issue.)

This account is consistent with Nisbett and Wilson's (1977) review of people's awareness of and ability to report on mental events. In all of the cases presented by Nisbett and Wilson, it would appear that the how becomes occluded through the use of automatic processes but the what is still available to awareness. The one exception to this statement is the possibility of processing based on subliminal stimuli, which have been shown to modify affective processes. Evidence for the role of subliminal stimuli on typical dynamic decision making, as opposed to affective processes, however, is less apparent.

In addition, the degree to which automatic processing occurs without any attention or awareness has been questioned. Reason (1984) argued that a minimum level of attention is required for all activity—even automatic processes—in order to bring appropriate schemata into play at the right times and to restrain unwanted schemata from interfering. At this very low level of attention, there would be no awareness (equated with consciousness) of the detailed procedures. Once a plan has been put into motion, it serves to execute scripts and process schema as instructed.

An example of the possibility of decision making without conscious SA is that of a person driving home from work who follows the same predetermined path, stops at stoplights, responds to brake lights, and goes with the flow of traffic, yet can report almost no recollection of the trip. Did this person truly operate with no conscious awareness? Or, is it that only a low level of attention was allocated to this routine task, keying on critical environmental features task, keying on critical environmental features. The low level of consciousness simply did not provide sufficient salience to allow that particular drive home to be retrieved from memory as distinguishable from a hundred other such trips. I would argue, in agreement with Reason (1984), that this latter alternative is far more likely. Several authors in support of this view have found that when effortful processing is not used, information can be retained in long-term memory and can affect subject responses (Jacoby and Dallas, 1981; Kellog, 1980; Tulving, 198S).

The major implications of the use of automatic processes are (a) good performance with minimal attention allocation, (b) significant difficulty in accurately reporting on the internal models used for such processing and possibly on reporting which key environmental features were related, and (c) unreliability and inaccuracy of reporting on processes after the fact. Based on this discussion, automaticity is theorized to provide an important mechanism for overcoming human information-processing limitations in achieving SA and making decisions in complex, dynamic environments.

The primary hazard created by automatic cognitive processing is an increased risk of being less responsive to new stimuli, as automatic processes operate with limited use of feedback. A lower level of SA could result in atypical situations, decreasing decision timeliness and effectiveness. For example, when a new stop sign is suddenly erected on a familiar route, many people will initially proceed through the intersection without stopping, as the sign is not part of their automatic process and is not heeded.

Goals

SA is not generally thought of as a construct that exists solely for its own sake. SA is important as needed for decision making regarding some system or task. As such, it is integrally linked with both the context and the decisions for which the SA is being sought: it is fundamentally linked with a person's goals. Goals form the basis for most decision making in dynamic environments. Furthermore, more than one goal may be operating simultaneously, and these goals may sometimes conflict (e.g., "stay alive" and "kill enemies"). In most systems, people are not helpless recipients of data from the environment but are active seekers of data in light of their goals.

In what Casson (1983) has termed a top-down decision process, a person's goals and plans direct which aspects of the environment are attended to in the development of SA. That information is then integrated and interpreted in light of these goals to form Level 2 SA. The observation of each of three parameters of a system is not in itself meaningful. When integrated and viewed in the context of what they indicate about the goal of operating the system in a given manner, however, they become meaningful. The decision maker then selects activities that will bring the perceived environment into line with his or her plans and goals based on that understanding.

Simultaneously with this top-down process, bottom-up processing will occur. Patterns in the environment may be recognized that will indicate that new plans are necessary to meet active goals or that different goals should be activated. In this way a person's current goals and plans may change to be responsive to events in the environment. The alternating of top-down and bottom-up processing allows a person to process effectively in a dynamic environment.

This process also relates to the role of mental models and schemata. The model in Figure 8.4 can be used to visualize the relationship. Mental models of systems can be seen to exist as set (although slowly evolving) memory structures. Independently, individuals form a set of goals that relate to some system. These goals can be thought of as ideal states of the system that they wish to achieve. The same set of goals may exist frequently for a given system or may change often. Conversely, a set of goals may relate to more than one

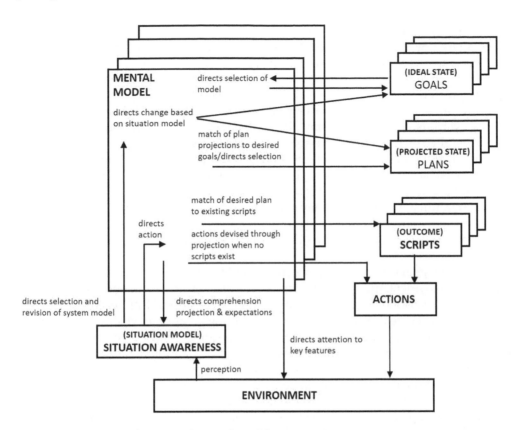

Figure 8.4 Relationship of goals and mental models to situation awareness.

system model. A person's current goal(s), selected as the most important among competing goals, will act to direct the selection of a mental model. The selected goals will also determine the frame (Casson, 1983), or focus, on the model that is adopted.

Plans are then devised for reaching the goal using the projection capabilities of the model. A plan will be selected whose projected state best matches the goal state. When scripts are available for executing the selected plan, they will be employed (Schank and Abelson, 1977). When scripts are not available, actions will have to be devised to allow for plan completion. Again, the projection capabilities of the system model will be used to accomplish this.

As an ongoing process, an individual observes the current state of the environment, with his or her attention directed to environmental features by the goal-activated model and interpreted in light of it. The model that is active provides a future projection of the status of key environmental features and expectations concerning future events. When these expectations match that which is observed, all is well. When they do not match because values of some parameter are different or an event occurs that should not, or an event does not occur that should, this signals to the individual that something is amiss and indicates a need to change goals or plans because of a shift in situation classes, a revision of the model, or selection of a new model.

This process can also act to change current goal selection by altering the relative importance of goals, as each goal can have antecedent rules governing situation classes in which each needs to be invoked over the others. When multiple goals are compatible with each other, several may be active at once. When goals are incompatible, their associated priority level for the identified situation class determines which shall be invoked. Similarly, plans may be altered or new plans selected if the feedback provided indicates that the plan is not achieving results in accordance with its projections, or when new goals require new plans. Through learning, these processes can also serve to create better models, allowing for better projections in the future.

To give a detailed example of this process in a military aircraft environment, a pilot may have various goals, such as stay alive, kill enemy aircraft, and bomb a given target. These general goals may have more specific subgoals, such as navigate to the target, avoid detection, avoid missiles, and employ missiles. The pilot would choose between goals (and subgoals) based on their relative importance and the existing situation classification. Staying alive is a priority goal, for example, which usually is active (except in extreme kamikaze circumstances). A pilot may alternate between the goals of bombing a target and killing an enemy aircraft based on the predetermined criticality of each goal's success to the current mission and the specifics of the situation (which would convey the likelihood of each goal's success).

The current goal would indicate the model and frame to be active. A model for the goal of missile employment might direct attention toward key environmental features, such as dynamic relative positions of own and threat aircraft (location, altitude, airspeed, heading, flight path), and current weapon selection, including weapon envelope and capabilities, current probability of kill, and rate of change of probability of kill. If this model was active, the pilot would be inclined to seek out and process those key elements of the environment. Use of the resultant situation model (SA), in conjunction with the missile employment model, would allow the pilot to determine how best to employ the aircraft relative to the enemy aircraft and missile launch timing (plans and actions).

While carrying out this goal, the pilot will also be alert to critical features that might indicate that a new model should be activated. If the pilot detected a new threat, for example, the activated goals might change so that the pilot would cease to operate on the missile employment model, and a threat assessment model would be activated consistent with that goal. The model selected, if detailed enough, would be used to direct situation comprehension, future projection, and decision making. A threat assessment model might include information as to what patterns of threats and threat movements constitute offensive versus defensive activities, for example. Future threat movements might be predictable from the model based on current threat movements and known tactics. Appropriate tactics for countering given threat actions might also be resident in the form of scripts, simplifying decision making.

Summary

To summarize the key features of SA in this model, a person's SA is restricted by limited attention and working memory capacity. Where they have been developed, long-term memory stores, most likely in the form of schemata and mental models, can largely circumvent these limits by providing for the integration and comprehension of information and the projection of future events (the higher levels of SA), even on the basis of incomplete information and under uncertainty. The use of these models depends on pattern matching between critical cues in the environment and elements in the model. Schemata of prototypical situations may also be associated with scripts to produce single-step retrieval of actions from memory. SA is largely affected by a person's goals and expectations which will influence how attention is directed, how information is perceived, and how it is interpreted. This top-down processing will operate in tandem with bottom-up processing in which salient cues will activate appropriate goals and models. In addition, automaticity may be useful in overcoming attention limits; however, it may leave the individual susceptible to missing novel stimuli that can negatively affect SA.

Task and system factors

A number of task and system factors can also be postulated to influence an individual's ability to achieve SA. Although a full list of these factors has yet to be determined, a few major issues would seem apparent.

System design

Figure 8.5 shows the sequence by which a person gains access to information from the environment (Endsley, 1990a). Some information may be acquired directly. In many domains of interest, however, an intervening system senses information and presents it to a human operator. In this process, transmission error, defined as a loss of information, can occur at each transition.

First of all, the system may not acquire all of the needed information (e1). Most aircraft systems, for example, even those with the latest radar, do not provide complete tracks on all aircraft. Nor do they provide everything the pilot would like to know about those aircraft that are detected. Similarly, most systems will acquire only certain information, based on the designer's understanding of what is required and technological limitations.

Of the information acquired by the system, not all of it may be displayed to the operator (e2). This may be because the interface is either not set up to display certain information or only subsets can be displayed at any one time. Frequently, the operator can determine to a certain degree which subset of data is displayed (and also in some systems in which data are acquired). Finally, of the information displayed by the system and that directly acquirable from the environment, there may be incomplete or inaccurate transmission to the human operator (e3 and e4) because of perceptual, attention, and working memory constraints, as discussed earlier.

The first external issue influencing SA, therefore, is the degree to which the system acquires the needed information from the environment. The second major issue involves the display interface for providing that information to the operator.

Interface design

The way in which information is presented via the operator interface will largely influence SA by determining how much information can be acquired, how accurately it can be acquired, and to what degree it is compatible with the operator's SA needs. Hence, SA has become a topic of great concern in many human factors design efforts. In general, one

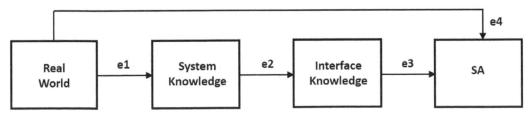

Figure 8.5 Situation awareness inputs.

seeks designs that will transmit needed information to the operator without undue cognitive effort. In this light, mental workload has been a consideration in design efforts for some time. At the same time, the level of SA provided (the outcome of that process) needs to be considered.

Determining specific design guidelines for improving operator SA through the interface is the challenge fueling many current research efforts. Several general interface features can be hypothesized to be important for SA, based on the model presented here.

1. As attention and working memory are limited, the degree to which displays provide information that is processed and integrated in terms of Level 2 and 3 SA requirements will positively affect SA. For instance, directly portraying the amount of time and distance available on the fuel remaining in an aircraft would be preferable to requiring the pilot to calculate this information based on lower-level data (e.g., fuel, speed, altitude, etc.).

2. The degree to which information is presented in terms of the operator's major goals will positively affect SA. Many systems provide information that is technology oriented-based on physical system parameters and measurements (e.g., oil pressure or temperature). To improve SA, this information needs to be SA oriented. That is, it should be organized so that the information needed for a particular goal is collocated and directly answers the major decisions associated with the goal. For example, for the goal of weapons employment, factors such as opening/closing velocity, weapon selected and firing envelope, probability of kill, target selected, and time to employment would be relevant elements that should be presented in an integrated form for this goal.

3. Considering that mental models and schemata are hypothesized to be key tools for achieving the higher levels of SA in complex systems, the critical cues used for activating these mechanisms need to be determined and made salient in the interface design. In particular those cues that will indicate the presence of prototypical situations will be of prime importance. Kaplan and Simon (1990) found decision making is facilitated if the critical attributes are perceptually salient.

4. Designs need to take into consideration both top-down and bottom-up processing. In this light, environmental cues with highly salient features will tend to capture attention away from current goal-directed processing. Salient design features, such as those indicated by Treisman and Paterson (1984), should be reserved for critical cues that indicate the need for activating other goals and should be avoided for non-critical events.

5. A major problem for SA occurs when attention is directed to a subset of information and other important elements are not attended to, either intentionally or unintentionally (Endsley and Bolstad, 1993). It is hypothesized that designs that restrict access to SA elements (via information filtering, for instance) will contribute to this problem. A preferred design will provide global SA-an overview of the situation across operator goals at all times, while providing the operator with detailed information related to his or her immediate goals, as required. Global SA is hypothesized to be important for determining current goals and for enabling projection of future events.

6. Although filtering out information on relevant SA elements is hypothesized to be detrimental, the problem of information overload in many systems must still be considered. The filtering of extraneous information (not related to SA needs) and reduction of data (by processing and integrating low-level data to arrive at SA requirements) should be beneficial to SA.

7. One of the most difficult and taxing parts of SA is the projection of future states of the system. This is hypothesized to require a fairly well developed mental model. System-generated support for projecting future events and states of the system should directly benefit Level 3 SA, particularly for less-experienced operators.
8. The ability to share attention between multiple tasks and sources of information will be very important in any complex system. System designs that support parallel processing of information should directly benefit SA. For example, the addition of voice synthesis or three-dimensional audio cues to the visually overloaded cockpit is predicted to be beneficial on this basis.

These recommendations may not appear that radically different from those that have been espoused, at least singularly, elsewhere. This is because the SA theory described here rests on various information-processing constructs that have been discussed for some years. The value added by the SA concept is as a means of integrating these constructs in terms of the operator's overall goals and decision behavior. As such, this provides several advantages in the design process.

1. The integrated focus of SA provides a means of designing for dynamic, goal-oriented behavior, with its constant shifting of goals. Traditional design approaches (Meister, 1971) have focused on task analysis, which works fairly well for fixed, sequential tasks but does not provide the mechanisms or flexibility necessary for dealing with dynamic tasks and fluctuating goals. By focusing at the level of operator goals, the degree to which multiple goals may be operating simultaneously can be considered and the precursors to goal activation represented. Thus, a more compatible representation of operator behavior can be generated for creating "user-centered" designs.
2. It provides a means of moving from a focus on providing operators with data to providing operators with information. When focusing on data, all of the integration, comprehension, and projection is still up to the operator. When focusing on information, the design focus is on presenting what the operator really needs to know in the format it is needed in, thus allowing the operator to achieve more SA at a given level of workload. By presenting the Level 1, 2, and 3 SA requirements associated with each goal or subgoal, this can be accomplished.
3. It provides a means of incorporating into the design a consideration of the interplay of elements, wherein more attention to some elements maybe at the expense of others. Many design guidelines are at the level of the specific component (e.g., a dial or audio signal's characteristics). Yet the real challenge in designing systems arises when the components must be integrated. The SA provided to the operator as a result of the combination of system components becomes the goal of the integration process.
4. Perhaps most important, this integrated level of focus provides a means for assessing the efficacy of a particular design concept that an examination of underlying constructs (attention, working memory, etc.) does not provide. As an integrated system, the degree to which a particular design provides SA (as a resultant state) can be determined after all these other factors, with their associated trade-offs and interactions, have come into play.

The number of possible display formats, technologies, and design concepts that have been or may be contemplated for improving SA are too numerous to mention. A few major design issues, however, pose a serious enough challenge to SA across numerous systems to warrant special consideration: stress, workload, complexity, and automation.

Stress

Several types of stress factors exist that may act to influence SA, including (a) physical stressors-noise, vibration, heat/cold, lighting, atmospheric conditions, drugs, boredom or fatigue, and cyclical changes; and (b) social psychological stressors-fear or anxiety, uncertainty, importance or consequences of events, aspects of task affecting monetary gain, self-esteem, prestige, job advancement or loss, mental load, and time pressure (Hockey, 1986; Sharit and Salvendy, 1982).

Mandler (1982) stated that these stressors "are effective to the extent that they are perceived as dangerous or threatening" (p. 91). That is, they are stressors only if the person perceives them as being stressing. A large interpretive component exists in the process. A certain amount of stress may actually improve performance by increasing attention to important aspects of the situation. A higher amount of stress can have extremely negative consequences, however, as accompanying increases in autonomic functioning and aspects of the stressors can act to demand a portion of a person's limited attentional capacity (Hockey, 1986).

Stressors can affect SA in a number of different ways. The first, and probably most widespread, finding is that under various forms of stress, people tend to narrow their field of attention to include only a limited number of central aspects (Bacon, 1974; Baddeley, 1972; Bartlett, 1943; Callaway and Dembo, 1958; Davis, 1948; Eysenck, 1982; Hockey, 1970). Under perceived danger, a decrease in attention has been observed for peripheral information (i.e., those aspects that attract less attentional focus; Bacon, 1974; Weltman et al., 1971). Broadbent (1971) found that there was an increased tendency to sample dominant or probable sources of information under stress. Sheridan (1981) has termed this effect cognitive tunnel vision.

This is a critical problem for SA, leading to the neglect of certain elements in favor of others. In many cases, such as in emergency conditions, it is those factors outside the operator's perceived central task that prove to be lethal. A United Airlines DC-8 crashed in Portland, Oregon, in 1978 when it ran out of fuel. It was reported that the captain, preoccupied with a landing gear problem, neglected to keep track of fuel usage (National Transportation Safety Board, 1979). Many similar incidents of attentional narrowing can be found.

Premature closure, arriving at a decision without exploring all information available, has also been found to be more likely under stress (Janis, 1982; Keinan, 1987; Keinan and Friedland, 1987). This includes considering less information (Janis, 1982; Wright, 1974) and attending more to negative information (Wright, 1974). Several authors have found that scanning of stimuli under stress is scattered and poorly organized (Keinan, 1987; Keinan and Friedland, 1987; Wachtel, 1967).

Complex tasks with multiple input sources appear to be particularly sensitive to the effects of stressors (Broadbent, 1954; Jerison, 1957, 1959). Woodhead (1964) found that performance decrements that occurred during intermittent noise stress took place during the information input stage. It would seem, then, that stress significantly affects the early stage of the decision-making process that is involved in the assessment of the situation. It is expected that stress will significantly influence SA on this basis, beginning with the initial perception of environmental elements (Level 1).

A second way in which stress may affect SA is through decrements in working memory capacity and retrieval (Hockey, 1986). Wickens et al. (1988) found that optimality of performance was negatively affected by stress only on decision tasks with a high spatial component, however, and not on those with purely a high working memory or long-term memory component.

The degree to which working memory decrements will affect SA depends on the resources available to the individual operator. In tasks in which achieving SA involves a high working memory load, a significant impact on SA Levels 2 and 3 would also be expected. In the Vincennes incident, the systems operators had to rely on working memory to calculate whether an incoming aircraft was ascending or descending. Their error in believing the incoming aircraft was descending could have been associated with reduced working memory capacity in a stressful combat environment. If long-term memory stores are available to support SA, less effect will be expected.

Workload

In many dynamic systems, high mental workload is a stressor of particular importance, so much so that at least one major approach to SA measurement combines workload features (supply and demand of operator resources) with information features (Taylor, 1989). Endsley (1993a), however, demonstrated independence between these two constructs across a wide range of values. That is, the following may exist:

1. Low SA with low workload: The operator may have little idea of what is going on and is not actively working to find out because of inattentiveness, vigilance problems, or low motivation.
2. Low SA with high workload: If the volume of information and number of tasks are too great, SA may suffer because the operator can attend to only a subset of information or may be actively working to achieve SA, and yet has erroneous or incomplete perception and integration of information.
3. High SA with low workload: The required information can be presented in a manner that is easy to process (an ideal state).
4. High SA with high workload: The operator is working hard but is successful in achieving an accurate and complete picture of the situation.

Thus, SA and workload are hypothesized to diverge because of characteristics of the system design, tasks, and the individual operator. If the operator is exerting effort at attaining SA and if the demands associated with this task and others exceed the operator's limited capacity, only then will a decrement in SA be expected.

Complexity

A major factor creating a challenge for operator SA is the increasing complexity of many systems. System complexity is hypothesized to negatively affect both operator workload and SA through factors such as an increase in the number of system components, the degree of interaction between these components, and the dynamics or rate of change of the components. In addition, the complexity of the operator's tasks may increase through the number of goals, tasks, and decisions to be made in regard to the system.

Each of these factors will increase the amount of mental workload required to achieve a given level of SA. When that demand exceeds human capabilities, SA will suffer. This complexity may be somewhat moderated by the degree to which the operator has a well-developed internal representation of the system to aid in directing attention, integrating data, and developing the higher levels of SA, as these mechanisms may be effective for coping with complexity.

Automation

A lack of SA has been hypothesized to underlie the out-of-the-loop performance decrement that can accompany automation (Carmody and Gluckman, 1993: Endsley, 1987a: Wickens, 1992b). System operators working with automation have been found to have a diminished ability to detect system errors and subsequently perform tasks manually in the face of automation failures as compared with manual performance on the same tasks (Billings, 1991; Moray, 1986; Wickens, 1992a: Wiener and Curry, 1980). Although some of this problem may result from a loss of manual skills under automation, SA is also a critical component.

Operators who have lost SA may be slower to detect problems and also will require extra time to reorient themselves to relevant system parameters in order to proceed with problem diagnosis and assumption of manual performance when automation fails. This has been hypothesized to occur for a number of reasons: (a) a loss of vigilance and increase in complacency associated with the assumption of a monitoring role under automation, (b) the difference between being an active processor of information in manual processing and a passive recipient of information under automation, and (c) a loss of or change in the type of feedback provided to operators concerning the state of the system under automation (Endsley and Kiris, in press). In their study, Endsley and Kiris found evidence for an SA decrement accompanying automation of a cognitive task that was greater under full automation than it was under various levels of partial automation. Lower SA in the automated conditions corresponded to a demonstrated outof-the-loop performance decrement, supporting the hypothesized relationship between SA and automation.

SA may not suffer under all forms of automation, however. Wiener (1993) and Billings (1991) have stated that SA may be improved by systems that provide integrated information through automation. In commercial cockpits, Hansman et al. (1992) found that automated flight management system input was superior to manual data entry, producing better error detection on clearance updates. Automation that reduces unnecessary manual work and data integration required to achieve SA may provide benefits to both workload and SA. The exact conditions under which SA will be positively or negatively affected by automation· need to be determined.

Errors in SA

From an operational point of view, there is major concern about situations in which the operator has poor SA, thus increasing the probability of undesirable performance. Errors in SA can be discussed in terms of the presented model. It is not the intention here to discuss all types of human error, for which several taxonomies exist (Norman, 1983; Rasmussen, 1986; Reason, 1987) but, rather, to investigate the factors that can lead to breakdowns in the SA portion of the decision-making process. These breakdowns can occur from either incomplete SA—knowledge of only some of the elements—or inaccurate SA—erroneous knowledge concerning the value of some elements. The discussion will be separated into those factors affecting SA at each of its three levels.

Level 1 SA

At the very lowest level, a person may simply fail to perceive certain information that is important for SA in the assigned task (incomplete SA). In the simplest case, this may result from a lack of detectability or discriminability of the physical characteristics of the signal in question, from some physical obstruction preventing perception (visual barrier, auditory masking, etc.), or from a failure of the system design to make the information available to the operator. Accurate, reliable weather information for aircrew is frequently lacking, for instance. The crew of a Northwest Airlines DC-9 attempted to take off from Detroit unaware that the aircraft flaps were retracted, leading to the death of 154 people (National Transportation Safety Board, 1988). A partial reason cited for this lack of knowledge was the failure of a takeoff warning system to alert the crew to the problem with the flaps. (In addition, the crew failed to fully execute a checklist, thus they did not directly check the flaps themselves.)

In extreme cases, the only cue a person will have regarding the presence of certain information will coincide with the occurrence of an error. Rasmussen (1986) gave the example of a person not realizing that it is icy until he or she slips. In this case, the condition could be discerned only in conjunction with the error and not sufficiently in advance to allow for behavior modification to prevent the error. In other cases, because of luck, no error may result from the lack of SA; however, the potential for error would rise significantly.

In many cases in which SA is incomplete, the relevant signals or cues are readily discernible but not properly perceived by the subject. The failure of the Northwest Airlines crew to manually check flap status would fall into this category. There can be several underlying causes for not perceiving available information. Many complex environments present an overabundance of information. Data sampling should maintain a fair degree of accuracy on each of the relevant variables (Wickens, 1992a), in which case errors in SA would be small (determined by the amount of change in each variable between successive samples) and distributed across the various variables of concern. Failures in information sampling are commonplace, however, and may result from the lack of an adequate strategy or internal model for directing sampling. Wickens (1992a) has also noted that humans have several general failings in sampling, including misperception of the statistical properties of elements in the environment and limitations of human memory (forgetting what has already been sampled). The phenomenon of visual dominance can act as a further limit; auditory information is less likely to be processed in some situations (Posner et al., 1976).

Furthermore, some people appear to be better than others at dividing their attention across different tasks (Damos and Wickens, 1980). Martin and Jones (1984) have found cognitive errors to be significantly correlated with capabilities in distributing attention across tasks. So, although environmental sampling can be an effective means of coping with excessive SA demands, human limitations in sampling, attention, and attention sharing can lead to significant Level 1 SA errors.

This problem is compounded by the addition of stress, which can affect the information input stage through premature closure, changes in factors attended to, and deterioration of the scanning process. The narrowing of attention brought on by stress or heavy workload can lead to a lack of SA on all but the factor at hand. In 1972 an L-1011 commercial airliner went down in the Florida Everglades because all of the crew members were so focused on a problem with the nose gear indicator that they failed to notice that the aircraft was descending. Ninety-nine lives were lost (National Transportation Safety Board, 1973). A major problem with attentional narrowing is that often a person will be sure he or she is attending to the most important information, but there is no way to know whether or not that assumption is valid without having some idea of the value of the other elements. In other cases, the normal sampling strategy has merely been interrupted and not reactivated in a timely manner. In either case, attentional narrowing can lead to serious errors in SA.

Inaccurate SA—the belief that the value of some variable is different from what it actually is—can also occur. In relation to Level 1 SA, this would occur through the misperception of a signal—for instance, seeing a blue light as green because of ambient lighting or seeing a 3 as an 8 on a dial. Exemplifying this problem is the instance in which a Boeing 737 hit power lines near Kansas City, Missouri, because the pilot misidentified lights north of the runway as the runway approach lights (National Transportation Safety Board, 1990). Erroneous expectations can be a major contributor to these misperceptions.

Level 2 SA

SA errors are most often the result of an inability to properly integrate or comprehend the meaning of perceived data in light of operator goals. Orasanu et al. (1993) described five National Transportation Safety Board (NTSB) aircraft accident reports. In all five cases, sufficient environmental cues were present, but the aircrew did not determine their relevance to important goals.

This misreading of cues can occur for several reasons. A novice will not have the mental models necessary for properly comprehending and integrating all of the incoming data or for determining which cues are actually relevant to established goals. Fischer et al. (1993) found that less effective crews lacked sensitivity to contextual factors, indicating a failure to recognize prototypical situations. In the absence of a good internal model, one must accept low SA and thus be compromised in decision making, develop a new model, or adapt an existing model to the task at hand. SA errors will exist in the form of inaccurate or incomplete Level 2 SA when the adapted or newly developed model fails to match the new environment.

In other cases, a person may incorrectly select the wrong model from memory, based on a subset of situational cues, and use this model to interpret all perceived data. Mosier and Chidester (1991) found evidence that aircrews made "recognitional, almost reflexive judgments, based upon a few, critical items of information; and then spent additional time and effort verifying its correctness through continued situational investigation." This strategy can be effective. Mosier and Chidester found that the best performing crews obtained a substantial portion of their information after making a decision.

However, if the wrong mental model is initially selected, based on a subset of cues, a representational error may occur. These errors can be particularly troublesome, as pointed out by Carmino et al. (1988). They noted that realizing that the wrong model is active can be very difficult because new data are interpreted in light of the model. Difficulties in recognizing the error may also be compounded by confirmation bias (Fracker, 1988). Thus data that should indicate one thing are taken to mean something quite different based on the incorrect model.

Klein (1993) reported on errors in medical decision making in which successive symptoms continued to be interpreted into an existing diagnosis even though they clearly pointed to a different diagnosis. Fracker also pointed out that an incorrect model may be selected initially because of representativeness and availability biases.

Even when a person has selected the correct model with which to interpret and integrate environmental stimuli, errors can occur. Certain pieces of data may be mismatched with the model or not matched at all, resulting in a failure to recognize a prototypical situation (Klein, 1989b; Manktelow and Jones, 1987). The National Transportation Safety Board (1981) noted that several aircraft conflicts were related to the fact that air traffic controllers received the same aural signal for both conflict alerts and low-altitude warnings. In this case, inadequate perceptual salience of the signals probably prevented an immediate correct match of cue to model.

In addition, SA errors could occur from over-relying on the default values embedded in a model (Manktelow and Jones, 1987). In general, when new situations are encountered in which the known default values are not appropriate, the model is modified to include the new class of situations. Before this occurs, or if cues received have not flagged the specific situation type, significant SA errors can occur by incorrectly assuming defaults for some variables. The newly developed French Airbus 320 crashed during a low flyover demonstration in 1988. The inquiry noted that the pilot may not have been adequately aware of effects on handling performance when flying near the angle-of-attack limits of the aircraft and may have been relying on the much-advertised envelope protection designed into the new aircraft (Ministry of Planning, Housing, Transport and Maritime Affairs, 1989). In terms of this paper, a refined model for the specific aircraft capabilities had not yet been developed, and the pilot had to rely on a general understanding of envelope protection.

When no model exists at all, Level 2 SA must be developed in working memory. An inability to perform this integration in an accurate, timely manner—resulting from insufficient knowledge or working memory limitations, particularly under stress—can also lead to inaccurate or incomplete SA.

Level 3 SA

Finally, Level 3 SA may be lacking or incorrect. Even if a situation is clearly understood, it may be difficult to accurately project future dynamics without a highly developed mental model. Klein (1989b) has noted that some people simply are not good at mental simulation. Lack of a good model and attention and memory limitations would account for this. Simmel and Shelton (1987) described the problems pilots have in determining potential consequences of assessed situations. Amalberti and Deblon (1992) and MacMillian, Entin, and Serfaty (1993) noted, however, that experts frequently determine possible future occurrences in order to plan ahead.

General factors

A few general underlying factors may also lead to SA errors at all three levels. Martin and Jones (1984) pointed out that people who have trouble with distributed attention may be having trouble in maintaining multiple goals. This could lead to considerable SA problems in complex systems, in which the ability to juggle goals on the basis of incoming information is a necessity. An inability to keep multiple goals in mind could seriously degrade an operator's receptivity to highly pertinent data related to the neglected goal, leading to significant errors.

A second major type of error affecting SA relates to the role of habitual schemata (or automaticity). In the normal course of events, habitual schemata will be automatically activated based on the presence of environmental cues. While the schema is active, environmental cues will be processed in a predetermined manner. When a change needs to be made, however, problems can occur. A person leaving work and getting into the car may automatically embark on the "drive home" schema. If on a particular day the person wishes to stop at the store, he or she must change or intempt the schema. Often, however, the person arrives home to realize the desired detour was completely forgotten.

Although this has been termed a slip of action (Reason, 1984), it can also be shown to be related to SA. Under normal circumstances, environmental cues (the store sign) will be processed in light of current goals (stop at the store). While habitual schemata are operating, however, the new, nonhabitual goal is suppressed, and seeing the store sign does not conjure the associated goal of stopping. While the habitual schema is operating, the person either is not receptive to the nonhabitual cues or does not generate the appropriate higher-level SA from the perception of the cues because the appropriate schema is suppressed.

Detection of SA errors

A real issue concerns how people know when their SA is in error. Very often they may be completely unaware of how much they do not know or of the inaccuracy of their internal representation of the situation. The main clue to erroneous SA will occur when a person perceives some new piece of data that does not fit with expectations based on his or her internal model. When a person's expectations do not match with what is perceived, this conflict can be resolved by adopting a new model, revising the existing model, or changing one's goals and plans to accommodate the new situation classification (Manktelow and Jones, 1987). The inappropriate choice could easily sabotage SA efforts for some time.

If the new data can be incorporated into the model, this may merely indicate that a new prototypical situation (state of the model) is present that calls up different goals and plans accordingly. If the new data cannot easily fit into the existing model, the model may be revised. A common problem is whether to continue to revise the existing model to account for the new data or choose an alternate model that is more appropriate. For the latter to occur, something about the data must flag that a different situation is present. Without this flag, the person may persist in a representational error whereby the data continue to be misinterpreted in light of the wrong model. Of course, if the inadequacy of the existing model is recognized but no appropriate new model exists, significant errors may still occur while a new model is being developed.

Conclusions

This paper presents a model of SA, including various mechanisms and factors hypothesized to be important for its generation. Based on this model, a taxonomy of SA errors was generated. The model also presents a means of conducting future research on SA.

Theoretical hypotheses

Several characteristics of individuals and systems have been presented that are believed to affect a person's ability to acquire and maintain SA. In terms of information-processing mechanisms available to individuals, the following key features affecting SA are hypothesized:

1. The way in which attention is directed across available information is critical to achieving SA (particularly in dynamic and complex systems in which attention is overloaded).
2. In the absence of long-term memory structures, SA will be constrained by the limitations of attention and working memory.
3. Schemata and mental models are presented as mechanisms for (a) directing attention in the perception process, (b) providing a means of integrating and comprehending

perceived information, and (c) projecting the future states of the environment. These mechanisms allow decision makers to develop SA when they have only limited information from the environment.

4. A person's expectations or preconceptions about future events and environmental features, as generated from mental models, instructions, and communications, will influence the perception process and the interpretation of what is perceived.

5. SA is viewed as being generated from a combination of goal-directed (top-down) and data-directed (bottom-up) processing. As such, it will be affected by both the operator's current goals and the presence of salient environmental cues.

6. The operator's current goals will act to direct the selection of a mental model and the focus (or frame) taken on the model.

7. Knowledge of critical cues in the environment is highly important for (a) directing the selection of active goals from among possible operator goals (and thus mental model selection) and (b) pattern matching with schemata of prototypical situations according to the current model.

8. Automaticity is presented as an additional mechanism for overcoming attention and working memory limitations. When operating with automaticity, it is expected that operators will have reduced awareness of environmental factors (lower SA) particularly for those elements outside the automated sequence, and thus will be more likely to make errors under novel circumstances.

In addition, several characteristics of systems and tasks are hypothesized to influence an individual's ability to achieve SA.

1. The degree to which relevant features of the environment are available to the operator either directly or through the system's displays fundamentally affects a person's ability to achieve SA.

2. The way in which information is presented via the operator interface will affect a person's ability to achieve SA. Specific features hypothesized to positively impact SA include: integrated and goal-oriented information presentation, salience of critical cues, support for parallel processing of information, elimination of unneeded information and reduction in salience of noncritical information, presentation of global information across goals and detailed information on current goals, and system support for projection of future events and states.

3. Although small amounts of stress may improve SA through an increase in arousal and attention, excess stress will negatively affect SA through disruptions in acquiring information and, in some cases, through reductions in working memory capacity.

4. SA and workload are hypothesized to be essentially independent across a wide range of these constructs. Only under high levels of perceived workload will decrements in SA be expected.

5. Increases in perceived system complexity are expected to negatively affect both workload and SA unless moderated by the presence of a mental model for dealing with that complexity.

6. Automation of human decision making and active system control is hypothesized to negatively affect operator SA, leading to out-of-the-loop performance problems. Automation of peripheral tasks (e.g., data integration) is expected to positively affect SA by reducing the load on limited working memory.

Directions for further research

The model presented provides an integrated framework for conceptualizing the SA construct, thus providing a common ground for moving forward. As such, it provides several capabilities.

SA requirements. The model can be used to generate a means of determining SA requirements (elements) for individual domains of interest. The criticality of operator goals in the SA process dictates that SA requirements (at all levels) are dependent on the operator's goals in relation to the system. Thus, a goal-directed task analysis methodology is indicated in which the requirements for system data, the comprehension and integration of that data, and the projection of future states are determined for each of the operator's major goals and subgoals.

A methodology for conducting this type of analysis has been developed and applied to airto-air fighter aircraft (Endsley, 1993c), advanced bombers (Endsley, 1989a), and air traffic control (Endsley and Rodgers, 1994). In many domains, designers are working with only simple information requirements, without determining how the information needs to be integrated to support operator goals. This methodology can be applied to these domains to determine the SA requirements for systems.

Individual abilities. Endsley and Bolstad (1994) found evidence of fairly stable differences between individuals in their ability to achieve SA given the same system. Based on the present model, variations in SA abilities were hypothesized to arise from individual differences in (a) spatial abilities; (b) attention sharing; (c) memory, including working memory capacity and long-term memory stores; (d) perceptual skills, including perceptual speed, encoding speed, vigilance, and pattern-matching skills: and (e) higher-order cognitive skills, including analytic skills, cognitive complexity, field independence, and locus of control. Testing these hypotheses on a group of experienced fighter pilots, Endsley and Bolstad found strong evidence for the importance of spatial skills and perceptual skills and partial support for the importance of attention-sharing and pattern-matching skills.

More studies are needed to expand these findings to a larger, broader population. In addition, the degree to which such capabilities generalize across different domains, indicating a general SA skill or ability, needs to be determined. The identification of basic human abilities that are important for SA may be useful for improving operator SA through either selection or training.

Training

Programs directed at improving operator training by making it "SA oriented" can also be generated from the model. (See Endsley, 1989b, for a detailed discussion.) They can be developed to instruct operators to identify the important characteristics of mental models in specific domains, such as the components, dynamics and functioning of the components and projection of future actions based on these dynamics. SA-oriented training would focus on training operators to identify prototypical situations of concern associated with these models by recognizing critical cues and what they mean in terms of relevant goals.

As SA is not a passive process, the skills required for achieving and maintaining good SA need to be identified and formally taught in training programs. Factors such as how to employ a system to best achieve SA (when and where to look for what), appropriate scan patterns, or techniques for making the most of limited information need to be determined and explicitly taught in the training process. This type of focus greatly supplements traditional technology-oriented training that concentrates mainly on the mechanics of how a system operates.

In addition, the role of feedback in the learning process may be exploited. It may be possible to provide feedback on the accuracy and completeness of operator SA as a part of training programs. This would allow operators to understand their mistakes and better assess and interpret the environment, leading to the development of more effective sampling strategies and better schemata for integrating information. Training techniques such as these need to be explored and tested to determine methods for improving SA with existing systems.

Design

Several general hypotheses and recommendations concerning how to design systems to enhance SA were generated by the model. More research is needed to apply, test, and expand on these recommendations in relation to the design of specific systems in various domains. Several factors need to be determined, including ways to determine and effectively deliver critical cues; ways to ensure accurate expectations; methods for assisting operators in deploying attention effectively; methods for preventing the disruption of attention, particularly under stress and heavy workload; and ways to develop systems that are compatible with operator goals.

Research is being conducted to investigate a host of new technologies and designs being considered for future systems, including three-dimensional visual and auditory displays, voice control, expert systems, helmet-mounted displays, and virtual reality. This model should be useful for generating hypotheses concerning the effect of new technologies on SA in the context of a particular domain and system interface. Through controlled testing and an objective determination of the impact of these concepts on SA, specific design guidelines for their implementation, alone and in conjunction with one another, can be established.

SA construct

Future research on the SA construct is greatly needed. Several major hypotheses have been formulated concerning underlying information-processing mechanisms. The role of each of the major components needs to be formally tested and explored. In addition, empirical data are needed on SA as a whole in order to better understand and validate the hypothesized interactions and integration of individual factors. SA has been presented as a three-level concept. The relative importance of these levels needs to be established. How critical of a role does projection play, for instance? How is higher-level SA generated from lower-level data? Mental models and goals are hypothesized here as critical mechanisms, but they need further exploration.

Research is also needed to better understand the processes operators use to achieve SA. The way in which information is acquired by individuals and teams needs to be determined to identify successful techniques for coping with complex, dynamic systems. Useful critical cues that may be vital to achieving good SA (or cues that lead to poor SA via the representational error) need to be determined. The degree and nature of individual differences in such processes are not widely known at this point, except anecdotally.

In addition, the concept of SA may be useful in researching other constructs. For instance, situation models (or SA), which are a virtual reflection of system models, may shed some light on the concept of a mental model. Problems with the nebulous use of the term and the need for more precise specification of mental models have been expounded by Wilson and Rutherford (1989). If mental models are truly "mechanisms whereby humans are able to generate descriptions of system purpose and form, explanations of system functioning and observed system states, and predictions of future systems states," as described by Rouse and Morris (1985, p. 7), then three of the four criteria (system functioning, states, and predictions) can be determined by examining situation models (SA) across various contexts or states of the model. This type of effort may help create a better understanding of the nature of mental models in specific domains.

SA measurement

The ability to objectively measure SA is seen as critical for future progress in this field. It provides a means of evaluating the efficacy of design concepts and technologies, providing diagnostic data for design iteration, and a means of evaluating and developing training concepts. It also provides a means of researching the SA construct, investigating the impact of various factors in SA, and explicitly testing the hypotheses concerning SA. Without this capability, no real progress in the area of SA design or theory can be made. Methodologies for measuring SA are discussed in the subsequent paper (Endsley, 1995), based on the model presented here.

Summary

A model of SA has been presented in relation to decision making in complex systems. Building on research in naturalistic decision making, a person's SA is viewed as a critical focal point of the decision process. In this role, SA is presented as a general construct, applicable across a wide variety of environments and systems.

SA is viewed as consisting of a person's state of knowledge about a dynamic environment. It incorporates the perception of relevant elements, comprehension of the meaning of these elements in combination with and in relation to operator goals, and a projection of future states of the environment based on this understanding. Using this knowledge, individuals with good SA will have a greater likelihood of making appropriate decisions and performing well in dynamic systems. By learning more about SA requirements and the SA construct as a whole, more effective interface designs and training programs can be established to support decision making in complex environments.

References

Adams, M. J., Tenney, Y. J., and Pew, R. W. (1995). Situation awareness and the cognitive management of complex systems. *Human Factors: The Journal of the Human Factors and Ergonomics Society*, 37(1), 85–104.

Amalberti, R., and Deblon, F. (1992). Cognitive modeling of fighter aircraft process control: A step towards an intelligent on-board assistance system. *International Journal of Man-Machine Systems*, 36, 639–671.

Ashby, F. G., and Gott, R. E. (1988). Decision rules in the perception and categorization of multidimensional stimuli. *Journal of Experimental Psychology: Learning, Memory and Cognition*, 14, 33–53.

Bacon, S. J. (1974). Arousal and the range of cue utilization. *Journal of Experimental Psychology*, 102, 81–87.

Baddeley, A. D. (1972). Selective attention and performance in dangerous environments. *British Journal of Psychology*, 63, 537–546.

Barber, P. J., and Folkard, S. (1972). Reaction time under stimulus uncertainty with response certainty. *Journal of Experimental Psychology*, 93, 138–142.

Bartlett, F. C. (1932). *Remembering: A Study in Experimental and Social Psychology*. London, UK: Cambridge University Press.

Bartlett, F. C. (1943). Fatigue following highly skilled work. *Proceedings of the Royal Society (B)*, 131, 147–257.

Bettman, J. R., and Kakkar, P. (1977). Effects of information presentation format on consumer information acqu tion strategies. *Journal of Consumer Research*, 3, 233–240.

Biederman, I., Mezzanotte, R. J., Rabinowitz, J. C., Francolin, C. M., and Plude, D. (1981). Detecting the unexpected in photo interpretation. *Human Factors*, 23, 153–163.

Billings, C. E. (1991). *Human-Centered Aircraft Automation: A Concept and Guidelines (NASA Technical Memorandum 103885)*. Moffet Field, CA: NASA-Ames Research Center.

Bowers, K. S. (1991). Knowing more than we can say leads to saying more than we can know: On being implicitly informed. In D. Magnusson (Ed.), *Toward a Psychology of Situations: An Interactional Perspective* (pp. 179–194). Hillsdale, NJ: Erlbaum.

Braune, R. J., and Trollip, S. R. (1982). Towards an internal model in pilot training. *Aviation, Space and Environmental Medicine*, 53, 996–999.

Broadbent, D. E. (1954). Some effects of noise on visual performance. *Quarterly Journal of Experimental Psychology*, 6, 1–5.

Broadbent, D. E. (1971). *Decision and Stress*. London, UK: Academic Press.

Callaway, E., III, and Dembo, D. (1958). Narrowed attention: A psychological phenomenon that accompanies a certain physiological change. *AMA Archives of Neurology and Psychiatry*, 79, 74–90.

Carmino, A., Idee, E., Larchier Boulanger, J., and Morlat, G. (1988). Representational errors: Why some may be termed diabolical. In L. P. Goodstein, H. B. Anderson, and S. E. Olsen, (Eds.), *Tasks, Errors and Mental Models* (pp. 240–250). London, UK: Taylor & Francis.

Carmody, M. A., and Gluckman, J. P. (1993). Task-specific effects of automation and automation failure on performance, workload and situational awareness. In R. S. Jensen and D. Neumeister, (Eds.), *Proceedings of the Seventh International Symposium on Aviation Psychology* (pp. 167–171). Columbus, OH: Ohio State University, Department of Aviation.

Casson, R. W. (1983). Schema in cognitive anthropology. *Annual Review of Anthropology*, 12, 429–462.

Cohen, M. S. (1993). Metacognitive strategies in support of recognition. In *Proceedings of the Human Factors and Ergonomics Society 37th Annual Meeting* (pp. 1102–1106). Santa Monica, CA: Human Factors and Ergonomics Society.

Damos, D., and Wickens, C. D. (1980). The acqu tion and transfer of time-sharing skills. *Acta Psychologica*, 6, 569–577.

Davis, D. R. (1948). *Pilot Error*. London, UK: HMSO, Great Britain Air Ministry.

Davis, E. T., Kramer, P., and Graham, N. (1983). Uncertainty about spatial frequency, spatial position, or contrast of visual patterns. *Perception and Psychophysics*, 5, 341–346.

deGroot, A. (1965). *Thought and Choice in Chess*. The Hague, Netherlands: Mouton.

Dreyfus, S. E. (1981). *Formal Models versus Human Situational Understanding: Inherent Limitations on the Modeling of Business Expertise (ORC 81–3)*. Berkeley, CA: University of California, Operations Research Center.

Endsley, M. R. (1987a). The application of human factors to the development of expert systems for advanced cockpits. In *Proceedings of the Human Factors Society 31st Annual Meeting* (pp. 1388–1392). Santa Monica, CA: Human Factors and Ergonomics Society.

Endsley, M. R. (1987b). *SAGAT: A Methodology for the Measurement of Situation Awareness (NOR DOC 87–83)*. Hawthorne, CA: Northrop Corp.

Endsley, M. R. (1988a). Design and evaluation for situation awareness enhancement. In *Proceedings of the Human Factors Society 32nd Annual Meeting* (pp. 97–101). Santa Monica, CA: Human Factors and Ergonomics Society.

Endsley, M. R. (1988b). Situation awareness global assessment technique (SAGAT). In *Proceedings of the National Aerospace and Electronics Conference* (pp. 789–795). New York: IEEE.

Endsley, M. R. (1989a). *Final Report: Situation Awareness in an Advanced Strategic Mission (NOR DOC 89–32)*. Hawthorne, CA: Northrop Corp.

Endsley, M. R. (1989b). Pilot situation awareness: The challenge for the training community. In *Proceedings of the Interservice/Industry Training Systems Conference* (pp. 111–117). Ft Worth, TX: American Defense Preparedness Association.

Endsley, M. R. (1990a, March). Objective evaluation of situation awareness for dynamic decision makers in teleoperations. *Presented at the Engineering Foundation Conference on Human-Machine Interfaces for Teleoperators and Virtual Environments*, Santa Barbara, CA.

Endsley, M. R. (1990b). Predictive utility of an objective measure of situation awareness. In *Proceedings of the Human Factors Society 34th Annual Meeting* (pp. 41–45). Santa Monica, CA: Human Factors and Ergonomics Society.

Endsley, M. R. (1990c). *Situation Awareness in Dynamic Human Decision Making: Theory and Measurement*. Los Angeles, CA: Unpublished doctoral dissertation, University of Southern California.

Endsley, M. R. (1993a). Situation awareness and workload: Flip sides of the same coin. In R. S. Jensen and D. Neumeister, (Eds.), *Proceedings of the Seventh International Symposium on Aviation Psychology* (pp. 906–911). Columbus, OH: Ohio State University, Department of Aviation.

Endsley, M. R. (1993b, February). Situation awareness in dynamic human decision making: Theory. *Presented at the First International Conference on Situational Awareness in Complex Systems*, Orlando, FL.

Endsley, M. R. (1993c). A survey of situation awareness requirements in air-to-air combat fighters. *International Journal of Aviation Psychology*, 3, 157–168.

Endsley, M. R., and Bolstad, C. A. (1993). Human capabilities and limitations in situation awareness. In *Combat Automation for Airborne Weapon Systems: Man/Machine Interface Trends and Technologies (AGARD-CP-520; pp. 19/1–19/10)*. Neuilly-Sur-Seine, France: NATO-Advisory Group for Aerospace Research and Development.

Endsley, M. R., and Bolstad, C. A. (1994). Individual differences in pilot situation awareness. *International/Journal of Aviation Psychology*, 4, 241–264.

Endsley, M. R., and Kins, E. O. (1995). The out-of-the-loop performance problem and level of control in automation. *Human Factors*, 37(2), 381–394.

Endsley, M. R., and Rodgers, M. D. (1994). *Situation Awareness Information Requirements for en Route Air Traffic Control (DOT/FAA/AM-94/27)*. Washington, DC: Federal Aviation Administration, Office of Aviation Medicine.

Eysenck, M. W. (1982). *Attention and Arousal: Cognition and Performance*. Berlin, Germany: Springer-Verlag.

Federal Aviation Administration. (1990). *The National Plan for Aviation Human Factors*. Washington, DC: US Department of Transportation.

Fischer, U., Orasanu, J., and Montalvo, M. (1993). Efficient decision strategies on the flight deck. In *Proceedings of the Seventh International Symposium on Aviation Psychology* (pp. 235–243). Columbus, OH: Ohio State University.

Fracker, M. L. (1987). *Situation Awareness: A Decision Model*. Dayton, OH: Unpublished manuscript.

Fracker, M. L. (1988). A theory of situation assessment: Implications for measuring situation awareness. In *Proceedings of the Human Factors Society 32nd Annual Meeting* (pp. 102–106). Santa Monica, CA: Human Factors and Ergonomics Society.

Fracker, M. L. (1989). Attention gradients in situation awareness. In *Situational Awareness in Aerospace Operations (AGARD-CP-478; pp. 6/1–6/10)*. Neuilly-Sur-Seine, France: NATO-Advisory Group for Aerospace Research and Development.

Gaba, D. M., Howard, S. K., and Small, S. D. (1995). Situation awareness in anesthesiology. *Human Factors: The Journal of the Human Factors and Ergonomics Society*, 37(1), 20–31.

Hansman, R. J., Wanke, C., Kuchar, J., Mykityshyn, M., Hahn, E., and Midkiff, A. (1992, September). Hazard alerting and situational awareness in advanced air transport cockpits. *Presented at the 18th ICAS Congress*, Beijing, China.

Herstein, J. A. (1981). Keeping the voter's limit in mind: A cognitive processing analysis of decision making in voting. *Journal of Personality and Social Psychology*, 40, 843–861.

Hinsley, D., Hayes, J. R., and Simon, H. A. (1977). From words to equations. In P. Carpenter and M. Just, (Eds.), *Cognitive Processes in Comprehension*. Hillsdale, NJ: Erlbaum.

Hockey, G. R. J. (1970). Effect of loud noise on attentional selectivity. *Quarterly Journal of Experimental Psychology*, 22, 28–36.

Hockey, G. R. J. (1986). Changes in operator efficiency as a function of environmental stress, fatigue and circadian rhythms. In K. Boff, L. Kaufman, and J. Thomas, (Eds.), *Handbook of Perception and Human Performance* (pp. 44/1–44/49). New York: Wiley.

Holland, J. H., Holyoak, K. F., Nisbett, R. E., and Thagard, P. R. (1986). *Induction: Processes of Inference, Learning and Discovery*. Cambridge, MA: MIT Press.

Humphreys, G. W. (1981). Flexibility of attention between stimulus dimensions. *Perception and Psychophysics*, 30, 291–302.

Jacoby, L. L., and Dallas, M. (1981). On the relationship between autobiographical memory and perceptual learning. *Journal of Experimental Psychology: General*, 110, 306–340.

Janis, I. L. (1982). Decision making under stress. In L. Goldberger and S. Breznitz, (Eds.), *Handbook of Stress: Theoretical and Clinical Aspects* (pp. 69–87). New York: Free Press.

Jerison, H. J. (1957). Performance on a simple vigilance task in noise and quiet. *Journal of the Acoustical Society of America*, 29, 1163–1165.

Jerison, H. J. (1959). Effects of noise on human performance. *Journal of Applied Psychology*, 43, 96–101.

Jones, R. A. (1977). *Self-Fulfilling Prophecies: Social, Psychological and Physiological Effects of Expectancies*. Hillsdale, NJ: Erlbaum.

Judge, C. L. (1992). Situation awareness: Modeling, measurement, and impacts. In *Proceedings of the Human Factors Society 36th Annual Meeting* (pp. 40–42). Santa Monica, CA: Human Factors and Ergonomics Society.

Kaempf, G. L., Wolf, S., and Miller, T. E. (1993). Decision making in the AEGIS combat information center. In *Proceedings of the Human Factors and Ergonomics Society 37th Annual Meeting* (pp. 1107–1111). Santa Monica, CA: Human Factors and Ergonomics Society.

Kahneman, D. (1973). *Attention and Effort*. Englewood Cliffs, NJ: Prentice-Hall.

Kaplan, C. A., and Simon, H. A. (1990). In search of insight. *Cognitive Psychology*, 22, 374–419.

Keinan, G. (1987). Decision making under stress: Scanning of alternatives under controllable and uncontrollable threats. *Journal of Personality and Social Psychology*, 52, 639–644.

Keinan, G., and Friedland, N. (1987). Decision making under stress: Scanning of alternatives under physical threat. *Acta Psychologica*, 64, 219–228.

Kellog, R. T. (1980). Is conscious attention necessary for longterm storage?. *Journal of Experimental Psychology: Human Learning and Memory*, 6, 379–390.

Klein, G. A. (1989a). Do decision biases explain too much?. *Human Factors Society Bulletin*, 32(5), 1–3.

Klein, G. A. (1989b). Recognition-primed decisions. In W. B. Rouse (Ed.), *Advances in Man-Machine Systems Research* (pp. 47–92). Greenwich, CT: JAI.

Klein, G. A. (1993). Sources of error in naturalistic decision making tasks. In *Proceedings of the Human Factors and Ergonomics Society 37th Annual Meeting* (pp. 368–371). Santa Monica, CA: Human Factors and Ergonomics Society.

Klein, G. A., Calderwood, R., and Clinton-Cirocco, A. (1986). Rapid decision making on the fire ground. In *Proceedings of the Human Factors Society 30th Annual Meeting* (pp. 576–580). Santa Monica, CA: Human Factors and Ergonomics Society.

Kuhn, T. (1970). *The Structure of Scientific Revolutions*. (2nd ed.). Chicago, IL: University of Chicago Press.

Kuipers, A., Kappers, A., van Holten, C. R., van Bergen, J. H. W., and Oosterveld, W. J. (1989). Spatial disorientation incidents in the R.N.L.A.F. F16 and F5 aircraft and suggestions for prevention. In *Situational Awareness in Aerospace Operations (AGARD-CP-478, pp. OV-E-1--OV-E- 16)*. Copenhagen, Denmark: NATO-Advisory Group for Aerospace Research and Development.

Lipshitz, R. (1987). *Decision Making in the Real World: Developing Descriptions and Prescriptions from Decision Maker's Retrospective Accounts*. Boston, MA: Boston University Center for Applied Sciences.

Logan, G. D. (1988). Automaticity, resources, and memory: Theoretical controversies and practical implications. *Human Factors*, 3D, 583–598.

MacMillian, J., Entin, E. B., and Serfaty, D. (1993). Evaluating expertise in a complex domain— Measures based on theory. In *Proceedings of the Human Factors and Ergonomics Society 37th Annual Meeting* (pp. 1152–1155). Santa Monica, CA: Human Factors and Ergonomics Society.

Mandler, G. (1982). Stress and thought processes. In L. Goldberger and S. Breznitz, (Eds.), *Handbook of Stress: Theoretical and Clinical Aspects* (pp. 88–104). New York: Free Press.

Manktelow, K., and Jones, J. (1987). Principles from the psychology of thinking and mental models. In M. M. Gardiner and B. Christie, (Eds.), *Applying Cognitive Psychology to User-Interface Design* (pp. 83–117). Chichester, UK: Wiley.

Martin, M., and Jones, G. V. (1984). Cognitive failures in everyday life. In J. E. Harris and P. E. Morris, (Eds.), *Everyday Memory, Actions and Absent-Mindedness* (pp. 173–190). London, UK: Academic.

Mayer, R. E. (1983). *Thinking, Problem Solving, Cognition.* New York: Freeman.

McCarthy, G. W. (1988). Human factors in FI6 mishaps. *Flying Safety,* 3(5), 17–21.

Meister, D. (1971). *Human Factors: Theory and Practice.* New York: Wiley.

Ministry of Planning, Housing, Transport and Maritime Affairs. (1989). Investigation commission final report concerning the accident which occurred on June 26th, 1988, at Mulhouse-Habscheim (68) to the Airbus A320, registered F-GFKC. Paris, France: Author.

Mintzburg, H. (1973). *The Nature of Managerial Work.* New York: Harper and Row.

Moray, N. (1986). Monitoring behavior and supervisory control. In K. Boff, L. Kaufman, and J. Thomas, (Eds.), *Handbook of Perception and Human Performance* (pp. 40/1–40/51). New York: Wiley.

Mosier, K. L., and Chidester, T. R. (1991). Situation assessment and situation awareness in a team setting. In Y. Queinnec and F. Daniellou, (Eds.), *Designing for Everyone* (pp. 798–800). London, UK: Taylor & Francis.

National Transportation Safety Board. (1973). Aircraft accident report: Eastern Airlines 4011L-1011, Miami, Florida, December 29, 1972 (NTSB/AAR-73–14). Washington, DC: Author.

National Transportation Safety Board. (1979). Aircraft accident report: United Airlines, Inc., McDonnell-Douglas DC-8–61, N8082U, Portland, Oregon, December 28, 1978 (NTSBI/AAR-79–07). Washington, DC: Author.

National Transportation Safety Board. (1981). Aircraft separation incidents at Hartsfield Atlanta International Airport, Atlanta, Georgia (NTSB/SIR-81–6). Washington, DC: Author.

National Transportation Safety Board. (1988). Aircraft accident report: Northwest Airlines, Inc., McDonnell-Douglas DC-9–82, N312RC, Detroit Metropolitan Wayne County Airport, August 16, 1987 (NTSB/AAR-99–05). Washington, DC: Author.

National Transportation Safety Board. (1990). Aircraft accident report: US Air Flight 105, Boeing 737–200, N282AU, Kansas International Airport, Missouri, September 8, 1989 (NTSB/AAR-90–04). Washington, DC: Author.

Neisser, U. (1967). *Cognitive Psychology.* New York: Appleton-Century, Crofts.

Nisbett, R. E., and Wilson, T. D. (1977). Telling more than we can know: Verbal reports on mental processes. *Psychological Review,* 84, 231–259.

Noble, D., Boehm-Davis, D., and Grosz, C. (1987). *Rules, Schema and Decision Making (NR 649–005).* Vienna, VA: Engineering Research Associates.

Norman, D. A. (1976). *Memory and Attention.* New York: Wiley.

Norman, D. A. (1983). Apsychologist views human processing: Human errors and other phenomena suggest processing mechanisms. In D. A. Norman (Ed.), *Five Papers on Human-Machine Interaction* (pp. 10–14). La Jolla, CA: University of California.

Orasanu, J., Dismukes, R. K., and Fischer, U. (1993). Decision errors in the cockpit. In *Proceedings of the Human Factors and Ergonomics Society 37th Annual Meeting* (pp. 363–367). Santa Monica, CA: Human Factors and Ergonomics Society.

Palmer, S. E. (1975). The effects of contextual scenes on the identification of objects. *Memory and Cognition,* 3, 519–526.

Posner, M. I., Nissen, J. M., and Klein, R. (1976). Visual dominance: An information processing account of its origin and significance. *Psychological Review,* 83, 157–171.

Posner, M. I., Nissen, J. M., and Ogden, W. C. (1978). Attended and unattended processing modes: The role of set for spatiallocation. In H. L. Pick and E. J. Saltzman, (Eds.), *Modes of Perceiving and Processing* (pp. 137–157). Hillsdale, NJ: Erlbaum.

Press, M. (1986). Situation awareness: Let's get serious about the clue-bird. Unpublished manuscript.

Rasmussen, J. (1986). *Information Processing and Humanmachine Interaction: An Approach to Cognitive Engineering.* New York: North-Holland.

Rasmussen, J., and Rouse, W. B. (Eds.). (1981). *Human Detection and Diagnosis of System Failures.* New York: Plenum.

Reason, J. (1984). Absent-mindedness and cognitive control. In J. E. Harris and P. E. Morris, (Eds.), *Everyday Memory, Action and Absent-Mindedness* (pp. 111–132). London, UK: Academic.

Reason, J. (1987). A framework for classifying errors. In J. Rasmussen, K. Duncan, and J. Leplat, (Eds.), *New Technologies and Human Error* (pp. 5–14). Chichester, UK: Wiley.

Roschelle, J., and Greeno, J. G. (1987). *Mental Models in Expert Physics Reasoning (OTIC AD-AI84106).* Berkeley, CA: University of California.

Rouse, W. B., and Morris, N. M. (1985). *On Looking into the Black Box: Prospects and Limits in the Search for Mental Models (OTIC AD-A159080).* Atlanta, GA: Georgia Institute of Technology, Center for Man-Machine Systems Research.

Salas, E., Prince, C., Baker, D. P., and Shrestha, L. (1995). Situation awareness in team performance: Implications for measurement and training. *Human Factors: The Journal of the Human Factors and Ergonomics Society,* 37(1), 123–136.

Sarter, N. B., and Woods, D. D. (1991). Situation awareness: A critical but ill-defined phenomenon. *International Journal of Aviation Psychology,* 1, 45–57.

Schank, R. C., and Abelson, R. P. (1977). *Scripts, Plans, Goals and Understanding.* Hillsdale, NJ: Erlbaum.

Sharit, J., and Salvendy, G. (1982). Occupational stress: Review and reappraisal. *Human Factors,* 24, 129–162.

Sheridan, T. (1981). Understanding human error and aiding human diagnostic behavior In nuclear power plants. In J. Rasmussen and W. B. Rouse, (Eds.), *Human Detection and Diagnosis of System Failures* (pp. 19–35). New York: Plenum.

Simmel, E. C., and Shelton, R. (1987). An assessment of nonroutine situations by pilots: A two-part process. *Aviation, Space and Environmental Medicine,* 58, 1119–1121.

Sundstrom, G. A. (1987). Information search and decision making: The effects of information displays. *Acta Psychologica,* 65, 165–179.

Sweller, J. (1988). Cognitive load during problem solving: Effects on learning. *Cognitive Science,* 12, 257–285.

Taylor, R. M. (1989). Situational awareness rating technique (SART): The development of a tool for aircrew systems design. In *Situational Awareness in Aerospace Operations (AGARD-CP-478 pp. 3/1–3/17).* Copenhagen, Denmark: NATO-Advisory Group for Aerospace Research and Development.

Treisman, A., and Paterson, R. (1984). Emergent features attention and object perception. *Journal of Experimental Psychology: Human Perception and Performance,* 10, 12–31.

Tulving, E. (1985). How many memory systems are there? *American Psychologist,* 40, 385–398.

Tversky, A., and Kahneman, D. (1981). The framing of decisions and the psychology of choice. *Science,* 211, 453–458.

US Air Force 57th Fighter Wing. (1986). Intraflight command, control, and communications symposium final report. Nellis Air Force Base, NV: Author.

VanDijk, T. A., and Kintsch, W. (1983). *Strategies of Discourse Comprehension.* New York: Academic Press.

Venturino, M., Hamilton, W. L., and Dvorchak, S. R. (1989). Performance-based measures of merit for tactical situation awareness. In *Situation Awareness in Aerospace Operations (AGARD-CP-478 pp. 4/1–4/5).* Copenhagen, Denmark: NATO-Advisory Group for Aerospace Research and Development.

Wachtel, P. L. (1967). Conceptions of broad and narrow attention. *Psychological Bulletin,* 68, 417–429.

Weltman, G., Smith, J. E., and Egstrom, G. H. (1971). Perceptual narrowing during simulated pressure-chamber exposure. *Human Factors,* 13, 99–107.

Wickens, C. D. (1984). *Engineering Psychology and Human Performance.* (1st ed.). Columbus, OH: Merrill.

Wickens, C. D. (1992a). *Engineering Psychology and Human Performance.* (2nd ed.). New York: HarperCollins.

Wickens, C. D. (1992b). Workload and situation awareness: An analogy of history and implications. *Insight: The Visual Performance Technical Group Newsletter,* 14(4), 1–3.

Wickens, C. D., Stokes, A. F., Barnett, B., and Hyman, F. (1988). Stress and pilot judgment: An empirical study using MIDIS, a microcomputer-based simulation. In *Proceedings of the Human Factors Society 32nd Annual Meeting* (pp. 173–177). Santa Monica, CA: Human Factors and Ergonomics Society.

Wiener, E. L. (1993). Life in the second decade of the glass cockpit. In R. S. Jensen and D. Neumeister, (Eds.). In *Proceedings of the Seventh International Symposium on Aviation Psychology* (pp. 1–11). Columbus, OH: Ohio State University, Department of Aviation.

Wiener, E. L., and Curry, R. E. (1980). Flight deck automation: Promises and problems. *Ergonomics*, 23, 995–1011.

Wilson, J. R., and Rutherford, A. (1989). Mental models: Theory and application in human factors. *Human Factors*, 31, 617–634.

Wirstad, J. (1988). On knowledge structures for process operators. In L. P. Goodstein, H. B. Anderson, and S. E. Olsen, (Eds.), *Tasks, Errors, and Mental Models* (pp. 50–69). London, UK: Taylor & Francis.

Woodhead, M. M. (1964). The effects of bursts of noise on an arithmetic task. *American Journal of Psychology*, 77, 627–633.

Wright, P. (1974). The harassed decision maker: Time pressures, distractions, and the use of evidence. *Journal of Applied Psychology*, 59, 555–561.

chapter nine

Globalization and defense manufacturing

Claude D. Vance

Contents

Preamble

> "It is of importance that the kingdom depend as little as possible upon its neighbors for the manufactures necessary for its defense" (Adam Smith, *The Wealth of Nations*)
>
> "WE THE PEOPLE of the United States, in Order to form a more perfect Union, establish justice, insure Tranquility, provide for the common defense, promote the general Welfare, and secure the Blessings of Liberty for ourselves and our Prosperity, do ordain and establish this Constitution of the United States of America." (Preamble, Constitution of the United States of America)

Introduction

Offshore manufacturing trends threaten the national defense of the United States of America. It is an area Americans worry about jeopardized security; both individual and national. It has become a topic of debate and political agendas.[1] Offshore manufacturing is a component of globalization. While globalization generates positive impacts on the domestic and world economies, concerns about defensive risks are prevalent.[2] Strategic decisions require an understanding of the impacts offshore manufacturing has on readiness.

What is defense readiness? The Department of Defense (DOD) considers military capability the aggregation of four major components: force structure, modernization, readiness, and sustainability.[3] It defines readiness as "the ability of US military forces to fight and meet the demands of the national military strategy."[3] A civilian version of readiness is "the state of having been made ready or prepared for use or action (especially military action)."[4] In the *National Response Plan*, the Department of Homeland Security (DHS) uses preparedness to describe capability. DHS and DOD are both involved in executing the national security strategy.[5] Because *readiness* and *preparedness* are synonyms, a distinction must be made for this research to alleviate potential confusion created by the disparity between definitions.

Capability is the interdependent relationship between preparedness and readiness. After evaluating definitions from DHS and DOD, the United States Marine Corps (USMC) and General James L. Jones, USMC (retired), a definition for defense capability was created.[6] Defense capability is the ability of a nation to deter aggression, protect sovereignty, deploy into areas of responsibility (AOR), sustain operations as situations warrant, expeditiously redeploy from AORs, and rapidly recover and reconstitute for future incidents.

Defense preparedness is the endeavor of planning, training, and equipping.[6] Decision-makers need to anticipate immediate and future operations, and identify vulnerabilities.[3] Personnel require initial, refresher, and up-grade training to maintain their knowledge and proficiency levels. The most effective assets—equipment, resources, and systems— must be available for an organization to successfully complete its objectives.[6]

Defense readiness is the acquisition, modernization, and sustainment of assets. In the acquisition process, organizations procure assets based on identified requirements and shortfalls.[7] Sustainment is the continuous maintenance of operational and reserve asset levels to support routine and incident operations efforts.[3] Through modernization, relevant assets remain viable through state-of-the-art upgrades, and advanced technologies replacing antiquated technologies.

Defense readiness ultimately supports national security. An old military adage states, "if you train the way you fight, you'll fight the way you're trained" [author unknown]. That piece of wisdom applies to all persons engaged in national defense. Defense preparedness depends on the assets made available through readiness. Defense readiness is less effective when personnel are unprepared to utilize assets, or assets are unavailable. That interdependency determines a nation's defense capability. When defense capabilities are degraded, national security vulnerabilities increase.

National defense organizations rely on manufactured goods to maintain readiness for daily and contingency operations. According to the United States Department of Commerce, manufacturing "is a cornerstone of the American economy."[8] It encompasses every aspect of finished goods from creation to consumption.[9] Although not all research literature focused on manufacturing overall, authors focused on one to four key areas either directly or indirectly. Those areas of people, innovation, production, and logistics.

Manufacturing depends on people. They are linked to every aspect of the value chain.[10] Some contribute directly through research, touch labor, goods delivery, or as customers. Other persons provide indirect labor such as administrative support and management. Internally, corporate success depends on the quality of employees and hierarchical relationships within the company.[11] Externally, customers are crucial to corporate success. A nation's defense capability relies on the manufacturing workforce to provide national security customers with finished goods for readiness and preparedness.

Innovation provides creative starting points for manufacturing. New goods, processes, and improvements come from "seeds of thought" through study and experimentation.[10,12]

The three elements of innovation are education, research and development (R&D).[10,13] Education is the continual attainment of knowledge. Studying concepts and experimenting with how to convert them into useful products, processes, and/or improvements is the purpose of R&D.[14] Relationships among the three spawn innovation. Innovation provides manufacturing with future vitality.[15] In turn, innovation supports a relevant and reliable national defense.

Production is the heart of manufacturing. By definition, it is the "combination of materials, parts, or subassemblies to increase their value" as finished goods.[9] Within the manufacturing supply chain, the workforce, innovation and logistics link together to create finished goods. Defense agencies acquire commercial and agency-specific goods to sustain asset levels, improve capabilities, and for modernization. In turn, preparedness can continually exist at acceptable levels. Therefore, production is linked to national security through readiness.

Logistics provides the means for delivering goods to agencies involved in national security. It is the management of how resources move through a supply chain for the creation and delivery of goods to customers.[16] Logistics pertains to material handling, distribution, storage, and information. Connections to innovation, production, and people make logistics the fourth facet of the economic pyramid (Figure 9.1).

The DOD specifically identifies "focused logistics" as a key part of the defense industrial base.[17] Logistics are essential for meeting customer demands and remaining competitive.[8] Major facets of logistics include procurement, provisioning, maintenance, movement, and planning.[18] Those facets are similar in definition to the aspects of defense preparedness and readiness.

Multinational connections and stakeholder perceptions cloud the definitions of domestic and foreign manufacturing. Domestic generally relates to the people, entities, ideas, and geography indigenous to a nation. Because this research addressed the national defense of the United States (US), domestic refers to anything associated with the US, being "American," or occurring within the sovereign boundaries of the United States of America (US) and its territories (US borders). Foreign alludes to all things indigenous to or occurring in other nations. For example, the Ford Motor Company (Ford) assembles Fusion automobiles in Mexico.[19] Because Ford is considered an "American" automotive company, those Fusions are foreign-produced domestic vehicles. Conversely, Nissan North American Manufacturing (NNAM) assembles the Titan pickup truck in Canton, Mississippi.[20] NNAM is owned by a Japanese company, Nissan Motor Company, Ltd. Thus, the Titan is a domestic-produced foreign vehicle.

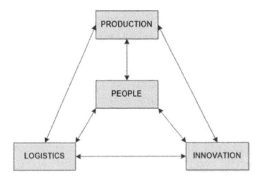

Figure 9.1 Economic pyramid.

Table 9.1 Defense manufacturer merger examples

Company	Former competitors acquired or merged
Boeing	McDonnell Douglas
Lockheed-Martin	General Dynamics Martin
McDonnell Douglas	McDonnell Douglas
Northrup Grumman	Northrup Grumman

Onshore and offshore refer to where activities take place or facilities are located. For this research, the US borders are the dividing lines for identification. Facilities and processes circumscribed by those borders were identified as being onshore. Therefore, Nissan Titans are assembled onshore. Offshore signifies that a location lies outside the US borders, including Canada and Mexico. Hence, Ford produces Fusions at an offshore facility. Offshore manufacturing is the innovation and/or production of goods outside US borders. A corporation may have onshore and/or offshore locations regardless of what consumers perceive as being domestic or foreign.

National defense depends on manufactured goods. As other nations experience growth in their manufacturing sectors, American manufacturing is in decline.[14] Domestic corporations compete with foreign competitors in commercial markets to remain viable. In the case of semiconductors, the military sector accounts for one percent (1%) of the market.[2] That number does not reflect the amount of non-military unique items used in daily defense operations. The reduction in defense contractors limits available options for acquiring domestically produced weapons systems (Table 9.1). In recent years, domestic-foreign partnerships have emerged as contractors attempt to satisfy the nation's defensive needs. The objective of this research is to show that offshore manufacturing trends are detrimental to the defensive readiness of the US.

Literature review

Federal legislation and policy affect national defense. Congressional concerns regarding procurement of offshore goods predate World War II.[21] Historically, domestic manufacturers have been inadequately prepared to supply and sustain military operations.[22] Under the Buy American Act, enacted in 1933, Federal agencies are required to purchase domestic goods for use in the US.[21] This statute applies to Federal agencies, outside the DOD, engaged in national defense and security. In 1941, enactment of the Berry Amendment mandated the use of domestic agricultural goods for national defense.[21] The Trade Act of 1962 authorized import limitations "for national defense purposes."[23] During the mid-1980s to early 1990s, foreign policy established under that authority successfully shored up declines in the domestic machine tool industry.[23] The Defense Authorization Act for Fiscal Year 2004 included legislation to incentivize defense contractors utilizing domestic capital assets.[23] Inclusion of such language was the result of concerns regarding domestic industries—particularly those supporting the defense industrial base—losing their market shares to foreign competition.[23] Legislation and policy, with respect to domestic production, will remain important national defense aspects as the nation moves into the future.

Activities within and without the domestic industrial base threaten national defense. The number of domestic companies with core competencies critical to national defense has diminished since the 1970s.[22,24] Reasons for the decreases include lack of work, leaving the defense industrial base completely, acquisition by domestic competitors or foreign interests, or ceasing to exist.[22,24] Foreign manufacturers exploit such opportunities to improve their capabilities and compete against the remaining domestic companies.[24] In turn, decreased domestic competencies and market shares translate into increased foreign dependency.[22]

As domestic companies move into systems integration and away from production, manufacturing is outsourced to lower tier suppliers.[25] At the start of Operation IRAQI FREEDOM, a European supplier withheld shipment of a munitions component critical for US military ordnance.[26] In the mid-1980s, an embargo against two Japanese electronics suppliers, for selling US technology to the Soviet Union, would have caused private industry layoffs and affected weapons production in the US.[22] Defense related issues can spill over into the private sector when procuring materials and capital assets from similar sources.

The innovation process is inherently linked to manufacturing, energy, and national security.[14,15] Innovation generates advanced manufacturing technologies. Domestic manufacturing invests the most capital into domestic R&D.[15] Domestic suppliers may be unable to implement advanced technologies (e.g., finances, labor with the necessary skills) while foreign competitors heavily invest in them.[25] As technology advancements push energy generation and distribution capabilities, strong innovation processes are needed to meet increasing requirements for clean, reliable sources.[14] The US relies on innovation to execute the national security strategy.[14] The percentages for government research spending have shifted in favor of meeting immediate needs via long-term basic R&D.[14] Innovative strength impacts the nation's economy, security, and prosperity.

Collocating manufacturing and innovation centers benefits both centers and their location. Close proximity facilities knowledge sharing, knowledge transfer, and innovation diffusion into adjacent centers.[15] Innovation centers proportionally grow with associated manufacturing centers.[2] The physical distance between centers and the amount of associated benefits are inversely related.[15] Globalization and the internet provide locations to virtually collocate centers in cyberspace.[13] Loss of domestic core competencies and dependence on offshore foreign products may create a reliance on offshore innovation for national defense.

Innovation relies on a well-educated people to conduct R&D. A large number of the domestic science and technology workforce, especially in defense, is close to retirement without a sufficient talent pool for passing on corporate knowledge.[2] Compared to primary and secondary students abroad, aptitude and interest in science and mathematics among US students is declining.[14] College students shun manufacturing careers based on negative stereotypes regarding work conditions, job uncertainty, and career growth.[11] As fewer domestic college students pursue science, technology, engineering, and mathematics (STEM) degrees, the number of foreign students earning STEM doctoral degrees is growing. Many DOD STEM occupations require US citizenship. A skilled labor shortage and knowledge gap could exist if the disparity between hiring and retirement continues to increase.

Domestic industries are able to compete globally. The terrorist attacks of 9/11 disrupted shipments from offshore. Before 9/11, manufacturers like Parker Hannifin competed against foreign competitors by improving processes and taking advantage of two important assets; people and superior logistics.[11] Companies can use people, logistics, and process improvement to stay competitive while supporting national defense. In 2001, Dr. Sheffi at the Massachusetts Institute of Technology presented four challenges companies should address for success. His challenges include preparing for future man-made and natural catastrophes, developing robust supply chains that take into account uncertainty and vulnerabilities across the entire value stream, establishing cooperative relationships with the federal government, and maximizing the strategic balance between corporate goals and security.[27] Meeting these challenges benefits the company and the nation.

Caveats

A vast amount of information related to offshore manufacturing exists. As a topic of debate and agendas, most literature was directed toward particular industries or contained biases. Protection and government involvement were recurring themes in those writings. Recommended solutions for industry called for means to bolster domestic manufacturing sectors and protection from foreign competition, including onshore foreign competitors. Political authors either addressed support for domestic industries onshore or contradicted policies of the current administration during their time.

Writings hailed the movement and/or establishment of overseas operations as "win-win" situations. Authors touted the benefits of offshore activities.[28] Their major selling point was the ability to remain competitive. Some studies lacked support from discussion and sufficient evidence.[29] One such study, by McKinsey & Company, was refuted as advertising consulting services for starting offshore operations.[29]

Negative connotations were attached to the term *protectionism*. The "protectionist" label implied a group or an individual was willing to prevent foreign competition or reluctant to embrace globalization. Pacific Rim and European nations also feel offshoring effects outside their borders.[29] The term and label perspectives were American. Authors favoring offshore activities applied both to discredit opposing views.

Publications directed toward national security or defense manufacturing were scarce. Literature related to national security mainly addressed two issues. The first issue was economic security in the private sector. The discussions concentrated on high technology, machine tool, and transportation industries. The second issue was policy. The writers recommended federal changes to processes or favoring domestic capabilities in the defense industrial base. With little literature available, parallels between the private and public sectors must be drawn.

The topic of offshore manufacturing contains much controversy. Healthy debate examines pros and cons. Filtering out bias serves as an injustice. Points brought forth from all sides illustrate the need to determine a balance between a nation's best interests and the benefits of its presence in the global economy. How does a nation protect national interests, and still be a "good" neighbor?

Model development

Discussions in the literature continually keyed in on various combinations of four elements. They were innovation, investment, production, and the workforce. Further reading supported the idea that strong bonds exist among those elements. Research kept returning to a challenge against Sir Walter Petty's stages of development theory. The dominance of a society's economic sectors evolves from agriculture through manufacturing to services (Figure 9.2).[30] This theory works well when assuming generated wealth remains in the society.[11]

Figure 9.2 Stages of development model.

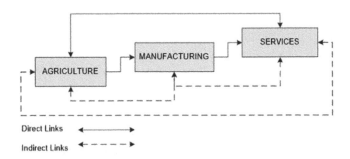

Figure 9.3 Modified stages of development model.

In challenging Petty's model, the Berkeley Roundtable on the International Economy argued prosperity, for a given economic sector, relies on direct and indirect interrelationships with the other two sectors.[30] Those relationships were added to the stages of development model (Figure 9.3). The modified model relationships indicate the potential for an economy to either regress toward a previously dominate sector or skip the manufacturing sector. Based on literature already researched, both scenarios were counterintuitive for Petty's theory. This generated two questions about relationships and benefits. First, is it possible that the three sectors are equally important? Second, do advancements in any one sector spill over as benefits to another?

The equal importance concept led to rearranging the modified model into an economic triangle (Figure 9.4). The economic sectors are in a horizontal relationship in this model. Growth in any sector is due to advances spilling over from the other two sectors.[11,30] A triangle was selected because of the relationships between fuel, heat, and air in the fire triangle. All three must be present, but lack a fourth element: combustion.[31] When all the

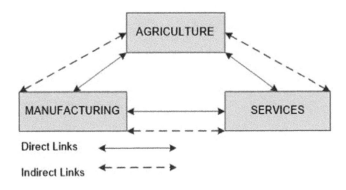

Figure 9.4 First economic triangle model.

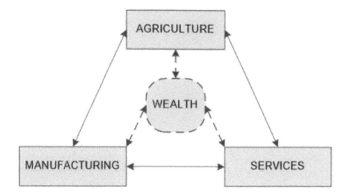

Figure 9.5 Second economic triangle model.

elements are present in correct proportions, something must occur for them to combine and generate growth.

The second iteration of the economic triangle depicts wealth as an indirect linkage among the three sectors (Figure 9.5). Wealth was viewed as a catalyst to spark an economy. Before providing wealth, a demand must be present. Therefore, the term *market* replaces *wealth* in the next model (Figure 9.6). This was because of the cyclical exchange of wealth and products between a market and the sectors which support it. For simplicity, direct and indirect linkages were combined in Figure 9.6 and subsequent models. Discussion about agriculture demonstrated significant sector growth was the result of advancements (e.g., technology) in related manufacturing and services. In turn, benefits spilled back into these sectors.[30]

As research moved toward manufacturing, technology became prevalent. It was seen as a driving force behind manufacturing advancements.[8,10,15] The benefits spawn more technology.[15,32] Further technology literature searches led to information concerning education and technology based service industries.[13,14] The importance of technology throughout the economy was the reason it replaced agriculture in the third economic triangle model (Figure 9.6).

With the market at the center of the third triangle, an examination of the four elements and their relationships took place. Services supply the other elements with

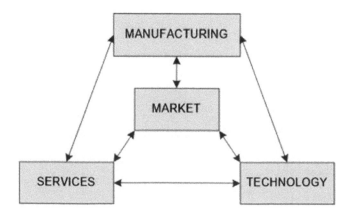

Figure 9.6 Third economic triangle model.

wealth and labor. Technology generates advancements. Manufacturing provides goods. The market is the exchange point. Questions quickly surfaced while determining interconnections. If services provide people and wealth to create technology, how does technology advance either one? What connections did technology have with the market? Because manufacturing was a source of wealth for laborers, was a direct link between goods and consumers more realistic than indirect links through the market? Why were exchanges circumventing the market? Erroneous interpretation caused the questions.

A new approach was needed as errors rapidly invalidated the triangle models. Market was incorrectly placed in the center. Forcing all transactions through the market meant no direct linkages among the other elements. Market was removed as a model element, and assumed to be the model itself. Using the term *wealth* created confusion. Its definition refers to anything of value. For this thesis, manufacturing wealth is capital. It may be working capital (i.e., money, investments), capital goods, or human capital (i.e., labor, knowledge). Replacing market with capital enabled connections to other markets.

Renaming market to capital initiated an epiphany. A recent change in the fire triangle added "the uninhibited chain reaction" of combustion.[31] It explained fire sustainment and the chemical reaction process. The relationship was changed to a pyramid; the fire tetrahedron.[31] Similarly, capital aided in the understanding of market processes. Hence, the economic triangle was converted to a pyramid.

The economic elements were perceived again in tetrahedral relationships. Manufacturing produces capital and consumer goods. It also generates wealth.[10] The agricultural and services sectors also produce goods. Assuming that an economic model should represent all sectors, manufacturing was renamed production (Figure 9.7). Services represent a sector instead of a market element. People and services were interchangeable during evaluations after the first economic triangle. People invest human and working capital into the other elements. Indirectly, they contribute goods through production. Hence, services was changed to people. Capital flows to the other markets in one or more of the previously stated forms. As a conduit to other markets, capital remained a separate element. Additional research about technology brought changes. Technology and innovation were tied to education, development, and research.[13,14] Research and development drive innovation.[13] Applying technology to manufacturing and education indicated it is a product of innovation.[10,15] Literature support for innovation, as an economic element, caused it to replace technology in the pyramid.

Capital, as an element, continued to present challenges. This is similar to the questions regarding market as an element. One or more types of capital served as direct links

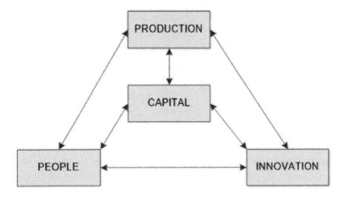

Figure 9.7 First economic pyramid model.

between other elements. That inferred that capital links elements rather than stands alone as one. Capabilities existed for people to provide human and working capital to other markets. Besides the flow of capital goods, the model no longer supported capital as a legitimate element.

Transferring capital between markets generated a replacement element: logistics. Logistics consists of two aspects. First, it involves the movement of capital.[18] Logistics links production to other markets.[27] Second, logistics focuses on operational process management.[16] Literature emphasized the importance of logistics in manufacturing.[8,27] Hence, logistics replaced capital in the economic pyramid (Figure 9.1).

Previous models were unstable because of erroneous justifications for market and capital. Innovation, people, and production were considered stable elements. Without all elements stable, those models collapsed. Could logistics support the model? It was necessary for indirect linkages between other elements. Producers need methods for shipping goods to consumers. In supply chains, they rely on suppliers for goods and technology from outside sources.[33] Logistics has direct links with the other elements. Growing complications mandate new innovations in logistics management.[27] Logistics requires labor, direct and indirect, and capital goods (e.g., tractor-trailers). Logistics shored up and completed the economic pyramid (Figure 9.1).

Security connection

While model development appeared to digress, it became the basis for further research. The initial hypothesis, effects of globalization on national security, was vague and unfocused. The unavailability of defense-specific information contributed to the broad approach. A holistic approach without the military was a failure. Civil agencies with national security responsibilities were identified during further reading and contemplation. Parallels between the pyramid, in the private sector, and those agencies were easier to draw.

Including civil agencies provided a better picture of defense readiness. The United States Coast Guard (USCG) has a dual role. During peacetime, it operates under DHS. In wartime, the United States Navy (USN) assumes authority of the USCG. Local law enforcement and fire departments are quasi-military organizations. These organizations are trained in (preparedness) and equipped for (readiness) emergency response (civil defense). Civil emergency management can be part of a military organization. The Tennessee Emergency Management Agency is part of the state's military department.[34] In recent years, new DOD doctrines incorporated civil support for homeland security.[35]

Readiness was a common theme in civil and military defense. While preparedness and readiness had similar definitions, both terms were used interchangeably. Distinctions were made to differentiate both terms. Preparedness pertains to people. Readiness refers to physical assets. Defense capability is dependent on both. Understanding the importance of all three—preparedness, readiness, and defense capability—to national security importance led to the definitions for each term.

Way forward

The economic pyramid elements were compared to capability, preparedness, and readiness. Direct links to people, innovation, and logistics were found. The defense industrial base provided connections to production. Literature for the initial hypothesis was reread several times. Those with overt national security and defense manufacturing connections were retained after the first read. Additional readings yielded four categories; indirect, inferred, supportive, and unrelated. Literature in the first two categories contained either indirect linkages or linkages that could be inferred. Supportive information and examples came from industry standards, dictionary definitions, and government reports. Literature unable to fit into any of the first three categories was classified as unrelated. Unrelated works were discarded.

Creating an annotated bibliography aided in narrowing the scope of this thesis. Six main topics exhibited a combination of relationships among manufacturing, government policy, and national security. The first topic was the effects of legislative actions on defense and the defense industrial base. Defense readiness strategy changes since World War I was the second topic. The third topic included attitudes toward and perceptions about manufacturing from society and legislation. The fourth topic was issues in domestic manufacturing giving rise to increased foreign competition. The fifth topic was concerned with increased dependence on offshore sources. Proximity relationships between manufacturing and innovation were part of the sixth topic. Researching onshore and offshore manufacturing led to identification of the topics.

The thesis contains five stages. First, a foundation was laid through research. Second, a hypothesis was built upon that foundation and finalized. Third, discussion will occur to either prove or disprove the hypothesis. Fourth, conclusions will be drawn and presented. Finally, recommendations for further research will be provided.

Perceptions and legislation

Negative perceptions concerning domestic manufacturing exist in the US. Plant closings and lost jobs are attributed to offshore competition.[1] Instead of associating manufacturing with high technology and innovation, college students used prison and slavery analogies to describe manufacturing careers in two major studies.[10] Consequently, few students pursue manufacturing related majors.[11] Traditional economic theories suggest that manufacturing has reached its apex and the service sector is becoming the dominant sector.[30] It is assumed that displaced manufacturing workers can simply retrain for new careers in other sectors.[13] Defense acquisition decisions, from 2001 to present, fuel notions about military dependence on components and subassemblies produced offshore (Table 9.2).[21] Lack of appreciation undermines the domestic industrial base.

Administrative and legislative connections to national security have existed since the US became a nation. Defense responsibilities and rights were included in the US Constitution and Bill of Rights.[36] Between World War I and the US officially entering World War II, important legislation supporting domestic production was enacted. In 1933, the Buy American Act (BAA) required the Federal government to procure assets, resources, and services from domestic [onshore] sources.[21] Even today, the BAA applies to civil agencies such as DHS.

The Berry Amendment is unique to the DOD. During World War II, concerns about the effects of war on the nation were brought forth in Congress. In 1941, the Berry Amendment was enacted to ensure service members were provided domestic agricultural goods (e.g., food, cloth for uniforms).[21] Since then, the number of items required to be produced onshore has grown. The Berry Amendment became part of the US Code in 2001.[21]

The Soviet launch of Sputnik ushered in new concerns for the US. The fear of the US losing its dominance in defense and scientific innovation prompted new legislation and programs. In 1958, Congress enacted the National Defense Education Act (NDEA).[14] The purpose of the NDEA has to bolster science and mathematics education in the US.[17] The Defense Advanced Research Projects Agency (DARPA) was established

Table 9.2 Federal acquisition competition examples

| Program | Agency | Competitor(s) | Base aircraft | | | Assembly location | |
			Model	Supplier	Nation(s)	Initial	Final
Deepwater[42]	USCG	Lockheed-Martin (LMA)	HC-235	CASA	Spain	Madrid, Spain	Unknown
UH-72A Helicopter[43]	US Army	American Eurocopter	EC-145	Eurocopter	European Union (EU)	Columbus, Mississippi	Columbus, Mississippi
Joint Cargo Aircraft[44]	US Army	L3 Communications	C-27J	Alenia North America	Italy	Turin, Italy	Jacksonville, Florida
	USAF	Boeing Raytheon	C-295	CASA	Spain	Madrid, Spain	Unknown
		LMA	C-130J	LMA	US	Atlanta, Georgia	Atlanta, Georgia
Law Enforcement[45]	DHS DEA FBI	American Eurocopter	EC-120 AS-350	Eurocopter	EU	Unknown Unknown	Unknown Unknown
Aerial Refueling Replacement Program[47]	USAF	Northrop Grumman	A330	Airbus	France	Toulouse, France	Mobile, Alabama
		Boeing	767	Boeing	US	Everett, Washington	Everett, Washington

within the time frame.[14] DARPA creates innovation advantages for defense.[37] Other than declared wars, the "Space Race" was a period in American history when substantial effort was placed on national defense.

Legislature and foreign policy can provide assistance to domestic manufacturing segments. The Trade Act of 1962 places importation limits for national security reasons.[23] The machine tool industry, a crucial defense industrial segment, has struggled.[25] Significant decline in domestic market share, thirty to fifty percent (30% to 50%), and possible loss of wartime surge capabilities caught the attention of the Reagan Administration.[23,26] Based on the Trade Act of 1962, Voluntary Restraint Agreements with Japan and Taiwan facilitated resurgence of domestic machine tools in the global market during the late 1980s and early 1990s.[26] By 2003, a report cited that twenty-five to thirty percent (25% to 30%) of machine tool skill and production moved offshore.[10] No sign of recovery since then was found.

Offshore dependence

The number of domestic defense manufacturers has decreased. Fifty years ago, an individual could count the number of prime defense contractors on more than two hands. That number has dwindled to a handful. Some companies, such as Vought, moved out of aircraft production. Those still in the aviation industry now supply components and subassemblies. Two of the remaining aircraft manufacturers, Boeing and Northrup Grumman, are combinations of former competitors (Table 9.1). Both companies have competed with foreign-based aircraft for major DOD contracts (Table 9.2). The increasing number of foreign-based aviation programs indicates two things. First, foreign dependence is present in US national security. Second, domestic defense manufacturing capabilities are declining.

A weak defense industrial base places the US at a political disadvantage. After a Japanese electronics firm provided the Soviet Union with military technology in 1986, Congress sought to prohibit sales of the company's products.[22] The negative effects on domestic electronics and weapons production countered the enactment of a ban.[22] Swiss-made components shipments, bound for onshore munitions production, were held up because of Swiss opposition to US policy and military actions in Iraq.[24,38] The US offered diesel submarines, a non-existent domestic production capability, to Taiwan.[24] Because Dutch and US foreign policies on China conflict, the Netherlands, a country with diesel submarine capabilities, refused assistance.[24] Dependence on foreign allies and companies shed doubt on defense readiness.

National security relies on foreign resources. The DOD defines essential industries as those "that transform the crude basic raw materials into useful intermediate or end products."[3] Raw materials are imported for defense and mobilization.[15,22] Berry Amendment waivers have been issued for Russian titanium in military aircraft.[21] As previously mentioned, the Berry Amendment pertains to the domestic procurement of military uniforms as well. The DOD requirements for wool berets create two problems. First, domestic wool sources are scarce.[21] Second, foreign companies with offshore production received beret manufacturing contracts to meet demand.[21] While offshore procurement and production of military uniform items may seem trivial, understanding the extent of foreign dependence is essential to readiness.

The energy powering security and the defense industrial base comes from natural resources.[14] In 2005, sixty-five percent (65%) of the petroleum consumed domestically came from overseas.[15] That does not include petroleum used by US military and government entities overseas. In response to the Energy Policy Act of 2005, the DOD is embracing domestic and renewable energy.[39,40] Such sources include scrap wood, wind, solar, and geothermal.[39] The unsuccessful attempt by a Chinese company to acquire Unocal, a domestic oil company, caused national security issues to surface.[17] Minimizing foreign energy dependence is necessary for national security.

Logistics risks

Logistics is vital to national security readiness. The word itself originated as a military term.[16] Events on September 11, 2001 (9/11) exposed domestic manufacturers to global supply chain vulnerabilities.[27] In keeping with commercial best practices, defense prime contractors are becoming systems integrators.[8] They manage the flow of information and products through supply chain networks.[27,32] As supply chains lengthen, both in distance and number of participants, control and security decrease.[13]

Supply chains depends on continuous and correct information flow. Commercial off-the-shelf information technologies, used by the government and private industries, are susceptible to cyberattacks.[13] Policies, procedures, and resources should be in place for data protection and disclosure prevention from those without a need to know.[13] Communications must be unambiguous and clearly understood. A sub-tier supplier, a few levels below a prime contractor, could produce parts based on an incorrect specification. The length of time between the error and detection may impact the production schedule or affect system performance. In the commercial environment, time loss and poor information translate into revenue losses. In national security, it means decreased readiness.

More risks are inherent to supply chains with offshore links than to domestic supply chains. A "black hole" was the description for transportation between suppliers and customers.[27] While advanced technologies and software may track the routes goods take, few updates are provided before goods arrive.[27] Offshore shipments are more vulnerable to disruptions than those from onshore suppliers.[13] The Swiss munitions component delays are an example. Another example came after 9/11 when the US government held up shipments coming from outside US borders.[27] Onshore automobile manufacturers experienced inventory shortages until offshore replenishment shipments could proceed.[27] Strategies utilizing onshore suppliers, enhanced tracking, and commercial-government partnerships may mitigate future risks.

Innovation

A network of related activities promotes growth and propels innovation processes forward. This combination of activities—such as R&D investment, knowledgeable workforce, proximity, etc.—is known as critical mass.[15] As innovations become more significant and frequent, the "spill-over" effects permeate the economy.[15] The manufacturing base acts as both a catalyst and a conduit. It feeds the innovation process while providing pathways to enlarge the critical mass. A competitive advantage exists where manufacturing capabilities and strong R&D are together.[14] Offshore manufacturing base growth and the establishment of new R&D bases place US R&D dominance at risk.[15] Foreign economies are gaining the critical mass necessary to support R&D offshore. Even if domestic manufacturing remains constant, its global share is shrinking with respect to increased foreign competition.[15] The US must enlarge its global R&D share to overcome the disparity.

Innovation requires two types of investment for prosperity and vitality. The first investment is education and training for the nation's workforce knowledge base.[1] Historically, the US federal government had led the charge when offshore competition threatened national prosperity and security—such as the Cold War. The US still faces offshore competition for global R&D dominance. Free tuition, since 1986, is part of a strong emphasis Ireland places on secondary and higher education.[14] Two Canadian innovation strategies, at the federal, provincial and local levels, seek to strengthen the national economy by "promoting R&D in the sciences and engineering..." and improving education, beginning at the elementary school level.[14] Determining how to tackle such challenges is vital to future innovation.

The future of domestic manufacturing depends on innovation. The second type of innovation investment is R&D investment.[1] R&D funding comes from public and private sources. In the decade prior to 2004, the largest manufacturing research funding source—federal funding—was reduced by half.[1] Manufacturing programs at DARPA and the DOD Manufacturing Technology program have also experienced elimination or significant reductions.[1] While overall funding was reduced in that period, priorities and allocations shifted from technologies to life sciences.[41] In contrast, South Korea sought to double R&D spending between 2002 and 2007.[14] In 2003, seventeen percent (17%) of global industrial R&D spending—about US$122 billion—came from domestic R&D for manufacturing in the US.[15] Between 1999 and 2003, domestic manufacturers' investments rose forty-two percent (42%) for offshore R&D and fell two and one half percent (2.5%) for onshore R&D.[15] These trends may indicate a weakening of the innovation critical to the US economy.

The domestic manufacturing sector employs large percentages of R&D essential personnel. Twenty-five percent (25%) of the science-related workers are in manufacturing; forty percent (40%) of them are from engineering fields.[15] Providing a sufficient source of educated workers begins at the elementary school level. Most domestic primary and secondary students lack a sufficient educational foundation for science-related fields. In 2003, less than one-third of fourth and eighth graders were proficient in math and science.[41] Proficiency among high school seniors was below twenty percent (<20%) three years earlier. When ranked internationally against peers from twenty other countries, domestic seniors were near the bottom; 19th for math and 16th for science.[41] Coincidentally, post-secondary and graduate science-related education experienced declines. During 1996 to 2003, the number of science and engineering doctoral degrees awarded to US citizens and permanent residents fell twenty percent (20%).[15] President John F. Kennedy expressed the relationship between prosperity and education when he stated, "Our progress as a nation can be no swifter than our progress in education. The human mind is our fundamental resource."

Global competition and dominance relies on an educated and literate domestic workforce.[11] R&D workers must come from other sources if US educational systems are unable to provide qualified domestic candidates. Such candidates may be foreign nationals studying either in the US or abroad. At the turn of the twenty-first century, domestic institutions awarded over fifty percent (+50%) of engineering and math doctoral degrees to foreign nationals.[41] Foreign graduate programs in nations like China and India are improving.[13] Engineering programs at Mexican and Turkish institutions were accredited by the Accreditation Board of Engineering and Technology during 2006–2007.[46] Although foreign talent bolsters US innovation, chances exist for foreigners to seek education overseas or return home with the knowledge they have obtained.[15,41]

Geographic proximity is important to manufacturing and innovation. Within the US, research centers like Silicon Valley and the Research Triangle came about through research spill-overs from academia nearby.[15] Collocating R&D and manufacturing capabilities improves competitive advantage.[14] Whether domestically or globally, corporations look to research conducted by sub-tiers within their supply chains.[15] Overseas, research occupational opportunities are expanding, and federal research investment is increasing.[13] The growth in manufacturing capabilities offshore is proportional with the increasing number of R&D centers overseas.[17] Reasons for relocating production and R&D offshore include reduced costs, highly skilled personnel who are eager to work, new technology, and proximity to growing markets.[14] Regional living standards are related more to productivity, fueled by innovation, than economics.[15] Offshore investment and public-private ventures abroad threaten the innovation critical to the US economy as R&D funding declines onshore.[41] Meeting these challenges requires a national effort.[17]

Besides economic security, innovation is important for national defense. Developing foreign economies challenge US technical competitiveness and leadership. Both are vital to economic health and military superiority.[17] Maintaining close proximity between R&D and manufacturing threatens the US military advantage. Throughout the Cold War, US federal policymakers understood that domestic innovation investment supported military superiority.[41] Since 1997, the following Chinese public law reinforces military support by the private sector[17]:

> *Combine the Military and Civil*
> *Combine Peace and War*
> *Give Priority to Military Products*
> *Let the Civil Support the Military*

China publicly acknowledges military dependence on civilian innovation and manufacturing. Another innovation threat comes from foreign acquisition of onshore capabilities.[24] It degrades the ability to maintain a strategic advantage by depending on foreign and offshore components and technologies.[1] Increasing private R&D reliance, insufficient DOD guidance, poor public-private coordination, and offshore R&D and manufacturing could place national defense at a disadvantage.[17]

Conclusion and recommendations

Offshore manufacturing has relationships with national security. Connections can be overlooked when evaluating innovation, logistics, people, and production separately. A fifth element, government, should be added to uncover additional linkages. As globalization continues, aggregated examination of these five elements becomes necessary. No single element is responsible for declining American dominance. The combination of activities, over several decades, fuels an impending "Perfect Storm." Trends directly and indirectly related to national defense, such as those in the private sector, will continue to threaten readiness and the ability to accomplish strategic objectives. For now, the risks appear minimal. The potential for risks to increase exists as other countries gain the critical mass necessary to grow their economies and gain defensive advantages. Retention and expansion of the gap between the US and foreign entities requires public and private actions to minimize the effects of offshore manufacturing on national defense. While foreign and offshore dependence do affect national defense, the full extent is unknown.

Confronting the effects of offshore manufacturing on national defense requires further research. This thesis focused on five elements individually: government, innovation, logistics, people, and production. How interactions among two or more elements affected relationships between offshore manufacturing and national defense were unclear. It is improbable to assume the full understanding of the effects on a macroscopic level. The following research recommendations are meant to better understand the effects of offshore manufacturing or reduce the threats through strategic decisions.

- Research federal policies; especially DOD acquisition policies. The US government publicly furnishes general policy and acquisition information. Surveying personnel directly involved in acquisitions may provide non-programmatic information regarding offshore manufacturing. Examine the effects of restricting/disallowing foreign competition in acquisitions and the effects on domestic competition in global and foreign markets.
- Evaluate the current defense manufacturing state. Publications directed toward defense manufacturing were scarce. Focusing on one or more aspects of the defense industrial base (e.g., fixed- and rotary-wing aviation) may generate insight into the effects of offshore manufacturing.
- Address domestic education. Research already focuses on the current state. Future planning needs strategies for improvement, supplying qualified talent, and flexibility to meet known and unanticipated challenges.
- Study the effects of art and its interconnections with STEM and innovation. Although STEM was the education focus for this thesis, what advantages does innovation gain from the arts (i.e., critical thinking, creativity, communication)? Art education includes language (i.e., writing, speaking), performance (i.e, music, theater), and visual arts such as painting. Just as steam drove industrialization, does the addition of art provide STEAM power for innovation?
- Analyze and compare three supply chain approaches for defense logistics. The first approach is sole domestic support. The second approach is primary domestic support with supplemental foreign sources. The third approach is an expedient logistics system where cost, schedule and performance drive selection of a domestic or foreign source.
- Study green alternatives to reduce dependency on foreign materials. Reusing and recycling could offset diminished and/or non-existent domestic sources. Advantages exist in alternative and renewable energy. The USN moved in this direction with Admiral Hyman Rickover's push for nuclear-powered naval vessels.

List of abbreviations

9/11	September 11, 2001
AOR	Area of Responsibility
BAA	Buy American Act
DARPA	Defense Advanced Research Projects Agency
DEA	United States Drug Enforcement Agency
DHS	United States Department of Homeland Security
DOD	United States Department of Defense
EU	European Union
FBI	United States Federal Bureau of Investigation
Ford	Ford Motor Company
LMA	Lockheed-Martin
NDEA	National Defense Education Act
NNAM	Nissan North American Manufacturing
R&D	Research and Development

STEM	Science, Technology, Engineering, and Mathematics
STEAM	Science, Technology, Engineering, Arts, and Mathematics
US	United States of America, or United States
USAF	United States Air Force
US Borders	Sovereign boundaries of the US and its territories
USCG	United States Coast Guard
USMC	United States Marine Corps
USN	United States Navy

References

1. Lieberman, J. (2003). *Making America Stronger: A Report With Legislative Recommendations on Restoration of US Manufacturing,* 2003. Retrieved September 25, 2006 from http://lieberman. senate.gov/documents/reports/ManufacturingReport.pdf.

2. Belt, D., Fellows, J., Kameru, B., Nazaroff, A., Pauroso, A., Schulz, F., Ballew, B., Bond, T., Demers, S., Kirkpatrick, S., & Thomas, W. (2005). *Electronics Industry Study Report.* Washington, DC: The Industrial College of the Armed Forces, National Defense University.

3. US Department of Defense (2007, October). *Department of Defense Dictionary of Military and Associated Terms,* Joint Publication 1-02. Washington D.C.: Government Printing Office.

4. Readiness (n.d.). *WordNet ® 3.0.* Retrieved July 20, 2007 from http://dictionary.reference.com/browse/readiness.

5. Office of the President of the United States of America (2006). *National Security Strategy of the United States of America,* March 2006. Retrieved September 25, 2006 from http://www.whitehouse.gov/nsc/nss/2006/nss2006.pdf.

6. Gongora, T. (2002, June). *The Meaning of Expeditionary Operations from an Air Force Perspective.* Paper presented at the Seapower Conference 2002, Dalhousie University, Halifax, Nova Scotia, Canada. Retrieved July 21, 2007 from http://wps.cfc.forces.gc.ca/papers/other/gongora.pdf.

7. US Department of Homeland Security (2006, May). *National Response Plan.* Washington, DC: Government Printing Office.

8. United States Department of Commerce (2004, January). *Manufacturing in America: A Comprehensive Strategy to Address the Challenges to US Manufacturers,* January 2004. Retrieved September 26, 2006 from http://www.ita.doc.gov/media/Publications/pdf/manuam0104final.pdf.

9. Institute of Industrial Engineers (1998). *Industrial Engineering Terminology.* Norcross GA: Engineering & Management Press.

10. National Association of Manufacturers (2003). *Keeping America Competitive: How a Talent Shortage Threatens US Manufacturing,* 2003. Retrieved September 27, 2006 from http://www.nam.org/s_nam/sec.asp?CID=201721&DID=230273.

11. Meredith, J. and Shell, R. (1988). *The Withering of US Manufacturing. Industrial Management,* 30(5), 13–18.

12. Innovation (n.d.). *WordNet ® 3.0.* Retrieved July 30, 2007 from http://dictionary.reference.com/browse/innovation.

13. Association of Computing Machinery (2006). *Globalization and Offshoring of Software,* 2006. Retrieved September 26, 2006 from http://www.acm.org/globalizationreport.

14. Committee on Prospering in the Global Economy of the 21st Century (2005). *Rising Above the Gathering Storm: Energizing and Employing America for a Brighter Economic Future.* Washington, DC: National Academies Press. Retrieved November 14, 2006 from http://www.hq.nasa.gov/office/oer/nac/documents/Gathering_Storm.pdf.

15. Popkin, J. and Kobe, K. (2006). *US Manufacturing Innovation at Risk.* Retrieved September 27, 2006 from http://www.nam.org/s_nam/bin.asp?CID20251&DID=236300&DOC=File.PDF.

16. Logistics (n.d.). *Investopedia.com.* Retrieved February 19, 2008 from http://dictionary.reference.com/browse/logistics.

17. US-China Economic and Security Review Commission (2005). *2005 Annual Report to Congress, 2005*. Retrieved September 27, 2006 from http://www.uscc.gov/annual_report/2005/annual_report_full_05.pdf.

18. Logistics (n.d.). *Dictionary.com Unabridged (v 1.1)*. Retrieved July 30, 2007 from http://dictionary.reference.com/browse/logistics.

19. Ford Motor Company (2007). *Ford Motor Company Sustainability Report 2006/7*. Retrieved July 28, 2007 from http://www.ford.com/en/company/about/sustainability/2006-07/overviewProfileGlobalDirectoryNAmerica.htm#mexico.

20. Nissan North American Manufacturing (2007, February 1). *At a Glance: 2007 Nissan Titan King Cab/Crew Cab*. Retrieved July 28, 2007 from http://www.nissannews.com.

21. Grasso, V. (2005). *Berry Amendment: Requiring Defense Procurement To Come From Domestic Sources, 2005*. Retrieved November 10, 2006 from http://www.fas.org/sgp/crs/natsec/RL31236.pdf.

22. Morrison, M. (1990). *The US Defense Industrial Base: Deterrence in Decline*. Marine Corps University Command and Staff College. Retrieved November 14, 2006 from http://www.globalsecurity.org/military/library/report/1990/MMR.htm.

23. Freedenberg, P. (2005, August 5). Are Machine Tools Important to Our National Security? *American Machinist*. Retrieved November 14, 2006 from http://www.americanmachinist.com/Classes/Article/ArticleDraw_P.aspx.

24. Hawkins, W. (2003, June 25). Homeland Defense: National-Security Resources Should not Be Manufactured in Foreign Lands. *National Review Online*. Retrieved November 14, 2006 from http://www.nationalreview.com/comment/comment-hawkins062503.asp.

25. Farrell, L. (2005). US Must Not Lose Manufacturing Edge. *National Defense Magazine*, 89(615), page 4. Retrieved November 14, 2006 from http://www.nationaldefensemagazine.com/issues/2005/Feb/US_Must.htm.

26. Freedenberg, P. (2005, January 1). Can Manufacturing Support a War? *American Machinist*. Retrieved November 14, 2006 from http://www.americanmachinist.com/Classes/Article/ArticleDraw_P.aspx.

27. Sheffi, Y. (2001). Supply Chain Management under the Threat of International Terrorism. *International Journal of Logistics Management*, 12(2), 1–11.

28. Chakrabory, K. and Remington, W. (2005). Offshoring of IT services: The impact on the US economy. *Journal of Computing Sciences in Colleges*, 20(4), 112–125.

29. Hira, R. *Implications of Offshore Outsourcing*. January 23, 2004. Retrieved September 25, 2006 from http://www.soc.duke.edu/sloan_2004/Papers/.

30. Hall, J. (1991). Sectoral transformation and economic decline: Views from Berkeley and Cambridge. *Cambridge Journal of Economics*, 5(2), 229–237.

31. National Fire Protection Association. *All About Fire*. Retrieved June 11, 2009 from http://www.nfpa.org/itemDetail.asp?categoryID=1330&itemID=30861&URL=Press%20RRoom/A%20Reporter's%20Guide%20to%20Fire%20and%20the%20NFPA/All%20about%20fire.

32. National Association of Manufacturers (2006, October). *The Facts About Modern Manufacturing, Seventh Edition*, October 2006. Retrieved October 4, 2006 from http://www.nam.org/s_nam/sec.asp?CID=210507&DID=229891.

33. Chopra, S. and Meindl, P. (2004). *Supply Chain Management: Strategy, Planning, and Operations, Second Edition*. Pearson, NJ: Pearson Education, Inc.

34. Tennessee Emergency Management Agency. *TEMA – The Agency, Its Missions and Goals*. Retrieved June 14, 2009 from http://www.tnema.org/index.htm.

35. US Department of Defense (2005, August). *Homeland Security*, Joint Publication 3-26. Washington, DC: Government Printing Office.

36. *The Constitution of the United States of America*. Article 1, Section 8. Article 2, Section 2. Second Amendment.

37. Defense Advanced Research Projects Agency (2008). *DARPA: Bridging the Gap*. Tampa, FL: Faircount LLC.

38. Freedenberg, P. (2004, January 1). Shoring Up the Defense Industrial Base. *American Machinist*. Retrieved November 14, 2006 from http://www.americanmachinist.com/Classes/Article/ArticleDraw_P.aspx.

39. United States Army (2009). *Army Energy Program: Renewable Energy.* Retrieved May 29, 2009 from http://army-energy.hqda.pentagon.mil/renewable/renewable.asp.
40. LOGAN Energy Corporation (2006, August 22). *US Army Schofield Barracks, HI PEM Demonstration Project Final Project Report.* Georgia: LOGAN Energy.
41. Kazmierczak, M. and James, J. (2005). *Losing the Competitive Advantage? The Challenge for Science and Technology in the United States,* 2005. Retrieved September 27, 2006 from http://aeanet.org/publications/idjj_CompetitivenessOverview0204.asp.
42. European Aeronautic Defense and Space Company (2007, May 2). *EADS CASA Will Supply Lockheed Martin Another Five (5) HC-235A Multi-mission Aircraft for the US Coast Guard's Deepwater Program.* Retrieved July 19, 2007 from http://www.eads.com/1024/en/pressdb/pressdb/Milita ry%20Transport%20Aircraft/20070502_EADS_CASA_HC-235A.html.
43. American Eurocopter (2006, December 2). *First UH-72A Light Utility Helicopter delivered to the US Army. New helicopter is a military version of the successful Eurocopter EC145.* Retrieved July 19, 2007 from http://www.eurocopterusa.com/Media/News/NewsDetail.asp?ID=409.
44. Putrich, G. (2007, June 14). *C-27J Tapped for Joint Cargo Aircraft.* Air Force Times. Retrieved July 19, 2007 from http://www.airforcetimes.com/news/2007/06/defense_JCA_070613/.
45. American Eurocopter (2007, July 12). *US Customs and Border Protection Orders Additional AS350 B3s for Homeland Security.* Retrieved July 19, 2007 from http://www.eurocopterusa.com/Media/News?NewsDetail.asp?ID=465.
46. Accreditation Board for Engineering and Technology (2010). *Search All Accredited Programs.* Retrieved March 31, 2010 from http://www.abet.org/AccredProgramSearch/AccreditationSearch.aspx.
47. Hedgpeth, D. (2008, March 4). A Foreign Air Raid? *The Washington Post.* Retrieved June 28, 2010 from http://www.washingtonpost.com/wp-dyn/content/article/2008/03/03/AR2008030303194_pf.html.

chapter ten

Is your organization ready for innovation?

Alfred E. Thal, Jr. and David E. Shahady

Contents

> "None of the most important weapons transforming warfare in the 20th century – the airplane, tank, radar, jet engine, helicopter, electronic computer, not even the atomic bomb – owed its initial development to a doctrinal requirement or request of the military."

The opening quote from Chambers (1999) suggests that the defense community has been at the forefront of innovation over the past century. Despite their success though, many organizations in the defense community struggle to explain specifically what they do to facilitate and implement innovation. To some, "being innovative" is interpreted as a means to empower employees to make decisions and solve problems at the lowest level possible. To others, "being innovative" is viewed as having open work spaces that lead to increased collaboration. However, innovation requires a much deeper understanding if it's to be successful. Beyond acknowledging the importance of innovation and inspiring the workforce though, what can leaders do to ensure their organizations are ready for innovation? To help answer that question, we think a good place to start is to review the organization's processes and dynamic capabilities. In many ways, these two concepts represent the DNA of the organization—and whether the organization is structured to facilitate innovation. We will then introduce a conceptual model that leaders can use to foster disruptive innovation. These three concepts—processes, dynamic capabilities, and the conceptual model—are equally applicable to organizations in both the public and private sectors.

Background

From the first powered flight by the Wright brothers in 1903 and the use of airplanes in the Army Air Corps to modern-day advances in military airpower, it's often said that innovation is a part of the Air Force culture. General Henry "Hap" Arnold alluded to this in 1945 when he suggested that *"... any air force which does not keep its doctrines ahead of its equipment, and its vision far into the future, can only delude the nation into a false sense of security."* Innovation has subsequently been highly touted by many of the Air Force's past and present leaders as being critical to the future success of the service. Furthermore, the vision statements for many Air Force organizations also acknowledge the importance of innovation. In fact, the Air Force's current vision statement is: The World's Greatest Air Force—Powered by Airmen, Fueled by Innovation.

Despite the importance placed on innovation though, two recent studies exploring the use of experimentation in innovation reported some sobering results. In the first study, the United States Air Force Scientific Advisory Board (SAB) concluded that the Air Force is very good at sustaining innovation but has "largely lost its ability to foster disruptive innovation" (United States Scientific Advisory Board, 2006). The SAB also concluded that Air Force organizations have not created an environment conducive to innovation. In the second study, the Air Force Studies Board (AFSB) expressed similar findings. Some of their key observations included a lack of space, time, and funding for experimentation-driven innovation, a fear of failure, a lack of appropriate processes, and a culture that is not supportive of innovation (AFSB, 2016). The results from both studies seem to indicate a stagnant environment in which the Air Force has lost momentum when it comes to technological innovation and is at risk of becoming irrelevant in the future battlespace.

Organizational processes

As W. Edwards Deming is fond of saying, "If you can't describe what you're doing as a process, then you don't know what you're doing." Let's put this into proper context for this chapter—if organizations are unable to describe their innovation efforts as a process, they're probably struggling with being innovative. This is consistent with Drucker (2002), who states that innovation is "capable of being presented as a discipline, capable of being learned, capable of being practiced." In other words, to make innovation more successful, it helps to view it as a process—a process which can be managed.

Processes are prevalent in organizations; they can be found in the way organizations operate, in their structures and cultures, and in the mindset of senior leadership (O'Reilly and Tushman, 2007). For those who may not have given it much thought, most organizations contain three general types of processes. Primary processes, also referred to as business processes, tend to be cross-functional. They often reflect the unique competencies of the organization and provide direct value to the customer; therefore, they are often considered mission essential. Support processes, on the other hand, usually do not provide direct value to the customer; instead, they are fairly standard and help sustain the organization. Common examples include management of information technology, infrastructure, capacity, and human resources. Finally, management processes provide direction and governance to ensure that the organization operates effectively and efficiently. They are generally conducted by senior leaders to develop and deploy strategy, manage the organizational structure, and establish organizational performance goals.

Regardless of its type, any process is "an organized group of related activities that work together to transform one or more kinds of input into outputs that are of value to the customer" (Hammer and Champy, 2001). Processes are thus designed to achieve a specific goal—a goal which, in turn, provides value to customers (either internal or external). This implies that processes are not random or ad hoc. Furthermore, every process in an organization should be viewed as either contributing to an organization's success or adding to its bureaucratic inefficiency—the key is being able to identify those processes that are a detriment to the organization and taking action to change them. When talking about processes and bureaucracy, an old adage often found in fortune cookies comes to mind: "People will do tomorrow what they did today because that is what they did yesterday." Ed de Bono refers to this as the "continuity of time sequence." Trapped by the sequence of our experiences, processes have a habit of developing almost arbitrarily yet becoming permanent.

It's human nature—and it explains a lot. It explains why many of today's practices are a reflection of "that's the way we've always done it." It explains how redundant processes develop and add to an organization's overhead. It shows how bureaucracy grows incrementally over time. Finally, it explains why few organizations run the way they should. The problem usually isn't about competence or effort—more often than not, the processes are the problem. Consider the following excerpt from Morison (1966).

> A time-motion expert … watched one of the gun crews of five men at practice in the field for some time. Puzzled by certain aspects of the procedures, he took some slow-motion pictures … A moment before the firing, two members of the gun crew ceased all activity and came to attention for a three-second interval extending throughout the discharge of the gun. He summoned an old colonel of artillery, showed him the pictures, and pointed out this strange behavior. What, he asked the colonel, did it mean. The colonel, too, was puzzled. He asked to see the pictures again. "Ah," he said when the performance was over, "I have it. They are holding their horses."

The earlier description relates to horse artillery units supporting the cavalry. However, as technology advanced and the process of firing artillery guns changed, part of the previous procedure remained intact. An argument can certainly be made regarding the importance of upholding tradition, especially in military organizations. In many other cases though, does the tradition provide value? Or is it simply a carryover from the past because that's the way it's always been done?

If we extend this line of reasoning to processes in general, how many processes in our organizations are simply carryovers from the past? To give this some critical thought, it might be helpful to evaluate the organization's core competencies. When organizations excel at an activity, they can easily become over-committed to it. If the organization holds on to them too tightly, those core competencies and their accompanying processes can easily become core rigidities (Leonard-Barton, 1992). In our own organizations, how many similar examples exist? How many processes do we have that were built in a different era and possibly for different purposes but continue to be blindly followed? Breaking away from these processes and the past requires conscious effort—it requires the will to question the existing processes and the inherent assumptions on which those processes were based.

As we review our organization's activities and processes, an important concept to consider is the value chain, which represents the primary and limited support processes that provide value to the customer (Porter, 1985). While organizations may have hundreds of work processes, they usually have very few business processes. As such, value-creating business processes begin and end with the external customer, tend to be large in scope, and commonly span multiple organizational components. Since this group of processes represents the core competencies of the organization, this is where performance improvement work is often focused. Furthermore, these processes must be aligned and integrated to enable effective performance of the organization.

With this brief introduction to processes, the question for most organizations is whether innovation is considered a core competency. If it is, does the organization treat innovation as a process that can be managed? And do other processes within the organization align with and complement the innovation process? An approach organizations might take to address these questions is to review their capabilities.

Organizational capabilities

When examining the success of organizations, a fundamental question that often arises is, "Why does a particular organization or group of organizations outperform other similar organizations?" To answer the question, two schools of thought have developed: the industry-based view and the resource-based view. The industry-based view assumes that success has something to do with the industry in which the organization operates; therefore, strategies are based on an external analysis (such as Porter's 5 Force model). On the other hand, the resource-based view assumes that success has something to do with the assets (or resources) the organization owns and controls; therefore, strategies are based on an internal analysis. Since empirical evidence suggests that organizational differences account for more variation in performance than industry differences (Rumelt, 1984), the resource-based view (RBV) has been increasingly referenced in the strategy literature. Although the RBV framework was initially developed to understand how businesses achieve and sustain competitive advantage (Prahalad and Hamel, 1990; Barney, 1991), its inward focus also makes it appealing to public sector organizations (Matthews and Shulman, 2005; Pablo et al., 2007).

Eisenhardt and Martin (2000) suggest that assets—physical, human, and organizational resources—are the foundation of the RBV approach. Furthermore, the "bundling" of these assets to perform specific business processes is often referred to as a capability. Organizational capabilities are thus the various routines (or patterns) and processes that transform inputs (i.e., resources) into outputs (i.e., goods and services that provide value to the customer). Routines represent sequences of actions for performing tasks in an organization. Institutionalized through technologies, formal procedures, and informal conventions or habits, they reflect "the way we do things around here."

Organizational capabilities can be characterized as either ordinary or dynamic as shown in Figure 10.1. Ordinary capabilities represent the routines and standard operating procedures within the organization. They tend to support the day-to-day operations of the organization and change little over time; in some cases, they are often referred to as "best practices." Dynamic capabilities are the real reason for an organization's long-term success; they represent a set of abilities that enable an organization to quickly build capability and affect change. Organizations with dynamic capabilities are thus better positioned to exploit opportunities by adapting organizational structures and routines. According to Teece (2006), strong ordinary capabilities are necessary but not sufficient for long-term

Ordinary Capabilities	**Dynamic Capabilities**
• Technical efficiency in basic business functions	• Strategic "fit" over the long run (evolutionary fitness)
• Operational, administrative, and governance	• Sensing, seizing, shaping, and transforming
• Relatively easy; imitable	• Difficult; inimitable
Doing things "right"	*Doing the "right" things*

Figure 10.1 Organizational capabilities. (Adapted from Teece, D., *Res. Policy*, 35, 1131–1146, 2006. With permission.)

success; they can be acquired (or "bought") from other organizations or through investments in training. However, strong dynamic capabilities are necessary and sufficient for long-term success; they cannot be bought and must be built. From an innovation perspective, this is a critical point—the ability to build and improve effective routines is often considered a necessary ingredient for successful innovation.

Teece et al. (1997) define dynamic capabilities as an organization's "ability to integrate, build, and reconfigure internal and external competencies to address rapidly changing environments." The term "dynamic" is meant to indicate an organization's capacity to establish new competencies in response to environmental conditions and the ability to reconfigure their assets and develop new routines (Lee and Kelley, 2008; Eisenhardt and Martin, 2000), while the term "capabilities" is meant to imply the importance of strategic management. Taken together, dynamic capabilities serve as the source of an organizations' competitive advantage. Additionally, they are often considered a necessary component of the innovation process (Lee and Kelley, 2008). To be specific, Lawson and Samson (2001) suggest three primary reasons dynamic capabilities align with innovation efforts: (1) the lack of a technology focus recognizes the importance of other resources; (2) the RBV basis makes it applicable to product, process, system, and business model innovation; and (3) asset heterogeneity reflects the expectation that there is no one generic formula. From a dynamic capability perspective, Tidd and Bessant (2009) describe the core abilities in managing innovation shown in Table 10.1.

Teece et al. (1997) describe organizational processes as routines of current practice serving three roles: coordination/integration, learning, and reconfiguration. These routines are used to integrate and exploit competencies. However, what the organization can accomplish with its dynamic capabilities is constrained by its asset positions and shaped by evolutionary and co-evolutionary paths (Teece et al., 1997). An organization's position reflects specific competencies in both tangible and intangible assets; these competencies may consist of technology capabilities, complementary assets, external relationships, etc. Paths represent options available to organizations based on core competencies, technology trajectories, and emerging opportunities. They tell us that the availability of current strategic choices is a reflection of past strategic choices (Teece et al., 1997). In other words, it is typically difficult for most organizations to ignore what has been done in the past and

Table 10.1 Managing innovation

Basic ability	Contributing routines
Recognizing	Searching the environment for technical and economic clues to trigger the process of change
Aligning	Ensuring a good fit between the overall business strategy and the proposed change—not innovating because it is fashionable or as a knee-jerk response to a competitor
Acquiring	Recognizing the limitations of the company's own technology base and being able to connect to external sources of knowledge, information, equipment, etc.
	Transferring technology from various outside sources and connecting it to the relevant internal points in the organization
Generating	Having the ability to create some aspects of technology in-house—through R&D, internal engineering groups, etc.
Choosing	Exploring and selecting the most suitable response to the environmental triggers which fit the strategy and the internal resource base/external technology network
Executing	Managing development projects for new products or processes from initial idea through to final launch
	Monitoring and controlling such projects
Implementing	Managing the introduction of change—technical and otherwise—in the organization to ensure acceptance and effective use of innovation
Learning	Having the ability to evaluate and reflect upon the innovation process and identify lessons for improvement in the management routines
Developing the organization	Embedding effective routines in place—in structures, processes, underlying behaviors, etc.

Source: Tidd, J. and Bessant, J., *Managing Innovation: Integrating Technological, Market, and Organizational Change,* John Wiley & Sons, Chichester, UK, 2009.

develop new ideas. Specifically in a research and development (R&D) environment, Cohen and Levinthal (1990) argue that an organization's innovative capability is a function of its prior related knowledge; without prior experience, the organization would not be in a position to recognize value and exploit it.

O'Reilly and Tushman (2007) suggest that capabilities are the result of senior leader actions to facilitate and ensure learning, integration, reconfiguration, and transformation; these processes thus dictate the paths (i.e., strategic choices) organizations take. They typically refer to this as the "sensing and seizing" of new opportunities to emphasize the key role of strategic management. Other researchers also include the role of "transforming" when referring to dynamic capabilities. As organizational leaders ponder their role, and the actions they take, to facilitate innovation through dynamic capabilities, a model may prove to be useful.

Disruptive innovation model

Numerous studies have been conducted regarding the importance of innovation. For example, the Council on Competitiveness (2005) concluded that, "Innovation will be the single most important factor in determining America's success through the twenty-first Century." In 2006, the American Management Association (AMA) commissioned a study on the emergence of innovation in global industries. The study concluded that "innovation is going to get considerably more important over the next decade;" therefore, it is essential for companies to eliminate the barriers of innovation and increase their innovative culture (American Management Association and Human Resource Institute, 2006). IBM Global Business Services conducted an innovation study focused on public and private sector senior leadership. According to the study, CEOs expected fundamental changes for their organizations and saw opportunities to be seized through innovation (IBM, 2006). The study concluded that business model innovation and external collaboration are extremely important, as well as the role of senior leadership, in fostering an innovative climate. A study by the Boston Consulting Group found that the leading innovative organizations were characterized by risk taking and investment in the long-term (*BusinessWeek*, 2007; McGregor, 2007). The study also found that gimmick-driven campaigns were not the deciding factor—companies became innovative through hard work.

Innovative organizations are revolutionary in that they aggressively take markets from competitors (Hamel, 2002). Furthermore, innovation helps good organizations become great organizations and equips strong companies to become long-lasting entities (Collins, 2001). Additionally, resilient groups embrace disruptive change (Hamel and Valikangas, 2003), and competitive organizations use breakthrough ideas to destroy the opposition (Foster, 1986). However, the difficult challenge for most groups is creating an environment to foster breakthrough innovation while marginalizing practices that stifle creativity. While many business scholars have articulated innovation as a key for survival, deriving a formula for success has proven to be a difficult challenge. Throughout the literature though, there is evidence that motivation, focus, barriers, and culture play a crucial role in the emergence of breakthrough and game-changing ideas. By examining these key elements with regard to innovation, a base model for the emergence of disruptive innovation can be formulated. After presenting the model, implications for the defense industry will be briefly discussed.

Motivations for pursuing innovation

The primary reason companies pursue innovation is to gain and/or maintain competitive advantage. Foster (1986) explained that competitive advantage can only be achieved by going on the attack and that companies can lose their markets almost overnight to faster-developing technologies. Based on recent research and literature, several consistent themes appear among both industry professionals and corporate CEOs. As illustrated in Table 10.2, the leading reasons for pursuing innovation are to increase profitability, respond to customer demand, and improve efficiency.

Increasing Profits: An increase in overall revenue and profit margins continues to be one of the primary motivations for companies to pursue innovation. The world's most innovative companies traditionally see greater revenue growth and margin growth compared to their less innovate counterparts (*BusinessWeek*, 2007). However, companies are finding it takes time to see profit growth and are often abandoning innovation investments for more short-term gains. Most decisions being made regarding innovation, and particularly the development of dynamic capabilities, would benefit from a long-term perspective.

Responding to Customer Demand: In today's marketplace, innovation is often seen as a primary means to acquire and hold onto customers. Peters (1997) explained this concept best: "If the other guy's getting better, then you'd better get better faster than the other guy's getting better, or you're getting worse." However, it is important to understand

Table 10.2 Reasons for pursuing innovation within industry organizations

The Quest for innovation (AMA, 2006)		Expanding the innovation horizon (IBM, 2006)	
Reasons	Rank	Reasons	Rank
To respond to customer demands	1	Profitable growth	1
To increase operational efficiency	2	Preempt business threats and create them	2
To increase revenues or profit margins	3	Drive needed efficiency	3
To develop new products and services	4	Develop multiple channels with different approaches for different customers	4
To increase market share	5		
To better use new technologies	6		

the level of customer interaction envisioned—while working closely with the customer provides great insight into their needs, it can also hinder the recognition of emerging needs and technologies (Francis and Bessant, 2005). Therefore, a high level of customer interaction seems to be more appropriate for sustaining/incremental innovation efforts, while disruptive/radical innovation typically requires less customer involvement.

Improving Efficiency: As shown in Table 10.3, companies need to reduce cycle-times and improve operational efficiency to survive. Hammer and Champy (2001) explain that because of customer power and customer choice, simply relying on acceptable process performance is no longer sufficient; furthermore, they state that conventional business remedies do not address the source of the problem, which is non-value added work resulting from fragmented processes.

Focus of innovation resources

While the need to focus resources on innovation is widely espoused, the optimal balance of investment is widely debated in the literature. Short-term investments necessitate close attention to detail, midterm investments demand capital and a willingness to take risks, and long-term investments require imagination and technological daring (Hayes and Abernathy, 1980). Innovation strategies by companies today are best described by looking at investments by functional area, innovation magnitude, and innovation type. The studies and literature indicate trends toward customer focus, reliance on business model innovation, and an emerging push toward new breakthrough products/services.

Table 10.3 Cycle-Time reductions in industry

Industry	Past	Recent	Goal
Automobile	84 months	24 months	<18 months
Commercial Aircraft	8–10 years	5 years	2.5 years
Commercial Spacecraft	8 years	18 months	12 months
Consumer Electronic	2 years	6 months	<6 months

Source: Defense Science Board, *2006 Summer Study on the 21st Century Strategic Technology Vectors: Volume IV Accelerating the Transition of Technologies into US Capabilities*, Defense Science Board, Washington, DC, 2007.

Table 10.4 Innovation within industry organizations

Functional areas of innovation		Focus areas of innovation	
Functional areas	Percent of responses	Areas	Percent of responses
R&D	27	Customer experience	15.2
Marketing	17.2	Service	11.6
Information Technology	12.2	Core processes	12.4
Sales	9.7	Product performance	12.2
Customer Service	8.9	Enabling processes	11.8
Manufacturing	6.5	Business models	10.6
Supply Chain	5.4	Brand	8.4
Planning	5.1	Networks and alliances	8.1
Human Resources	3.9	Product systems	4.7
Finance	2.4	Channel	3.6

Source: American Management Association and Human Resource Institute, *The Quest for Innovation: A Global Study of Innovation Management 2006–2016*, American Management Association, New York, 2006.

Customer Focused Innovation: According to the AMA study (2006) results outlined in Table 10.4, more than 25% of the innovation resources in participating companies were focused on supporting customer experience and service. In addition, the study found that while innovation occurs across various functional areas, the areas directly related to customer relationships are receiving the highest degree of focus. Marketing, sales, customer service, and supply chain functions accounted for over 41% of the functional areas of innovation.

Emphasis on Business Model Innovation: Companies are finding with greater certainty that business processes and organizational innovation are important. The IBM (2006) study found that "four out of every ten companies were afraid that changes in a business competitor's business model would upset the competitive dynamics of the entire industry." It's no wonder then that the CEOs of outperformers are placing nearly twice as much focus on business model innovation than the CEOs of underperformers.

Product/Service Migration toward Disruption: While competition has pushed companies to consider process innovation, the most popular type of innovation focus continues to be in the area of products/services. The recent industry shift is toward new products/services with "fewer companies focusing on incremental innovation or making minor changes to existing products" (*BusinessWeek*, 2007). This further solidifies the importance of understanding the emergence of disruptive innovation.

Barriers of innovation

Innovation can be a difficult and daunting challenge—one of the reasons for this is that most innovation experts agree that barriers hampering innovation are abundant. Many companies invest considerable resources into fostering ideas only to have their innovation efforts squelched by internal and external barriers (Kelley and Littman, 2001). Table 10.5 summarizes the most common barriers found in companies today. Although the semantics of obstacles varies from study to study, several common themes are consistent throughout the research: unsupportive culture, insufficient resources, lack of strategic vision, and poorly developed processes.

Table 10.5 Study findings in barriers of innovation

The quest for innovation (AMA, 2006)	Expanding the innovation horizon (IBM, 2006)	The world's most innovative companies (BusinessWeek, 2007)

• Insufficient resources
• Lack of formal strategy for innovation
• Lack of clear goals and priorities
• Unsupportive organizational structures
• Short-Term mindset
	Internal
	  • Unsupportive culture and climate
	  • Limited funding for investment
	  • Workforce issues
	  • Process immaturity
	  • Inflexible physical and IT infrastructure
	  • Insufficient access to information
	External
	  • Government and other legal restrictions
	  • Economic uncertainty
	  • Inadequate enabling technologies
	  • Workforce issues arising externally
		• Lengthy development times
		• Lack of coordination
		• Risk-averse culture
		• Limited customer insight
		• Poor idea selection
		• Inadequate measurement tools
		• Lack of ideas
		• Marketing or communication failure

Unsupportive Culture: The research findings summarized in Table 10.5 found unsupportive organizational cultures to be significant obstacles to innovation growth. This is consistent with Kelley and Littman's (2001) observation that company mindset is one of the biggest barriers to innovation. Risk-adversity, inflexibility, communication failures, workforce issues, and lack of ideas are all common symptoms of a poor innovative culture. Overcoming these barriers can best be addressed by cultivating a positive innovative culture. The characteristics of innovative culture are addressed in more detail later in the chapter.

Insufficient Resources: Innovation is not merely about financial investments—it also involves investments in people, facilities, markets, training, and technology. Many organizations are falling into the "performance" trap where the company is doing well and fails to explore other opportunities because of the time, money, and personnel required (AMA, 2006). Other organizations are opting to sacrifice long-term stability for short-term gains. With reductions in discretionary dollars and pressures from stockholders, many CEOs are forced to divert R&D resources to low-risk investments with guaranteed returns (IBM, 2006). According to the *BusinessWeek* (2007) assessment, "More than half of all CEOs, chairmen, and presidents of companies were happy with how they'd spent on growth initiatives. CFOs, not surprisingly, were among the least satisfied: A full 63% were unhappy with their results." This mindset clearly defines the difficulties faced by innovators attempting to gain access to needed resources.

Lack of Strategic Vision: Although it is debated in the literature whether companies can "direct" innovation, it is commonly acknowledged that innovation strategy plays a role in fostering new concepts. Based on the AMA (2006) research highlighted in Table 10.6, most companies fall dramatically short in developing a well understood strategy for innovation and a shared vision on how to execute a plan for innovation.

Poorly Developed Processes: Long development times, insufficient access to information, poor idea selection, ineffective organizational structures, and communication failures are all indicative of poorly developed processes. Hammer (1996) contends that

Table 10.6 Industry lack of innovation strategy

People in my company...	Percent of respondents
Have a shared definition of what innovation is.	41.3
Regularly review the progress of innovation.	22.4
Have a shared agenda to execute the innovation strategy.	12.3
Have a well-understood strategy for innovation.	12.1
Have well-defined roles and responsibilities.	11.3

Source: American Management Association and Human Resource Institute, *The Quest for Innovation: A Global Study of Innovation Management 2006–2016*, American Management Association, New York, 2006.

"it is not uncommon to find less than 10 percent of the activities in a process to be value-ending, with the rest mostly non-value-adding overhead." Process improvement is based on a commitment to optimize value through a process view of accomplishing work. It is not surprising that companies with inefficient processes struggle with innovation given that it takes creative and radical thinking to develop effective processes.

Characteristics of innovative culture

Organizational culture is defined as "a system of shared meaning held by members that distinguishes the organization from other organizations" (Robbins and Judge, 2007). An innovative culture is therefore a shared organizational environment designed to foster innovation. Many companies even specialize in teaching organizations to become more innovative. IDEO, ranked as the 28th most innovative company in the world (*BusinessWeek*, 2007), is considered a premiere leader in the development of the breakthrough spirit. With the recent emphasis being placed on innovation throughout the business world, it is not surprising that hundreds of articles and publications have been written on the characteristics of an innovative culture. Several common threads appear within the leading studies, summarized in Table 10.7, that help define the key characteristics: strong customer focus, collaboration, efficient processes, creative people, inspiring leadership, risk-taking, and motivation/reward systems.

Strong Customer Focus: The research suggests that organizations who place their existing and future customers at the forefront tend to be more innovative. Strong customer focus does not just mean delivering what customers ask for but rather "capturing

Table 10.7 Characteristics of innovative culture

The quest for innovation (AMA, 2006)	Expanding the innovation horizon (IBM, 2006)	The world's most innovative companies (BusinessWeek, 2007)
• Customer focus • Teamwork and collaboration with others • Appropriate resources • Organizational communication • Ability to select the right ideas for research • Ability to identify creative people	• Orchestration from the top • Collegial culture with individual rewards • Consistent business and technology integration	• Right organizational structures • Right processes • Right people • Inspired leadership

their ideas or actually allowing them to innovate on their own behalf" (AMA, 2006). According to Kelley and Littman (2001), co-founder of IDEO, true understanding comes not by talking to customers, but by watching them and becoming immersed in their environment. As a result of this strong customer focus, organizations are in a better position to implement disruptive product and process innovations that transform the marketplace and decimate the competition. Demonstrating this point, Christensen and Raynor (2003) reviewed the extensive market analysis conducted by a quick-service restaurant chain with regard to milkshake sales. The group examined not just what the customers wanted, but why they wanted it, when they wanted it, who they were with, and what they would be doing if they were not there buying a milkshake. They essentially focused on the job the customer was trying to get done.

Collaboration: External and internal collaboration is a common characteristic found in studies on innovation. According to Hargadon (2003), most significant innovations come from collaborative groups of people and not brilliant lone individuals. Collaborative innovation can be defined using the organizational Garbage Can Model (Cohen et al., 1972). The theory articulates that many solutions to problems can often be found by sifting through garbage in which ideas, or the ideas of others, have been tossed out as being irrelevant. Similarly, innovative cultures are best characterized by broad and often unrelated people that simply interact to make breakthroughs happen. Organizations that collaborate to a large extent typically perform better than the competition and receive strong benefits from the innovate spirit that is generated.

Efficient Processes: Efficient processes are streamlined and provide the appropriate level of performance to the organization. In addition, efficient processes undergo an endless cycle of improvement in which performance is measured, benchmarks are established, gaps are identified, and modifications are implemented (Hammer, 1996). According to the AMA (2006) assessment, innovative cultures are strongly tied to how efficiently organizations can capitalize on ideas. Innovative organizations know how to balance resource investments, select the right ideas, mobilize the right resources, and measure results. The level of disruptive innovative is directly related to an organization's ability to get funding and manpower required to cultivate new idea proposals (Christensen, 1997).

Creative People: Creative people, a key element in creating an innovative culture, solve problems by examining the world from different perspectives (Glover and Smethurst, 2003). Innovators are able to look beyond the status quo and visualize the realm of the possible while not allowing risk and adversity to hamper their progress. Henry Ford reportedly once said, "Failure is the only opportunity to begin again, this time more intelligently." Not everyone is naturally creative and many companies like IDEO have developed a series of innovation roles that allow people to contribute to the innovative culture. Although business scholars believe that innovation comes from groups of creative people, breakthrough teams are composed of individual characters and diverse personalities deliberately recruited to generate energy and ideas (Kelley and Littman, 2001).

Inspiring Leadership: Collins (2001) found that successful leaders, those who blend extreme personal humility with intense professional will, were the catalyst in building great companies. Supportive leadership has been shown to be an equally important characteristic in building an innovative culture. The extent to which the leader reflects on organizational objectives, strategies, and processes, and implements changes accordingly, is directly related to the organizational climate for innovation. In organizations with more reflective leaders, employees rated the innovative climate higher, organizational practices were more non-traditional, and there was a greater amount of change (Kazama et al., 2002).

Risk-Taking: "Innovation demands adherence to two fundamental principles: a willingness to accept risk and a willingness to wait for the return on investment" (Council on Competitiveness, 2005). While most scholars agree that innovation is a risky venture, only 20% of global companies actually recognize and reward intelligent risk-taking (AMA, 2006). Innovative cultures are made stronger by embracing failure as an option and taking the time to experiment. IDEO describes this innovation characteristic with the slogan, "Fail often to succeed sooner" (Kelley and Littman, 2001). Encouraging risk-taking helps create an environment where employees are willing to take chances with radical ideas.

Motivation and Reward Systems: Rewards for innovative behavior were a common characteristic cited in several publications on innovative culture in industry. Most companies use non-financial rewards as a means to promote innovation (AMA, 2006). Companies that "reward individual [innovation] contributions achieved 2 percent higher operating margins on average and grew nearly 3 percent faster than those who did not" (IBM, 2006). Motivation and reward systems are closely tied with organizational willingness to accept risk.

> How you encourage and reward innovative activities will ultimately determine whether your employees undertake them. Innovation starts with employees willing to take risks. Employees will be apprehensive of these activities if they perceive the upside to be limited and the downside to be significant. A truly innovative culture needs to make employees feel secure enough to believe that failure itself will not affect their position within the firm. (Deloitte, 2003)

Putting it all together

Christensen and Raynor (2003) propose that building an organization capable of disruptive growth requires a careful balance of resources, processes, and values. Combining these thoughts with previous studies of organizational innovation provides a model for fostering disruptive innovation. The model proposes the following: an increase in the right motivation, plus an increase in the right focus of innovation resources, plus a decrease in the barriers of innovation, plus an increase in the characteristics of innovative culture, will foster an increase in the emergence of disruptive innovation. This model, illustrated in Figure 10.2, is not intended to be an equation for guaranteed success but rather a conceptual formula to ensure that critical elements in the emergence of disruptive innovation are considered. While the interpretations, applications, and considerations will be domain dependent, the basic model is a universal framework for innovation improvement. Understanding the model is not sufficient though—to link it back to the previous discussion, fostering innovation also requires a full understanding of the organization's processes and its capabilities.

Defense implications

The model presented in Figure 10.2, along with the concepts of processes and dynamic capabilities, are applicable to all organizations in both the public and private sectors. To show the applicability to the defense community, each element of the model is briefly discussed in this section of the chapter.

Figure 10.2 Model for fostering disruptive innovation.

Motivations for pursuing innovation

Within the defense community, the reasons for pursuing innovation may be quite different and caution must be exercised. For example, defense organizations could be motivated by a desire to be viewed as state-of-the-art and capable of using new technologies more effectively. From a dynamic capabilities perspective, this could reflect a perceived need to enhance the organization's assets and improve its technological positioning. However, it could also be an indication of focusing too much on incorporating the newest technologies to create a "wow" factor. It could also indicate a reliance on technology, and perhaps a focus on invention instead of innovation, to meet the customers' needs. Depending on the situation, a better approach may be to focus on the job the customer is performing and strive to help the customer perform that job better (e.g., more quickly, more effectively, less costly, etc.).

Relying on policies to encourage innovation may not be very effective. Tidd (1993) found that policies often do not support technology strategies; instead, organizations tend to either follow industry trends or act in an ad hoc manner in response to a near-term need. This type of reactive approach may be due to existing learning mechanisms (or the absence thereof); therefore, defense organizations interested in innovation might consider examining their organizational processes to ensure appropriate structures and polices are in place to develop congruent strategies. Recognizing the impact of past strategic choices, coupled with critical thought about the impact of current decisions on future opportunities, could also be helpful. Since Chesbrough and Rosenbloom (2002) consider the business model to be a mediator between technology and value, the defense community might consider placing more emphasis on the business model aspect of innovation and developing appropriate value propositions. Finally, processes should be in place to facilitate cross-functional teamwork and integration, as well as to introduce employees to new technologies and make them aware of their potential uses and benefits; this learning process thus affects the evolutionary path of the organization.

The key factor is whether processes are in place to address the motivations for pursuing innovation. To facilitate this desire, it is important that organizations consider all components of the dynamics capabilities framework and develop commensurate strategies. With a narrow focus instead of a broader perspective, organizations may be overlooking opportunities to improve their dynamic capabilities and be more innovative.

Focus of innovation efforts

The functional and focus areas of innovation were shown in Table 10.3. The low percentages shown in the table could indicate an overall weak application of innovation efforts; however, it could also reflect a lack of focus. When this happens, Francis and Bessant (2005) suggest that innovation efforts often develop without any coherent strategy and are often inefficient and sometimes contradictory. They also suggest that systematic analysis and comparative benchmarking might help facilitate more alignment between incongruent innovation efforts. A more structured approach to the development of organizational processes could also be helpful.

Networks and alliances are a key source of innovation (von Hippel, 1988) in which the primary reason for collaborating is to access either complementary technologies to support innovation activities or new markets (Tidd, 1993; Greis et al., 1995). However, too much focus on collaborations (i.e., networks and alliances) could reflect a reliance on external entities to drive innovation efforts instead of developing organic capabilities. Additionally, collaborations can affect an organization's evolutionary path by potentially shaping, sometimes positively and sometimes negatively, future strategic choices (Teece, 2006). Therefore, defense organizations are encouraged to develop a healthy strategy towards the use of collaborations.

When it comes to the customer experience, business models and brand areas are important components. Since business models help convey the organization's value proposition (Chesbrough and Rosenbloom, 2002), the choice of business model will influence the organization's processes, positions, and paths (Francis and Bessant, 2005). Therefore, more emphasis on innovative business models could potentially provide new benefits for defense organizations. Although branding may not be very applicable in the defense community, it may be helpful in establishing effective communication channels with customers to provide a better understanding of what innovation can do for them.

Barriers to innovation

Consistent with the resource-based view of the organization and other research (e.g., Blumentritt and Danis, 2006), defense organizations often indicate that insufficient resources is a primary barrier. However, Liao et al. (2009) found that the primary constraint hindering innovation is the lack of integrative capabilities (e.g., routines for integrating external knowledge and identifying opportunities). What this tells us is that organizations tend to lack processes to perform the coordination/integration, learning, and reconfiguration roles Teece et al. (1997) claim are necessary to develop new competencies quickly.

Furthermore, a lack of guidance from the organization's leadership may suggest that innovation is accomplished in an ad hoc manner. Employees may feel they are getting adequate support from their immediate supervisors but not receiving clear guidance from the organization's senior leaders. The defense community may thus benefit from examining strategies and guidance since the ability of senior managers to "sense and seize" opportunities while overcoming organizational inertia and path dependencies is at the core of dynamic capabilities (O'Reilly and Tushman, 2007). This is especially important since organizational constraints are often "hidden" in everyday activities and processes.

An often overlooked constraint is the organization's history and the path-dependent nature of capabilities created by the organization's routines (Rindova and Taylor, 2002). As previously mentioned, these core capabilities can easily become core rigidities (Leonard-Barton, 1992).

Finally, a "fear of failure" culture is a potential innovation barrier. Employees may not relate individual attitudes to barriers; however, when viewing culture as a barrier, they may be thinking of the organization's processes, policies, and procedures. This may be why factors related to organizational culture—threat of new ideas, lack of rewards, and short-term mindset—are often rated higher than the "fear of failure" barrier. In some ways then, culture may be viewed in terms of bureaucracy, which Francis and Bessant (2005) characterize as unfriendly to innovation.

Characteristics of an innovative culture

Of primary concern to the defense community may be the freedom to innovate, which may be because of the bureaucratic and structured nature of most government organizations. This is consistent with the SAB's (2006) finding that the Air Force relies too much on technology demonstrations instead of experimentation. To be truly disruptive, Christensen and Raynor (2003) suggest the use of discovery-driven planning, to include experimentation and learning. An innovative culture also requires appropriate organizational processes and leadership ability to reconfigure assets and "sense and seize" opportunities. This may be lacking in government organizations, thus making the culture not as conducive as it could be in terms of facilitating innovation. Additionally, defense leaders may want to ensure there is a clear understanding, shared definition, and strategy for innovation in their organizations.

Industry considers the best way to establish an innovative culture is to focus on the customer. Although the defense community may consider customer focus to be important, it may struggle with the degree of "connectedness" to the customer and efforts to develop an appropriate value proposition and business model. Another important factor for industry is effective organizational communication, which requires effort and supporting processes. Therefore, poor communication may contribute to it being seen as a barrier to innovation. It may also imply more of a team approach to developing innovative solutions as compared to the typical "stovepipes" in more bureaucratic organizations.

Senior leader involvement

Although defense organizations may consider innovation to be extremely important, they may find that it is not integrated very well into the overall organizational strategy. Blumentritt and Danis (2006) have suggested that "strategic orientation may be a powerful explanatory variable that accounts for important differences in how innovation is managed." In fact, de Jong and Marsili (2006) found that there is a correlation between the presence of a documented innovation strategy and the level of innovative activity in an organization. Similarly, O'Reilly and Tushman (2007) recommend that leaders articulate a vision and strategic intent, along with identifying specific complementary organizational processes. Furthermore, Lawson and Sampson (2001) found that innovation often requires visionary leadership; coordination between innovation, business, and technology strategies; and a commitment to results. Therefore, the defense community may want to consider using strategy to facilitate the integration of innovation. They may also find it helpful to develop new value propositions and business models.

Final thoughts

Managing innovation creates a dilemma for organizations. A loose organizational structure is often perceived as flexible and thus preferred if one wants to foster innovation, creativity, and adaptability. However, a formal structure and key management controls are required to coordinate and communicate innovation efforts. The key is to have a broadly structured framework within which employees have the freedom to make decisions about the best approach to take for a specific effort. At a minimum, each organization should have a tailored version of the innovation funnel. The intent of the funnel is to generate ideas, narrow the list of ideas to those that are most promising, and then implement the ideas that are selected to increase the value provided to the customer.

If an organization wants to become more innovative, the following principles are offered for consideration.

1. Create a strategic vision that establishes innovation as a priority.
2. Inspire the workforce by clearly identifying the organization's challenges and discussing how innovation will help address those challenges. Keep in mind that innovation is not required in every organization.
3. Evaluate the organization's dynamic capabilities and determine the changes required to align them with the strategic vision. Successful innovation depends on two key factors—resources and capabilities. Does the organization have the appropriate resources? Does the organization have the appropriate dynamic capabilities?
4. Review the organization's existing processes and create/change processes as required. This includes the innovation process itself, as well as complementary processes within the organization. Determine how innovation will be integrated with other processes in the organization.
5. There's an old adage in organizations—"you get what you measure." Therefore, spend some time developing an effective set of metrics to measure innovation and communicate the results.
6. Innovation is accomplished through people. Therefore, provide training to the workforce in terms of product and/or process innovation tools, managerial tools, and general problem-solving skills.
7. Recognize innovative behavior and reward innovative results.
8. Promote experimentation and prototyping as a way to develop a "fail early and often" mindset.

Although innovation is rooted in curiosity and discovery, it's not free-wheeling and void of structure—it's driven by a system of principles and practices which support and encourage people to solve problems. Therefore, and as previously mentioned, innovation should be considered a process—a process which can be managed. It's ultimately a management and leadership question involving choices to be made about resource allocation and coordination. With the right choices and the proper approach to developing dynamic capabilities, the military can position itself to fulfill the following vision expressed by Gen "Hap" Arnold at the end of World War II.

> "The next war may be fought by airplanes with no men in them at all...Take everything you've learned about aviation in war, throw it out of the window, and let's go to work on tomorrow's aviation. It will be different from anything the world has ever seen."

References

Air Force Studies Board. (2016). *The Role of Experimentation Campaigns in the Air Force Innovation Life Cycle.* Washington, DC: The National Academies Press.

American Management Association and Human Resource Institute. (2006). *The Quest for Innovation: A Global Study of Innovation Management 2006–2016.* New York: American Management Association.

Barney, J. (1991). Firm resources and sustained competitive advantage. *Journal of Management,* 17(1), 99–120.

Blumentritt, T., & Danis, W. (2006). Business strategy types and innovative practices. *Journal of Managerial Issues,* 18(2), 274–291.

BusinessWeek. (2007). Special report—2007 most innovative companies. Retrieved 7/25/07, from http://www.buisnessweek.com/innovate/di-special/20070503mostinnovative.htm.

Chambers, J. (1999). (Ed.). *The Oxford Companion to American Military History.* New York: Oxford University Press.

Chesbrough, H., & Rosenbloom, R. (2002). The role of the business model in capturing value from innovation: Evidence from xerox corporation's technology spinoff companies. *Industrial and Corporate Change,* 11(3), 529–555.

Christensen, C.M. (1997). *The Innovator's Dilemma.* Boston, MA: Harvard Business School Press.

Christensen, C.M., & Raynor, M.E. (2003). *The Innovator's Solution.* Boston, MA: Harvard Business School Press.

Cohen, M.D., March, J.G., & Olsen, J.P. (1972). A garbage can model of organizational choice. *Administrative Science Quarterly,* 17(1), 1.

Cohen, W., & Levinthal, D. (1990). Absorptive capacity: A new perspective on learning and innovation. *Administrative Science Quarterly,* 35, 128–152.

Collins, J. (2001). *Good to Great.* New York, NY: HarperCollins Publishers.

Council on Competitiveness. (2005). *Innovate America: Thriving in a World of Challenge and Change.* Washington, DC: Council on Competitiveness.

de Jong, J., & Marsili, O. (2006). The fruit flies of innovations: A taxonomy of innovative small firms. *Research Policy,* 35, 213–229.

Defense Science Board. (2007). *2006 Summer Study on the 21st Century Strategic Technology Vectors: Volume IV Accelerating the Transition of Technologies into US Capabilities.* Washington, DC: Defense Science Board.

Deloitte Touche Tohmatsu. (2003). Fostering and innovative culture: Sustaining competitive advantage. *Growth: The Executive Series for Dynamic Companies,* 10(1), 7–24.

Drucker, P. (2002). The discipline of innovation. *Harvard Business Review,* 80(8), 95–102.

Eisenhardt, K., & Martin, J. (2000). Dynamic capabilities: What are they? *Strategic Management Journal,* 21(10/11), 1105–1121.

Foster, R. (1986). *Innovation: The Attacker's Advantage.* New York: Simon & Schuster.

Francis, D., & Bessant, J. (2005). Targeting innovation and implications for capability development. *Technovation,* 25, 171–183.

Glover, C., & Smethurst, S. (2003). Creative license. *People Management,* 9(6), 1.

Greis, N., Dibner, M., & Bean, A. 1995. External partnering as a response to innovation barriers and global competition in biotechnology. *Research Policy,* 24, 609–630.

Hamel, G. (2002). *Leading the Revolution.* Boston, MA: Harvard Business School Press.

Hamel, G., & Valikangas, L. (2003). The quest for resilience. *Harvard Business Review,* 81(9), 52.

Hammer, M. (1996). *Beyond Reengineering: How the Process-centered Organization is Changing Our Work and our Lives.* New York: Harper Business, A Division of HarperCollins Publishers.

Hammer, M., & Champy, J. (2001). *Reengineering the Corporation; A Manifesto for Business Revolution.* New York: Harper Business, A Division of HarperCollins Publishers.

Hargadon, A. (2003). *How Breakthroughs Happen: The Surprising Truth About How Companies Innovate.* Boston, MA: Harvard Business School Press.

Hayes, R.H., & Abernathy, W.J. (1980). Managing our way to economic decline. *Harvard Business Review,* 58(4), 67.

IBM Global Business Services. (2006). *Expanding the Innovation Horizon: The Global CEO Study 2006.* Somers, NY: IBM Corporation.

Kazama, S., Foster, J., & Hebl, M. (2002). Impacting culture for innovation: Can CEOs make a difference? *17th Annual Conference if the Society for Industrial and Organizational Psychology,* Toronto, Canada.

Kelley, T., & Littman, J. (2001). *The Art of Innovation.* New York: Random House.

Lawson, B., & Samson, D. (2001). Developing innovation capability in organizations: A dynamic capabilities approach. *International Journal of Innovation Management,* 5(3), 377–400.

Lee, H., & Kelley, D. (2008). Building dynamic capabilities for innovation: An exploratory study of key management practices. *R&D Management,* 38(2), 155–168.

Leonard-Barton, D. (1992). Core capabilities and core rigidities: A paradox in managing new product development. *Strategic Management Journal,* 13, 111–125.

Liao, J., Kickul, J., & Ma, H. (2009). Organizational dynamic capability and innovation: An empirical examination of internet firms. *Journal of Small Business Management,* 47(3), 263–286.

Matthews, J., & Shulman, A. (2005). Competitive advantage in public-sector organizations: Explaining the public good/sustainable competitive advantage paradox. *Journal of Business Research,* 58, 232–240.

McGregor, J. (2007). The world's most innovative companies. *BusinessWeek,* Retrieved 7/25/07 from http://www.businessweek.com/innovate/ content/may2007/ id20070504_051674. htm?chan=innovation_special+report+---+2007+most+ innovative+companies_2007+most+innovative+companies

Morison, E.E. (1966). *Men, Machines, and Modern Times.* Cambridge, MA: The MIT Press.

O'Reilly, C., & Tushman, M. (2007). Ambidexterity as a Dynamic Capability: Resolving the Innovator's Dilemma. Working paper 07-088. Cambridge, MA: Harvard Business School.

Pablo, A., Reay, T., Dewald, J., & Casebeer, A. (2007). Identifying, enabling, and managing dynamic capabilities in the public sector. *Journal of Management Studies,* 44(5), 687–708.

Peters, T. (1997). *The Circle of Innovation: You Can't Shrink Your Way to Greatness.* New York: Vintage Books: A Division of Random House.

Porter, M.E. (1985). *The Competitive Advantage: Creating and Sustaining Superior Performance.* New York: Free Press.

Prahalad, C., & Hamel, G. (1990). The core competencies of the corporation. *Harvard Business Review,* 68(3), 79–91.

Robbins, S. P., & Judge, T. A. (2007). *Organizational Behavior.* Upper Saddle River, NJ: Pearson/Prentice Hall.

Rumelt, R. (1984). Towards a strategic theory of the firm. in Lamb, R.B. (Ed.) *Competitive Strategic Management,* Englewood Cliffs, NJ: Prentice Hall, pp. 557–570.

Teece, D. (2006). Reflections on "Profiting from Innovation." *Research Policy,* 35, 1131–1146.

Teece, D., Pisano, G., & Shuen, A. (1997). Dynamic capabilities and strategic management. *Strategic Management Journal,* 18(7), 509–533.

Tidd, J. (1993). Technological innovation, organizational linkages, and strategic degrees of freedom. *Technology Analysis & Strategic Management,* 5(3), 273–284.

Tidd, J., & Bessant, J. (2009). *Managing Innovation: Integrating Technological, Market, and Organizational Change.* Chichester, UK: John Wiley & Sons.

United States Scientific Advisory Board. (2006). Report on System Level Experimentation. No. SAB-TR-06-02. Headquarters, United States Air Force. Retrieved 7/25/07 from http://www. dtic.mil/dtic/tr/fulltext/u2/a463950.pdf.

von Hipple, E. (1988). *Sources of Innovation.* New York: Oxford University Press.

chapter eleven

Human monitoring systems for health, fitness and performance augmentation

Mark M. Derriso, Kimberly Bigelow, Christine Schubert Kabban, Ed Downs, and Amanda Delaney

Contents

Introduction

The Department of Defense (DoD), industry and academia have been investing in system monitoring technologies for man-made machines over the past several decades. In fact, the DoD established a policy in 2007 called Condition-Based Maintenance Plus (CBM+) for incorporating automatic monitoring technologies across its weapon systems [1]. The goal of CBM+ is to increase weapon systems availability and reliability while reducing downtime and operational cost. Many advances have been and are continuously being made for monitoring machines such as automobiles, locomotives and aircraft. However, the most valuable asset in the DoD arsenal is not being monitored at all during operations, the human machine.

The DoD toyed with the idea of utilizing human monitoring technologies in the past for health and performance monitoring but no official policies ever materialized. Recently, there has been an explosion in the development of wearable technologies in the commercial industry due to the advances in flexible electronics, wireless communications and miniaturized computing technologies. Wearable technologies are currently being used in different applications such as healthcare, fitness, athletic and military domains to some degree. In most cases, these technologies provide feedback of physiological and/or biomechanical parameters continuously in real time. Professionals can use these parameters to develop an appropriate intervention plan to prevent injuries and for achieving the desired outcomes.

Because of the investments and advancements industry is making in wearable technologies the DoD renewed its interest in human monitoring. This was evident in the 2012 article published in the Armed Forces Journal titled "The Quantified Warrior" [2]. In this article, Jack Blackhurst et al. discussed the motivation, need, and how DoD should lead in developing human performance monitoring and augmentation technologies. Furthermore, the article presented a foundational framework, "sense-assess-augment," for implementing human performance monitoring and augmentation technologies in military applications. Similarly, the athletic community also recognize the benefits of human monitoring technologies for improving performance. In February, 2017 Harvard University held a forum entitled "The Rise of the Quantified Athlete" hosted by the Sports Innovation Laboratory [3]. Participants included business executives, scientists, professional athletes, military personnel and entrepreneurs from across the country. The purpose of this gathering was to explore the future of sports technology with all stakeholders and discuss the advances and issues related to quantifying athletic performance. Although there have been significant advancements made in wearable technologies over the past few years, there are still some critical challenges that must be addressed before human performance monitoring technologies reaches its full potential.

Currently, most human monitoring systems are based on measuring physiological and/or biomechanical parameters during training or operations as a means to evaluate health, fitness and performance. These approaches are useful for assessing cardiorespiratory, musculoskeletal and cognitive systems' health and fitness but falls short of truly quantifying a person's ability/capability to perform operationally. However, if used appropriately wearable technologies could enable the ability to identify, measure and quantify key parameters needed to assess and predict operational performance. Such a capability would revolutionize healthcare, athletics and military domains by the improved analytical ability to recruit, train, rehabilitate and enhance personnel for optimal operational performance.

This chapter will first discuss the components of human monitoring systems. The chapter will then describe the state of the art in human performance sensing and modeling. Next we will discuss applications for human performance monitoring. Then we will conclude with some current research we are performing in this area.

Human monitoring systems

A typical human monitoring system is comprised of three major elements: (1) sensors, (2) data acquisition and communication, and (3) data processing and analytics. Sensors are used to measure desired parameters from the individual being evaluated. The most popular type of sensor systems used for continuously monitoring human parameters are wearables. Wearables are non-invasive, light-weight body worn devices that have the ability to sense, collect and store body measured data locally and/or transmit it to a remote location. Figure 11.1 depicts a diagram of a typical human monitoring system using wearables.

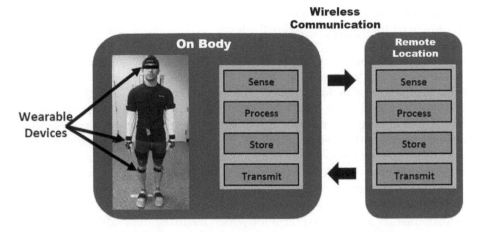

Figure 11.1 A typical human monitroing systems.

The type of sensing, data collection and analytical techniques used in a human monitoring system depends upon the application. Most human monitoring systems are employed as a means to assess a person's health, fitness and performance levels. The parameters used for performing these assessments are primarily from two categories, biomechanical and physiological. Biomechanical sensors are used to capture movement of specific parts of the human body while in motion. This data could then be used to estimate the body kinematics (i.e., motion) of the individual being assessed. Motion information are useful for evaluating a person's functional movements, body coordination and flexibility.

Physiological sensors are also used in human monitoring for measuring physiological parameters such as respiration, cardiopulmonary, electroencephalography (EEG), body temperature and stress. One of the most popular parameters that professionals use to assess their patients'/clients' health and fitness is heart rate. Heart rate measurements can be used to detect potential irregularity within the cardiorespiratory system. Additionally, heart rate could also serve as a means to measure an individual's recovery time by measuring their resting heart rate and heart rate after performing a physical task. There are different approaches currently being investigated for assessing a person's cognitive state. One common method involves measuring and monitoring electrical activity of the brain via EEG signals. By analyzing changes in EEG signals under known cognitive conditions, models are being developed for estimating a limited number of cognitive states.

Using wearable devices for collecting biomechanical and physiological parameters to assess a person's health and fitness conditions is easy to comprehend since physicians have been exploiting this type of data for years to diagnose patients. Conversely, using these same parameters to assess and estimate an individual's performance is not as conceivable. The next section will discuss the state of the art in human performance sensing.

Human performance sensing

The DoD introduced the term "human performance sensing" around 2014 as a way to communicate and capture research associated with their human monitoring and augmentation "sense-assess-augment" framework mentioned earlier. Although no formal definition exists for "human performance sensing" the overarching goal is to monitor a person's performance in real time via sensing technologies and if needed, apply the appropriate corrective action to obtain an acceptable level of performance. In order to realize this capability two critical questions must be answered: (1) What is performance? and (2) What needs to be sensed to measure it? Today the term "performance monitoring" is used liberally regardless of what is being sensed and measured. For example, applying wearables to an athletic team for monitoring heart rate is probably not the most effective parameter to measure for estimating game performance. Heart rate measurements could provide some insight into game performance if the reason that an individual's performance is below normal is due to a decline in health or fitness. In this scenario, the term "performance monitoring" would be appropriate because health/fitness is directly correlated with game performance; however, that is not always the case. In fact, an athlete could be completely healthy and fit but perform subpar during a game because of other factors that were neither considered nor measured. In this situation, heart rate would not be a suitable predictor of game performance since there was no direct correlation between heart rate and game performance.

"Performance" is an overused term for which meaning depends on the situation and objective of the task being performed. Therefore each task could require different parameters and/or parameter values in order to execute a task successfully. For instance, a college program may accept a student based upon a grade point average (GPA) of 3.0 or higher. The same college program may only accept a GPA of 3.5 or higher into their honors program. The differences in the threshold values of this performance parameter (i.e., a GPA of 3.0 versus 3.5) is task-specific, determined by the work required for each program. Both the standard and honors programs established "GPA" as a performance indicator for the success of the student in their respective curriculums. The athletic and military domains alike have been searching for the factors needed to become elite athletes and combat warriors for many years. More recently, diverse groups of researchers from academia, government and industry have been collaborating in search of this elusive answer. Trainers, instructors and coaches continue to express the need for improved methods to measure and analyze an athlete's performance. That is, a methodology to truly sense, measure and quantify improvements in factors that have a direct relationship to performance. Since wearable devices are incapable of sensing performance directly, human performance models must be utilized for providing meaning to the sensed data.

Human performance modeling

Of the three major elements of the human monitoring system, human performance modeling is directly concerned with data processing and analytics. However, the usefulness of a proposed model in assessing and predicting performance is directly influenced by all three major elements and begins with the two critical questions stated earlier. As a general analytical framework for human performance or any modeling, the first step is to define the research questions and the outcome of interest. Performance, as an outcome, must be defined for the specific application and task. For example, consider developing a model that uses body mass index (BMI) to predict performance in an endurance-related task. Although modeling can demonstrate an association or correlation of BMI to success in an endurance task, this correlation may be more indicative of confounded effects. Since it is known that those with high cardiovascular health and fitness usually have low BMI, directly measuring a parameter related to health/fitness might be more appropriate. Therefore, a formal definition of "what is performance" and parameters or features that directly quantifies it are paramount to sense and eventually build performance models.

Once a definition of performance is obtained and parameters established, the next critical question is how to sense these features of performance. Typically, performance is not measured as part of the human monitoring system. Instead, features are used via the human monitoring system captured by biomechanical and physiological sensors. In addition, other human features and characteristics are exploited in order to predict performance. Performance outcomes are often more tangible such as mission success (e.g., scoring a touch-down). What's critical here, is that human monitoring systems are used to sense phenomenon that can predict performance such as mission success; however, performance is not known until either the task or mission is completed. In order to build models to predict performance, data must be observed, collected and analyzed using a known performance outcome. Without the true, gold-standard, performance outcome, it is likely that no models using any human monitoring system data would be able to accurately predict performance.

Therefore, to build a model for performance, a known, true measure of performance is needed as the outcome of interest (i.e., the dependent variable) and candidate features or parameters from the human monitoring systems are needed as predictor variables (i.e., the independent variables). These predictor variables result from the data acquisition and processing of the phenomenon sensed by the human monitoring system, and are often also referred to as "the data." Data could be the measurements observed from the variables to be used as predictors in an eventual model or that which the sensors collect and then process into features (i.e., variables). Similarly, careful thought to what is required for an outcome should be employed. Consideration should be given to the following: (1) What are the important, human-based features that are related to the application and task-specific performance? (2) How might these features be measured, or sensed? and (3) How do these features relate to performance?

Determining which features would be important to include as potential predictor variables requires planning and often collaboration across multi-disciplined experts well-versed in the application and task. Frequently, experts contain, either explicitly or intrinsically, theoretical knowledge of what features relate to the performance outcome of interest. Heart rate may not be indicative of being able to score in a basketball game; therefore, to improve performance an expert trainer may not work to monitor and improve heart rate in a particular athlete as much as he/she would work to improve short-term speed or mobility. On the other hand, the trainer may use heart rate to monitor and improve the performance of a long-distance track athlete. As such, these experts contain theoretical knowledge as to what phenomenon relates to performance, and therefore, what potential features could be used to capture and measure those phenomena of interest. In a multi-disciplined collaboration, engineers can take the information from the experts to design, augment, or derive sensors to collect the data that the application experts believe are related to the performance outcome. Although many biomechanical and physiological sensors exist, it does not make sense to use a physiological sensor, either because it is easy, cheap or "the latest state of the art," if physiological features are theoretically not directly linked to the performance outcome. As in the example earlier, for the athlete needing to improve his/her scoring ability in a basketball game, a biomechanical sensor from which direction and speed can be ascertained may be more important than a physiological sensor that measures heart rate and blood pressure.

In fact, despite current social trends encouraging the collection of data continuously, from the fitness-based watches to cell phone apps, it is instead advisable to not include measures that are theoretically unfounded. In order to model a particular phenomenon such as performance, the investigative team should take time and care to understand their definition of performance and what has been theoretically, either via literature or expert opinion, linked to that performance. Collecting data using high- or low-technical sensors that measure features not related to the outcome induces unnecessary cost in the processing and storage of data as well as burden to the human subject. Using features that are expected to be related to the performance outcome should make a very informed, and highly predictive model.

Once the appropriate features to be used as the potential predictor variables are determined, it becomes important to hypothesize a model that, through the use of data collection, can be fit, augmented, and analysed to determine how well performance can be predicted. There are many common approaches to human performance modeling, some of which are data-driven techniques whereas others may be physics-based, or a combination of the two. However, the vast majority of these models are rooted in standard, statistical and data analytic methods used for applications in which prediction of an outcome is of primary interest.

Data-driven models

In a data-driven model, data is used to understand the relationships between the potential predictor variables and the performance outcome. Although such models are driven by data, it is important to realize that the data, and therefore, the variables considered should be used because these represent the correct, or theorized, variables related to the performance outcome. Variables and data should not be included for reasons such as (1) the sensor computes these additional variables anyway, (2) justification for excessive or previous data collection efforts on the same subjects, and (3) the effort to include additional sensors that are currently assessable is minimal. Instead, data and variables collected should be hypothesized in a relationship to the performance outcome. The hypothesized relationship defines the form of the model to be fit to the data.

Common data-driven models include statistical linear and non-linear regression techniques in which hypothesized relations between the predictor variables and the outcome are explicitly expressed. Machine learning techniques such as neural networks in which relations are often implicit, and non-parametric modeling techniques such as classification trees and random forests. In all of these techniques, an association is established between the predictor variables and the performance outcome of interest. Ideally, this association is considered strong and can be measured through common statistics such as (1) the R-squared value in a regression analysis which measures the proportion of variation in the performance measure that can be explained by the combination of predictor variables used, (2) the Mean Squared Error (MSE) which measures the squared deviations of the performance measure and the model predicted performance measure, or (3) classification rate for a dichotomous performance measure (e.g., such as "touch-down" or "no touchdown") that provides the percent of correctly predicted outcomes. However, as is true for all data-driven models, the outcome of interest is virtually always predicted with some error. That is, the collection of variables used to predict the outcome are not mathematically able to express and predict the outcome without some error. In very good models with carefully chosen variables and hypothesized relationships, this error can be quite small and the relationships between the predictors and the outcome can be quite strong.

It is important to note that when a strong association is established and a model predicts well, most of these data-driven models provide only this associative relationship and not a causal relationship. As an associative relationship, it is recognized that there is a relationship between the predictor variables and the performance outcome (i.e., dependent variable). It does not necessarily imply that the performance outcome is caused by the various levels of the predictor variables. In fact, when predictor variables are not chosen with extreme care, they can act as surrogates and mask the variables that contain the true relationships to the performance outcome. A lower game day temperature may be related to a lower total score in the football game; however, it is not the game day temperature that causes the low score but the lack of thermal-dynamic compensation and muscle inelasticity of the offensive players. Similarly, including variables from data that is not previously theorized to predict the performance outcome may induce spurious relationships in mathematical models. Especially if models are over-fit, they contain more variables for which relations had to be estimated than the number of subjects could mathematically allow. When such items are of concern, standard modeling techniques such as regression, contain methods to assure that the fitted model has adequate power for estimation given the sample size considered. In such a case, walking through the mathematical exercise of computing power for the proposed model is warranted and there are many software packages, online calculators and expert statisticians who can accomplish such calculations.

When using data-driven models, it is important to know what your predictor variables can and cannot tell you about performance. Well-chosen variables should have reasonable relationships to the performance outcome. Such relations are easy to see when using techniques such as linear or non-linear regression models as the model parameters establish the relationships between the predictor variable and the outcome. Other modeling techniques, such as neural networks, require a range of inputs to establish these relationships. However, as a data-driven method, the extent of the relationships are determined by the data, yet from the resulting model, it is possible to discover, quantify and test hypothesized relationships.

Physics-based models

In a physics-based model, relationships between predictor variables and the outcome are known and can be expressed with mathematical precision. $E = mc^2$ is an example of a physics-based model in which equivalent energy (E) can be expressed with mathematical precision as a function of the mass of an object (m) times the square of the speed of light in a vacuum (c^2). Unfortunately, performance as an outcome may not be so easily expressed. Yet, it is possible, as a result of data-driven models and theoretical findings, to be able to begin expressing performance in a model as a hybrid of physics-based and possibly data-driven variables. The advantage of striving for physics-based modeling is in the precision with which performance can be predicted. In Einstein's famous equation above, energy is predicted without error as a function of mass and the speed of light. Although context is important, the ability to predict in a physics-based model removes uncertainty, replacing unexplained error through mathematical expressions proven to estimate a phenomenon. Balance can be assessed through a variety of tests such as the Balance Error Scoring System or Star Excursion Balance Test. Although the measures that result from these tests are not physics based and are recommended for specific applications and tasks, it is conceivable that in the near future, physics-based expressions of force, mass and motion can be combined to specifically measure the extent to which weight is evenly or not evenly distributed (i.e., the definition of balance).

Verification and validation

Whether a model of performance is constructed using data-driven techniques or is physics-based, one critical element of the process is an evaluation of the model. "Verification" and "validation" are often used as terms to ensure that a system, process, or in this case, a model, performs to internal and external standards. With respect to human system modeling, verification is used to assure that the model meets a particular specification. In most prediction models, this is often equated to a minimum level of unexplained error often measured through either MSE or the R-square statistic for continuous performance outcomes and correct classification rates for group-level outcomes such as "success" or "no success." During a model building process, a final model is usually not presented until verification is established, usually by meeting a pre-specified level of error (i.e., minimum MSE or error rate). Validation is used to assure that the model meets a particular, and often similar, level of prediction when applied to independent sample(s) from the population—this is considered the external standard. For well-trained data analysts and statisticians, validation is usually of primary concern, as meeting a similar level of prediction in an independent sample from the population provides assurance that the model may estimate consistently or suitably when applied to alternate data.

In human performance modeling, as in any other modeling task, models should both be verified and validated. Typically, verification has occurred when an appropriate model, one that perhaps minimizes MSE to a particular standard, is developed. However, just because an appropriate model is found, there is no guarantee that it will predict just as well in an independent sample. In fact, models may produce larger error when applied to an independent sample for many reasons such as (1) the selection of predictor variables missed an important feature that was not as apparent in the model building data, (2) time, alternate settings or subpopulations invoke different behavioror, (3) the model was mathematically overly parameterized, fitting too closely to the data used to build it and not adequately modeling the expected or natural variation among subjects. For example, the data used to build the model may have been collected from an elite pool of professional athletes, whose variability with respect to, say, reaction time is not as large as that found in elite college athletes or the general population. Therefore, an aspect of performance that is captured by reaction time could not be incorporated into the model. Demographic and personal characteristics may also come into play. If the sample of subjects who were measured and whose data was used to generate the human performance model is not indicative of the population for which this model was constructed to represent, the model will not predict well when applied to the population at large. This is an error that validation methods seek to prevent.

There are several common ways to validate a model through either comparison to expected theoretical results, collection of new data or hold-out of existing data [4]. Resulting model coefficients and level of error may be compared to that expected from theory or expert opinion. This may include examining the size and direction of model coefficients for reasonable and expected values. Most commonly, a second collection of data, or a hold-out of data from the model building, is used for comparison. In these methods, the model is built upon the original data collection or data that was not withheld. Then, the second data set or hold-out data is used to check that the model has a similar level of prediction in addition to fitting a model with the same set of predictor variables to determine if the coefficients in the model are similar. When possible, a second collection of data is encouraged. As an independent sample gathered separately from the original collection it behaves as a pure estimate of whether or not the model will still be applicable to data beyond that which was used to generate the model. If feasible, during the study design phase, the second data collection should be included in budget, time and analysis development.

With proper planning, the second data collection is quite achievable. Unfortunately, cost, lack of foresight, and other constraints often hinder the ability to perform a second data collection. Therefore, the hold-out method may be used to mimic the findings of a validation. It is imperative when holding out data, that the data is divided carefully into a hold-out data set and a model building data set prior to beginning any model building. The division of data may occur using random number generators to produce two data sets of randomly divided subjects or, depending on the application, a matching scheme may be more appropriate. In either case, such action seeks to maintain the independent nature of the hold-out data and retain similarity between the subjects in each data set. Limitations of this method include an inability to determine if timing of the data collection has an effect on model prediction since the model building data and hold-out data were collected from the same cohort and the need for data splitting which reduces the number of samples available for model building. In general, 6 to 10 observations per predictor variable should be retained in any model building data set for linear regression models. Sample size requirements for neural networks, classification trees, and non-linear regression models vary by the number of weights and/or components and are specific to the modeling method chosen.

In order for human monitoring systems to successfully transition to commercial and military applications, the reliability of the system performance must be confirmed using some of the techniques discussed earlier. The next section will discuss potential applications of human monitoring systems in the fields of healthcare, rehabilitation, fitness and athletics.

Human monitoring applications

Healthcare and rehabilitation

In a nation that is seeing a surge in aging Americans and an increase in disability, the need for advanced healthcare and rehabilitation technologies only becomes more critical. This is coupled with more stringent requirements relative to documenting and demonstrating needs and progress for purposes of insurance reimbursement, and often shorter than desired length of financially covered care. This has led clinicians and researchers to already begin envisioning innovative approaches that include a focus on wellness and activity, preventative care, regular monitoring and screening, telemedicine and at-home or group rehabilitation. Sensor development has begun to integrate into these areas, but the potential for further enhancements in the sensor monitoring industry are endless.

Wellness has become a major focus in healthcare as of late. If people can stay well, their own health outcomes are better, the healthcare system is not overburdened and there is a significant financial benefit for all. Often a key component of this focus is keeping individuals fit and active. The emergence of commercially available physical activity monitors, such as the popular Fitbit, has been one step in this direction. It is estimated that over 23 million individuals actively use a Fitbit [5]. Research has found that individuals who use pedometers and other activity monitors are more physically active than those who do not [6]. Additionally, individuals who began using a pedometer increased their physical activity on average 26.9 percent over baseline and took 2,491 steps per day compared to controls [6]. Because of this impact, and the positive health effects associated with the increase in physical activity, there has been increased focus on leveraging this technology especially in those who could benefit; for example, older adults with osteoarthritis of the knee, those with type 2 diabetes and sedentary workers.

Considering this, one key takeaway that emerges and is seen filtering into other health applications is that individuals seem to become more motivated when they receive real-time feedback on their *performance*. In the particular example of the pedometer or physical activity monitor, this is fairly simplistic: a device that is worn records the number of steps taken and individuals walk more. This is either because they know it is recording how many steps they take, or because they regularly look at that number of steps taken and make corrective action. In either case, as they walk more, improved health outcomes may be seen (e.g., decrease in weight, lower blood pressure). The commercial devices that have emerged provide additional platforms to try to encourage more activity through apps that track and display trends over time, provide motivation and virtual rewards and allow individuals to compete against friends or use social media to share accomplishments. People can and have begun to recognize and dream up other means of using such technology within the health and rehabilitation field.

However, it should also be recognized how limited this technology is in its current state. Similar to the issues outlined earlier, while the technology is emerging there is so much more that needs to be incorporated and learned before this technology can be fully realized. For example, in the application of the pedometer or activity monitor, only number of steps is counted. There is nothing to be said about the quality of those steps. There is nothing to be said about the quality of the rest of the body's movement. There is nothing indicating whether that number of steps at that quality is affecting the body in some direct way. There is nothing to indicate a prediction as to whether that number of steps taken will lead to reduced disability, likelihood of being taken off medication or other tangible outcomes. There is nothing to indicate how this relates to better "performance," the intangible word that might mean improved ability for community ambulation and participation, ease of getting around or more. There is nothing to indicate whether the steps taken indicate some underlying disease state or physical concern that should be addressed. So while a particular technology may indicate progress and be beneficial in doing its intended function of counting steps, these non-trivial questions would need to be addressed in order to keep innovating and paving critical ground in this area.

While the University of Dayton (UD) Wellness through Biomechanics Laboratory has not yet sought to answer these questions and advance this work in the area of wellness and physical activity, we have sought to address the related areas of preventative care, regular monitoring and screening. In particular, we have focused on the prevention of falls in older adults. It is estimated that more than 1 in 4 adults aged 65 and older fall annually [7]. These falls are associated with serious physical, emotional and financial repercussions. Falls are the leading cause of unintentional injury-related deaths in this age group, and even for those who sustain non-injurious falls, the fear of falling again often becomes so overwhelming that individuals self-restrict activity and community participation. The costs associated with falls, to Medicare alone, totaled over $31 billion in 2015 [8]. Therefore, preventing individuals from falling in the first place is a critical need. To do this, individuals at risk must first be identified, and once identified there are a number of evidence-based practices and clinical recommendations that can be implemented in an attempt to lower fall risk.

The complication of doing this is the multi-faceted nature of fall risk. A comprehensive review of the literature identified the top risk factors for falls as including: muscle weakness, history of falls, gait deficits, balance deficits, use of an assistive device, visual deficits, arthritis, impaired ability to complete Activities of Daily Living, depression, cognitive impairment, and being over the age of 80. These are risks that even relatively healthy older adults may possess and yet not be aware of their increased risk of falls. Therefore, regular screening is critical. Current screening, however, tends to be quite limited in its sensitivity. The Berg Balance Scale (BBS) is one often used tool, where individuals complete a number of balance-related tasks and are scored by a trainer-observer on a 0–4 rating scale. The test, however, has a ceiling effect for higher performing individuals and takes approximately

15 minutes to complete, as well as significant space (15 ft) which precludes it from widespread use during regular clinical exams. The Timed Up and Go (TUG) is another commonly used assessment. Individuals stand from a chair, walk a distance, turn, walk back to the chair and sit down while being timed. A stopwatch is used to measure the total time duration. Normative databases have been established to allow the recorded time to be compared to the normative data to establish likely fall risk. This test too faces ceiling effects, as well as space constraints.

As wearable sensors emerged, the clinical and research communities recognized an opportunity: inertial measurement units (IMUs)—small wearable sensors—could be placed on an individual during the completion of the TUG. This quickly emerged as a new commercial product—the instrumented Timed Up and Go (iTUG). While the task is essentially the same, the presence of the sensors has enabled clinicians and researchers to move from making clinical decisions based solely on total time duration to now considering objective data obtained from each aspect of the task—the trunk angle upon standing, the speed of the turn, the time it takes to go from standing to sitting. Clinically each of these outcomes is important—if an individual takes more time to go from a sitting-to-standing position, but is able to complete the walk within normative ranges, perhaps quadriceps muscle strength needs to be improved and sit-to-stand transition practiced. In this way, the instrumentation of the TUG can pinpoint specific areas of challenge for an individual and identify individualized needs that physical therapy should work to address. How much more informative! In addition, because the assessment is now more sensitive, groups exhibiting subtle differences that could not be differentiated by the original TUG have demonstrated significant differences identifiable during iTUG performance [9]. This is especially important as questions have arisen as to the effectiveness of the TUG, with conflicting results of whether differences between fallers and non-fallers are actually detected. From this example we see these two important benefits and opportunities that the sensors allow: (1) the identification of individualized deficits that can be addressed through physical therapy and (2) better sensitivity in detecting subtle underlying differences (that may be a result of pathology). This takes a screening assessment and makes it something more, though it is recognized that this may not always be appropriate. It also provides better documentation, especially as the iTUG runs through commercially available software. While the iTUG is one example, the "instrumentation" made possible by wearable sensors opens up possibilities for all of the other many screenings and assessments that are used across the clinical/rehabilitation, athletics and occupational fields.

As another example within this same area, the UD have focused extensively on measuring and monitoring balance as an indication of fall risk and for noting disease progression, changes due to interventions and therapies and other reasons. Because balance is dependent on contributions of the visual, vestibular and proprioceptive systems, assessing balance provides an overview of the individual's function, with deficits in any of these individual systems often being able to be identified by doing the testing under manipulated sensory conditions. To measure balance, we utilize a force platform and ask individuals to stand as still and as quietly as they can while standing on the plate. The sensors within the plate capture data indicative of how much the individual is swaying. Exhibiting larger amounts of sway while trying to stand as still as possible is undesirable and is often a marker of an underlying problem and fall risk. These measurements are made under four different testing scenarios: (1) eyes open on a flat plate (all sensory systems contributing), (2) eyes closed on a flat plate (visual input removed), (3) eyes open on a foam pad (proprioceptive feedback reduced), and (4) eyes closed on a foam pad (primarily an outcome of the vestibular system's function). While this technology has been standard in balance research in academia for decades, it has yet to make the transition to everyday clinical practice. To address this, in collaboration with Dr. Necip Berme of the Ohio State University, the UD envisioned that if a force platform could be embedded into a traditional clinical-style scale, balance could be measured during a doctor's appointment at the same time weight was taken. However, as the reoccurring problem with sensors—just because we know an individual's balance, this did not on its own indicate anything about the outcomes that was most important (i.e., fall risk, fall history, etc.). As such, we conducted a study to collect balance data from our 150 older adults. A statistical analysis of this data enabled us to develop a prediction model that indicated one's likelihood of being a faller. In developing the model, this helped identify which of the many outcomes were the best, yet non-redundant, indicators of fall risk. At the end of the effort we had developed a method where within 30 seconds, in an everyday clinical environment, individuals could have their probability of being a faller determined and documented for comparison visit-to-visit. While the application has not yet caught on, there has been interest in it, though there are also challenges especially due to the natural variation in the postural sway of older adults. It would be exciting to see what others can do within this area and to what other applications this type of approach could extend to.

Advancing these efforts is a growing interest in human movement variability. Whereas interest generally lies most strongly in quantifying the "amount" of the outcome of data (e.g., the amount of sway an individual exhibits, the size step that they take), there is new research emerging that the "structure" (or pattern) of the outcome data might be even more meaningful. For example, an individual may exhibit a pattern of movement that demonstrates a highly repeatable pattern. In the past, it was thought that this was very desirable (e.g., always taking the exact same size step). Now it has been found that this is generally not advantageous because it does not allow for the adaptability necessary to respond to unexpected external perturbations. On the other hand, exhibiting too much variability is also undesirable and has been described as a "drunken walk" where getting to the end point is questionable and utilizes excess energy. Therefore, the *Optimal Movement Variability* perspective suggests that individuals should exhibit a level of variability somewhere between these extremes; furthermore, this exact level is likely task specific and influenced by age, health and other factors [10]. Researchers who have quantified the level of variability through Detrended Fluctuation, Sample Entropy, Lyapunov Exponents or other analysis methods have identified between-group differences that traditional measures had not revealed, while also providing additional insight that has been informative for physical therapists [10,11]. Research has also shown, for example, that while an individual's electrocardiogram (EKG) may not appear differently through traditional observation, an examination of the underlying patterns demonstrates life-threatening changes [12]. Similar phenomenon has been identified across other physiological measures, suggesting other critical changes in health state [13]. All of this additional insight that is gained and usable in any number of ways starts with wearable sensor monitoring. As development in both wearable sensor monitoring and the way in which the data is analyzed and considered continues, innovative approaches to healthcare will only continue to soar.

These advances in clinical practice are mirrored by the rehabilitation field. Physical therapists, occupational therapists and other clinical partners continue to strive to be innovative in how they can maximize their time and impact in the face of growing patient workloads and reduced face-to-face time as reimbursable. A wide range of approaches already exist, some using wearable monitoring. Group classes have emerged such as: the group kickboxing class modified for individuals with multiple sclerosis that we developed and evaluated, at-home therapy programs, as well as telemedicine/teletherapy options especially useful for individuals in rural communities or others who have a difficult time routinely accessing clinical care. It is envisioned, and researchers have begun to explore, the use of instrumented casts or braces that include sensors that monitor weight-bearing and alert the patient and/or clinician if thresholds have been exceeded. Similarly sensors have been proposed as a way to study whether an individual is completing their at-home therapy exercises, perhaps even giving real-time feedback based on the quality of those movements and/or deviation from the desired movement patterns that should be being practiced.

The possibilities are endless. Applications that are sensor-based are emerging rapidly, often with each iteration attempting to do more and become more sophisticated in their abilities. While known challenges exist, we see consistent areas of opportunity and benefit. And while all of these have the ability to affect any individual, we can also envision all of these catered specifically for the purposes of military medicine and rehabilitation. The further development, testing and application of sensors can aid in the preventative care, diagnosis and treatment of service men and women. Applications of human monitoring as it relates to fitness and athletics will be discussed next.

Fitness and athletics

Fitness is a key attribute of active duty DoD personnel. Each military service (i.e., Army, Navy, Marines, Coast Guard and Air Force) creates their own fitness program to ensure their respective service members stay within the desired physical condition. Physical tests are performed at least once a year to assess the fitness level of each military personnel. For example, the Air Force's fitness program evaluates fitness based on cardiorespiratory endurance, muscle fitness and body composition. Cardiorespiratory endurance is assessed by measuring the time it takes an Airman to complete a one and a-half mile run. Furthermore, muscle fitness tests are evaluated by counting the number of push-ups and sit-ups/crunches an Airman can accomplish in a one minute period per exercise. Lastly, body composition is assessed by measuring the abdominal circumference of an Airman, and comparing that measurement to a standard body composition point chart. A quantified fitness score is computed for every Airman based on their age and outcome of each test. These scores range from 0 to 100 and are assigned a fitness rating in the following matter (0–70 poor; 70–74.9 marginal; 75–89.9 good; 90–100 excellent) [14]. For Airmen who receives a rating of "good" or above are retested only annually. However, for those who score in the "marginal" or below range are retested every three months. In addition, those in the marginal category are required to attend a healthy living workshop, while those in the poor category are required to attend the workshop and participate in a fitness improvement program. The DoD has successfully established quantitative assessment methods for evaluating general health and fitness for its military population. However, techniques for measuring and assessing performance of specific military career fields (e.g., Special Operations Forces) which require advanced physical and mental abilities have yet to be developed.

Elite athletes and combat warriors possess very similar attributes. They are capable of combining cognitive and physical abilities to enable effective decisions and task execution in uncertain environments. Additionally, they are strategic and adaptive, which empowers them to perform beyond typical human expectations, especially in dynamic situations. These advanced levels of "open skills" are what separate the elite performers from average athletes/warriors. Open skills can be defined as "a series of movement patterns performed in an unpredictable environment such that the individual must adapt movements in response to changes within the environment" [15]. For many years, professional coaches and trainers have been measuring performance factors such as speed, agility and vertical jump with devices such as radar guns, stop watches and vertecs, respectively. These devices are used to measure different aspects of an athlete's physical abilities; for example, how fast a football player runs the 40 yard sprint, how quick a basketball player change directions and how high a volleyball player jumps. These types of information are still being used to help coaches recruit players. However, using this information alone to predict an individual's game performance has proven to be unreliable on numerous occasions. An athlete may have a faster than average 40 yard sprint time (e.g., 4.4 seconds) but then may decide to run in the wrong direction during a game. A volleyball player may have a 38 inch vertical jump but may jump at the wrong time. Therefore, if all of these performance factors are not used properly and at the time needed the mission has a greater chance of failing.

Pro-Trainer Ed Downs is a highly sought after personal trainer in the professional sports arena that is amongst fitness experts in search of reliable performance factors. From his 25 years of experience working with elite athletes and combat warriors he created the ProTERF methodology as a means to improve transferability of training to game performance. ProTERF stands for Professional Training Endurance Response-time and Functionality. The theoretical basis for ProTERF training was built upon the American Council on Exercise Fitness (ACE) Integrated Fitness Training (IFT) model (See Figure 11.2) [16]. The IFT model is a training framework consisting of four progressional phases: (1) Stability and Mobility, (2) Movement Training, (3) Load Training and (4) Performance Training. ProTERF extends the IFT model by integrating cognitive training throughout the phases and adding two additional phases: (5) Multi-Angular Movement, (6) Multi-directional/Multi-planar Movement. The uniqueness in the ProTERF methodology is how each phase not only builds off the previous phase but also incorporates it into the next phase. Table 11.1 shows the six phases of ProTERF.

The most powerful factor when using ProTERF is the implementation of cognitive training. An athlete without the proper cognitive ability is not as successful as one with the appropriate cognitive skills in his/her sport or position. Cognitive flexibility is a critical component needed to successfully perform in dynamic and uncertain environments. Cognitive flexibility can be defined as "the ability to adapt behaviors in response to changes in the environment" [17]. Observing the environment, orienting and analyzing

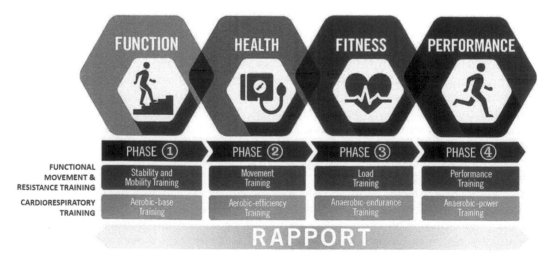

Figure 11.2 ACE integrated fitness training model.

Table 11.1 ProTERF 6 phases

Phase	Component		
	Physical	Cardiorespiratory	Cognitive
1	Stability, mobility, static balance	Aerobic base	Cognitive working memory
2	Static movement, dynamic balance, Intro resistance training	Aerobic efficiency	
3	Strength training, dynamic movement Dynamic coordination		Cognitive functional movement
4	Dynamic strength training, speed Explosive power	Anaerobic endurance	Cognitive patterned sequencing
5	Closed skills agility, multi-angular dynamic movement, response time	Anaerobic power	Cognitive dynamic patterned sequencing Cognitive random sequencing
6	Multi-planar dynamic movement Open skills agility		Cognitive decision making

the situation, making proper decisions and then acting on those decisions are all part of "open skills." When a decision is made to act, all the physical features must be accessible to execute the desired decision. One may need to slow down their speed, increase speed, control their balance or even use agility to evade a defender. Many factors must be integrated to successfully execute a single move or a sequence of movements (e.g., National Football League (NFL) running back having to evade multiple defenders). This methodology of training calls for a development of exercises that mimic actual movements the athlete would execute in a game. One of the challenges associated with this training approach is the lack of equipment/devices available to effectively train and measure particular functional movements. Wearable devices have the potential to fill this critical gap for the professional sports arena from rehabilitation to game-time operational performance.

For several years ProTERF has shown impressive performance outcomes for elite athletes and combat warriors. Both ACE and ProTERF believe that there are six primary factors that could be used for estimating an individual's performance ability; however, these factors have not been scientifically proven. These factors include: Speed, Balance, Coordination, Agility, Response-time, and Explosive Power. In order to validate these factors the following question needs to be addressed: What needs to be sensed to measure them? ProTERF teamed up with the Air Force Research Laboratory (AFRL), University of Dayton (UD) and the Air Force Institute of Technology (AFIT) to investigate this problem.

Current research

In order to examine this problem, a thorough study protocol was designed. Selecting the proper wearables for sensing became the first concern. To obtain both the biomechanical and physiological measurements, the team selected the Xsens 3D motion tracking system and the Zephyr bioharness band, respectively. The next step involved determining the cognitive and physical drills needed to test the six primary performance factors recommended by ACE/ProTERF as well as additional factors required to predict operational performance. Lastly, in an attempt to measure performance in both a closed and an open environment for comparison, two final drills were added.

The human subject study took place in the UD's Engineering Wellness through Biomechanics Laboratory. A total of twenty males between the ages 19 and 23 (mean age: 21.0 ± 1.1) were recruited to complete the study over two data collection sessions, each an hour and half in length. Recruitment methods included flyers and emails posted and sent around campus. Rational for the two testing sessions can be explained by attempts to increase recruitment and decrease fatigue. Furthermore, all subjects completed both testing sessions approximately one week apart. All individuals recruited were screened to ensure that their physical activity prior to the study included exercises more intense than walking for over 150 minutes per week. Additionally, participants were required to have participated in a sport at the club or collegiate level within the past four years. As a result, the participant population was as follows: 2 collegiate athletes (golf), 3 previous collegiate athletes (2 football, 1 cross country), 11 currently involved in intramural sports, and 4 currently involved in a club sport (3 soccer, 1 basketball). Consent as well as a personality survey was given to all participants for measuring conscientiousness and neuroticism. The study breakdown completed by all 20 participants can be seen in the following with a brief explanation of the method behind the chosen tasks.

Cognitive battery assessments

To independently measure the cognitive capabilities of each athlete recruited, three tests were administered on a tablet. For assessing one's spatial visualization, cognitive flexibility, and decision making skills the Manikin Test, Wisconsin Card Sorting Test, and the Rapid Decision Making Test were used, respectively [18]. These tests were selected because they assess the cognitive attributes we believe are essential for successfully performing in an open environment.

Traditional physical assessments

Three standard training exercises were tested to evaluate their effectiveness when training for operational performance: (1) 300 yard shuttle run, (2) Pro Agility 5-10-5, and (3) 30 second pushups. The 300 yard shuttle run was used to test anaerobic capacity, the 5-10-5 to test agility, and pushups to assess muscular endurance. We hypothesize that these factors are essential for assessing health and fitness but will not be as strongly correlated to performance. All of these tests were administered during day 1 of testing.

ProTERF assessments

With the emergence of performance training such as ProTERF, there has been minimal research completed to determine its efficacy to transfer trained skills to "open skills" needed for effective game performance. However, ProTERF training has shown to produce impressive performance enhancements, possibly due to the drills that mimic real game-like movements. Therefore, we took six drills corresponding to the six primary factors, to implement during testing. These drills consisted of: (1) Lock and Load, (2) First Step, (3) Landmine 30 second press, (4) Bosu Ball Squat with problem solving, (5) Medicine Ball Toss, and (6) Agility Random Lights. Most drills integrated at least two primary factors of performance as seen in Table 11.2.

Table 11.2 ProTERF drills

Exercise	Primary performance factors
Lock and load	Balance, coordination
First step	Explosive power, speed, coordination, balance
Landmine 30 seconds press	Explosive power, speed,
Bosu ball squat with problem solving	Balance, coordination, response time
Medicine ball toss	Explosive power
Agility random lights	Agility, response time

D4 closed and open

The setup seen in Figure 11.3 was used for the Agility Random Lights and both the D4 open and closed drills. D4 stands for "Dynamic, Detection, Diagnose and Decide." These drills were designed using a programmable lighting system (i.e., Fitlight) to simulate the dynamics and uncertainties present in actual open environments. For the closed sequence, the participant was given directions to run and hit equally spaced lights 1, 3, and 6, returning to the initial position after each touch. For the open sequence, all 6 lights were programmed to display 2 or 3 different colors that were turned on simultaneously. One of the lights always had a distinct color, where the rest had a matched pair. The participant was told all 6 lights would turn on and they must run to and hit the light that had no color pair, returning to the start after each light, as before. To add another factor into the "open skills" drill participants were asked to remember the color order of the lights they hit. None of the participants in this study were colorblind. Testing both static and dynamic sequences enabled a comparison to be made between training techniques for open and closed environments.

Variables

For each task that was performed over the two sessions, three trials of each test were administered with the appropriate recovery time between trials. Variables collected during

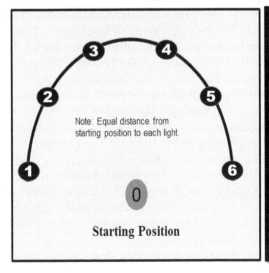

Figure 11.3 Diagram of agility random lights/D4 setup and participant performing D4 test.

Table 11.3 Qualitative and quantitative measurements

Variable	Test(s)
Reaction time	Manikin test, rapid decision making, and the Wisconsin Card Sorting test
Number correct	Manikin test, rapid decision making, the Wisconsin Card Sorting test, and the Dual Task done during Bosu Ball Squats
Time to complete	300 yd shuttle run, Pro Agility 5-10-5, Bosu Ball Squats, Agility Random Lights, D4 closed, and D4 open
Number of repetitions	Pushups and landmine
Distance	Medicine ball toss
Expert scoring of form	Lock and load, first step, Landmine, Bosu Ball Squats, and medicine ball toss

active study sessions consisted of both qualitative and quantitative results. A breakdown of which variables were recorded for each test can be seen in Table 11.3. These immediate response variables were used for further analysis. The complexity of the Xsens system and Zephyr band data called for evaluation of all scopes of the project to be set aside for now; however, full project effort will be resumed in the near future as extensions, in hopes of accurately finding means to a solution for accurate predictions of performance.

Results

Preliminary analysis of the immediate responses to build a predictive model is currently underway, but several interesting findings have been noted while reviewing the results. For the first trial of the D4 open skills test, 40 percent of the participants incorrectly recalled the color sequence. Of the 40 percent, it was noted that 2 of the 8 recorded lower scores, on average, for the rapid decision making test. Additionally, 1 participant displayed scores in the bottom half of the spatial awareness tests, and 1 displayed slow reaction times on his spatial awareness test, while the others appeared to score around average for all cognitive tests. By the third trial of the D4 open skills test, all subjects correctly recalled the color order and increased their time to completion. When looking at the D4 open versus closed tests, the average time to completion for the D4 open and closed trials differed by about 50 percent, with 12.79 versus 6.46 seconds for the open versus closed respectively. Formally, this is a statistically significant increase in average time of 6.32 seconds for the D4 open skill trial over that of the D4 closed skill drill (p-value < 0.0001) with a 95 percent confidence interval between 5.91 and 6.74 seconds. By design, the functional difference between the D4 open and closed trials was the inclusion of decision-based tasks. It appears that the inclusion of such cognitive needs during a physical task significantly increases the time to task completion. Speed alone cannot account for these differences. Correlations between the 300 yard shuttle run and the average completion time for the D4 open and closed skills were approximately 0.13 and 0.24, respectively, both demonstrating faster times in the 300 yard shuttle were related to faster times in the D4 test. However, in the open skill test, the correlation between the 300 yard shuttle time and the open skill test (0.13) was about half that for the closed skill test (0.24). Further, cognitive skills may be a critical link in the performance of an open skill task. How these physical and cognitive skills interrelate to performance, though, is still unknown.

Although we can study and relate individual measures of physical or cognitive ability to that of performance via models describing correlation between these factors, a model describing how all these factors co-relate to performance may provide better insight into which factors are most directly or indirectly related to performance and how performance may be modified through enhanced physical and cognitive skills. Moving forward, we are looking towards mathematical and statistical modeling techniques that can incorporate all these factors and model their relationships simultaneously. For instance, structural equation modeling is a statistical modeling technique in which the theoretical relationships between a number of variables may be tested simultaneously in a causal manner. That is, based upon a hypothesized model of performance, we may determine if measures of, say, anaerobic capacity (such as the 300 yard shuttle) are directly related to performance, indirectly related through other measures of endurance, or possibly moderated by cognitive ability. Looking specifically at First Step, the exercise incorporates explosive power, speed, coordination and balance. To measure explosive power, analyzing the maximum trunk acceleration of the individual is needed to determine the take-off force, but this is not something current wearables record as a direct response. Additionally, to calculate balance, analysis of the Center of Mass (CoM) in the anterior-posterior and medial-lateral direction is required to determine the overall resultant sway. When looking at values of CoM, smaller values correlate to better balance. Sensor data from the Xsens can be used for these calculations, but here is where integrating factors require models.

Therefore, this analytical method, structural equation modeling, allows estimation of both direct and indirect effects to investigate the processes underlying the relationships between the factors which are known as constructs [19]. In addition, effects and relations established in the structural equation model are casually related, that is, causal inference is now possible. The fitting of the structural equation model includes estimating the coefficients of the pathways between each of the constructs as well as computations and tests to modify the pathways. Model fit is assessed through fit indices [19–24].

Constructs, in structural equation modeling, may be composed of directly measured variables such as push-ups, or may represent an unmeasured (latent) construct such as "agility." In order, then, to estimate the effect of agility, we use physical assessments which are believed, via the literature and through expert consultation, to be related to each of these constructs. The measurements of these assessments (the variables) are combined using a statistical technique called factor analysis which determines which of the variables from these assessments are measuring the same (latent) construct such as agility. Formally, this is accomplished through a sequence of calculations starting initially by examining inter-task correlations to assure variables are truly measuring the same construct [20,25–26]. Then, exploratory factor analysis is used to examine the variables thought to represent the latent construct, such as agility. As an additional insight, inter-task correlations and factor analysis can also illuminate both variables that are and those that are not related to the constructs, allowing researchers to refine the sensing technology required for modeling. Finally, the structural equation model is fit and adjustments made as necessary to the variables retained. The results of such an analysis provide understanding into how all the constructs and factors relate to performance and how some constructs, such as cognitive ability, may moderate or mediate these relationships.

With the goal of modeling performance, structural equation modeling may provide the means to understand how fitness, health, cognitive ability and other pertinent related factors inter-relate to performance. Additionally, theorizing the constructs and factors that should be included in such a model allows researchers to identify the proper technology and tests to measure these variables as accurately as possible. With this current study, we aim to show that the use of the right technology for the measurement of the right (theoretically supported) variables in the right model can produce a way to not only predict, but understand how human factors merge and integrate into the performance of a specific task.

Conclusion

As discussed throughout this chapter monitoring systems for humans have broad and many uses for commercial and military applications. These applications include areas such as healthcare, rehabilitation, fitness and athletics. Significant investments are being made in wearable technologies and the popularity of using human monitoring devices on a daily bases is growing rapidly. One of the most common applications of wearable devices lies within the health and fitness realm. Many users are exploiting the step-tracking capability of devices such as Fitbits regularly to stay motivated and active; however, the potential benefits for human monitoring systems far exceeds basic activity monitoring.

Human monitoring technologies can provide greater insights into human performance across many domains. Today, the term "human performance" is used liberally with no concrete definition and therefore no standard way to analyze performance. Utilizing wearable technologies strategically could aid trainers, instructors, scientists and engineers in developing a theory and methodology for quantitatively assessing human performance. However, in order to realize the full capability of human performance monitoring, the two critical questions proposed earlier in this chapter need to be addressed: (1) What is performance? and (2) What needs to be sensed to measure it?

Addressing these two questions requires combined research amongst performance specialists, sensors developers, mathematicians and engineers for identifying the right performance factors, developing the right sensors, building the right models and computing the right performance assessments. Ideally when individuals are training, monitoring systems could provide professional trainers and/or users with identified areas of improvement needed in order to increase their performance readiness level. Furthermore, monitoring systems could also provide real-time feedback of individuals during games/operations to determine what adjustments/interventions are needed to maintain an acceptable level of performance.

References

1. Department of Defense. (2008). *Condition-Based Maintenance Plus DoD Guidebook*, DoDI 4151.22.
2. Blackhurst, J. L., Gresham, J. S., & Stone, M. O. (2012). The quantified warrior. How DoD should lead human performance augmentation. *Armed Forces J.*
3. NFLPA. (2017). The Rise of the Quantified Athlete. Retrieved from March 14 2017, https://i-lab.harvard.edu/event/quantified-athlete/ retrieved Mar 14, 2017.
4. Michael, H. K., Christopher, J., & Nachtsheim, J. N. (2004). *Applied Linear Regression Models*, 4th ed., McGraw-Hill/Irwin.
5. Smith, C. By the numbers: 37 Amazing Fitbit Statistics. *Digital Company Statistics*, Digital Marketing Ramblings, 28 July 2017, expandedramblings.com/index.php/fitbit-statistics/.
6. Bravata, D. M., Smith-Spangler, C., Sundaram, V., Gienger, A. L., Lin, N., Lewis, R., & Sirard, J. R. (2007). Using pedometers to increase physical activity and improve health: A systematic review. *JAMA*, 298(19), 2296–2304.
7. Stevens, J. A., Ballesteros, M. F., Mack, K. A., Rudd, R. A., DeCaro, E., & Adler, G. (2012). Gender differences in seeking care for falls in the aged Medicare Population. *Am J Prev Med* 43, 59–62.
8. Burns, E. B., Stevens, J. A., & Lee, R. L. (2016). The direct costs of fatal and non-fatal falls among older adults—United States. *J Safety Res* 58, 99–103.
9. Salarian, A., Horak, F. B., Zampieri, C., Carlson-Kuhta, P., Nutt, J. G., & Aminian, K. (2010). iTUG, a sensitive and reliable measure of mobility. *IEEE Transactions on Neural Systems and Rehabilitation Engineering*, 18(3), 303–310.

10. Stergiou, N., Harbourne, R. T., & Cavanaugh, J. T. (2006). Optimal movement variability: A new theoretical perspective for neurologic physical therapy. *J Neurol Phys Ther*, 30(3), 120–129.
11. Stergiou, N., & Decker, L. M. (2011). Human movement variability, nonlinear dynamics, and pathology: Is there a connection?. *Human Movement Science*, 30(5), 869–888.
12. Goldberger, A. L., Amaral, L. A., Hausdorff, J. M., Ivanov, P. C., Peng, C. K., & Stanley, H. E. (2002). Fractal dynamics in physiology: Alterations with disease and aging. *Proceedings of the National Academy of Sciences*, 99(suppl 1), 2466–2472.
13. Lipsitz, L. A. (2002). Dynamics of stability: The physiologic basis of functional health and frailty. *The Journals of Gerontology Series A: Biological Sciences and Medical Sciences*, 57(3), B115–B125.
14. Military.com, (2018), Air force fitness, Retrieved from February 10 2018, https://www.military.com/military-fitness/air-force-fitness retrieved Feb 10, 2018.
15. Medical Dictionary for the Health Professions and Nursing. *S.v. "open skill."* Retrieved from January 25 2018, https://medical-dictionary.thefreedictionary.com/open+skill retrieved Jan 25, 2018.
16. ACE, (2018), Integrated Fitness Training (IFT) Model, https://www.acefitness.org/certifiednewsarticle/684/ace-integrated-fitness-training-ift-model-for
17. Science Direct, Cognitive flexibility, Retrieved from January 3 2018, https://www.sciencedirect.com/topics/neuroscience/cognitive-flexibility.
18. O'Donnell, R. D, Moise, S., & Schmidt, R. M. (2005) Generating performance test batteries relevant to specific operational tasks. *Aviation, Space and Environmental Medicine*, 76, C24–C30.
19. Bollen, K. A. (1989). *Structural Equations with Latent Variables*. New York: John Wiley & Sons.
20. Hatcher, L. (1994). *A Step-by-step Approach to Using SAS for Factor Analysis and Structural Equation Modeling*. Cary, NC: SAS Institute.
21. Marsh, H. W., Balla, J. R., & McDonald, R. P. (1988). Goodness-of-Fit Indexes in confirmatory factor analysis: The effect of sample size. *Psychological Bulletin* 103, 391–410.
22. Bentler, P. M., & Bonnett, D. G. (1980). Significance tests and goodness-of-fit in the analysis of covariance structures. *Psychological Bulletin* 88, 588–606.
23. Bentler, P. M., & Chou, C. (1987). Practical issues in structural modeling. *Sociological Methods and Research* 16, 78–117.
24. Bentler, P. M. (1989). *EQS Structural Equations Program*. Los Angeles, CA: BMDP Statistical Software.
25. Churchill, G. A. (1979). A paradigm for developing better measures of marketing constructs. *Journal of Marketing Research* 16(1), 64–73.
26. DeVellis, R. F. (2010). *Scale Development: Theory and Applications* (Vol. 26). Newbury Park, CA: Sage.

chapter twelve

Enhancing innovation
Methods, cultural aspects, ideation approaches, and box busters

Daniel D. Jensen and Cory A. Cooper

Contents

Introduction

Innovation can be either "incremental" or "disruptive." "Incremental" innovation describes the small changes to an idea, process or technology that allow it to progress up the innovation S-curve toward maturity. "Disruptive" innovation describes the implementation of new ideas, processes or technologies that cause a jump to a new S-curve. In reality there

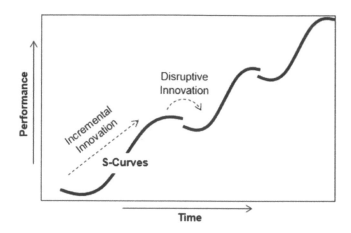

Figure 12.1 Innovation S-curves, incremental verses disruptive innovation. (From Foster, R. N., *Research Management*, 29(4), 17–20, 1986; Christensen, C. M., *The Innovator's Dilemma: When New Technologies Cause Great Firms to Fail*, Harvard Business Review Press, Cambridge, MA, 2013.)

is a continuum between incremental and disruptive innovation. However, in general, disruptive innovation occurs when developments allow us to meet new customer needs or provide new capabilities [1,2] (Figure 12.1).

The following three sections cover major ways to improve innovation in an organization and achieve innovative results. The descriptions and approaches are based on decades of study on the art and science of innovation and observations in top companies. The methods have been applied with award-winning results at the US Air Force Academy, other Department of Defense organizations, and top international companies.

In the following, a set of five methods that organizations use to strengthen their innovative ability are described. This is followed by a discussion of a number of aspects of an organization's culture that promote creativity and the resulting disruptive innovation. Finally, a set of concept generation (or ideation) methods which can be used to improve individual or group creativity and innovation are described.

Methods for increasing an organization's capability to innovate

Organizations may enhance their ability to innovate in these five ways:

1. Acquire innovative organizations
2. The "innovation guru" method (ex: Steve Jobs)
3. Skunk Works® model
4. Research and Development (R & D)
5. Integrate a culture that fosters innovation throughout the organization

Each of these five methods is described briefly in the following.

Acquire innovation organizations: Corporate acquisitions can be a very productive way to enhance the innovation capabilities of a company. One positive aspect of this method is that the acquired company has already proven its ability to innovate. In addition, the acquired organization may be able to integrate their innovation abilities into the larger company. This model was used successfully by Amazon to support and develop the integration of Alexa (artificial intelligence and automation control) to Echo and Dot technology platforms. The acquisition of smaller, innovative companies such as Alexa, a2z, Lab126, and Brilliance Audio set the stage to be able to leverage, integrate, and innovate. This acquisition approach to innovation does not always work though. If the larger (acquiring) company does not have certain cultural/structural features, it may inadvertently crush the innovation capability of the acquired company.

The "innovation guru" method: Some examples of this method are Steve Jobs of Apple Inc., Larry Page of Google LLC., Marissa Mayer of Yahoo!, and Ed Catmull of Pixar Animation Studio. Clayton Christensen has done significant work in characterizing the attributes that a leader must possess in order to play the role of the innovation guru (see *The Innovator's DNA*) [3]. Companies that have a "guru" seem to be able to retain the capability to innovative despite the tremendous pressure to default to safe, non-innovative, organizational structures or culture. However, innovative gurus are expensive, difficult to find and often create significant turmoil as they fight to create and maintain an innovative culture. Also, past success can be a predictor for future success, but there is no guarantee that the guru can create and maintain an innovation capability. This is especially true when trying to create disruptive (as opposed to incremental) innovation, as disruptive innovation is inherently risky and relatively unpredictable. Finally, the characteristics of the innovation guru are often not the same as those needed to facilitate a company's incremental growth. Therefore, the organization may struggle to maintain its ability to sustain incremental innovation or growth. A company's self-awareness of its place on the innovation S-curve is important to know which leaders it should be utilizing.

Skunk Works® model: The term "Skunk Works®" was originally coined by Lockheed Aircraft Corporation as a name for a design/development group tasked to create disruptive innovation in the aircraft industry [4]. The group had several key characteristics that apparently led to their success. First, they had significant support from top organizational leadership. This support took the form of finances, personnel, infrastructure and rewards. In addition, the group was insulated from the normal constraints and requirements of the larger corporate culture. Normal constraints on purchasing, prototyping and risk taking were removed. Also, requirements for reports, documentation and many other normal "checks and balances" were drastically changed. This model has been used successfully by many organizations. However, it is quite difficult to keep the mid- and top level managers from encroaching on the group and inhibiting the innovation. Also, the fact that the Skunk Works® group is insulated from the normal organization is critical to its ability to be innovative, but it also largely prohibits it from spreading its enthusiasm and strategies for innovation throughout the rest of the organization. Finally, as with all of these innovation strategies, there is no guarantee of success for this particular strategy.

Research and development: Research and development is not the same as innovation. While many organizations communicate that they are facilitating innovation through their funding of research and development (R & D) budgets, this may not be the case. R & D investment leads to one of three outcomes: (1) incremental advancement, (2) disruptive advancement, or (3) no advancement at all. The culture of the R & D group, which most often flows from the culture of the larger organization, will determine the likelihood of the three different outcomes. The manner in which the reward system is integrated into the R & D culture determines much of the outcome of the R & D efforts. If R & D developments that prove to have immediate application to a company's product line see the biggest rewards, then that will obviously shift the focus of the R & D to incremental and applied research. If research that results in new products, S-curve jumps in product development or original contributions to a research area receives significant rewards, then the R & D focus will shift toward disruptive innovation. One caveat to this is that for a culture of disruptive innovative to flourish, risk must be handled carefully. New ideas, products or inventions that do not show immediate effect to the corporate bottom line must still be rewarded appropriately. As a balance to this however, these new ideas must still receive some level of scrutiny to avoid ideas that are intellectually sloppy.

Integrate a culture that fosters innovation throughout the organization: This is probably the most challenging and the most rewarding of the five methods. The challenges come from the tremendous inertia in the culture of most organizations coupled with the need to align the organization's culture in a way that facilitates innovation. Also, it is quite possible that it is unwise to realign the entire organization's culture to foster innovation. This is true if the core productivity of a group is coming from (and possibly should continue to come from) incremental innovation as opposed to disruptive innovation. Note that there may be some incremental cultural changes that could be made to facilitate an organization's transition from a Skunk Works® model to a full innovation culture model. For example, a larger acceptance of risk may be communicated throughout the organization, but aggressive risk taking may only be fully implemented in certain segments of the organization. There are numerous aspects of an organization's culture that must be considered in order to access that organization's capacity for disruptive innovation. These will be discussed in detail in the following.

Organizational cultures that foster innovation

There are many aspects to an organizational culture that facilitate disruptive innovations. Although there appear to be general principles regarding this culture, it is likely that the details of many of these aspects differ for different organizations depending on the organization's goals, resources, and people. In broad strokes, the aspects of the culture can be broken down into the following eight categories:

1. Top level leadership's commitment to change
2. Organizational stability and resources
3. Reward structure
4. Risk tolerance
5. Physical environment
6. Communication culture
7. Characteristics of key personnel
8. Creativity/innovation training

Some comments in each area are provided in the following. Organizations desiring disruptive innovation are encouraged to self-assess their current organizations for potential adoption of the following aspects.

Top level leadership's commitment to change

Cultural change is most often quite difficult. Resistance to change is prevalent in all levels of most organizations. This general resistance to change can result in a tremendously hostile environment for those attempting to facilitate the changes. If the cultural changes intended to support disruptive innovation are not supported by the top levels of leadership of an organization, the odds of successful implementation are low. Exceptions to this occur if a group is able to function for a period of time outside of the oversight of the main leadership (similar to an unapproved Skunk Works® model) or if they are able (miraculously) to produce successful disruptive innovation that facilitates the company's bottom line in a short enough period of time to avoid corporate intervention.

Organizational stability and resources

In most cases, disruptive innovative takes significant time and resources (people and money) to be successful. It requires an S-curve jump that places the organization back on the lower slope area of a new S-curve. In addition, it is far more difficult to predict how long it will take to produce disruptive innovation than it is for incremental innovation. If a company is driven by the quarterly stock price, then the long wait and relative uncertainty of success when pursuing disruptive innovation will likely cause abandonment of the strategy. Unfortunately, many organizations address their need to facilitate innovation only when they are experiencing substantial financial struggles. If a cultural shift toward production of innovation necessitates significant, long term expenditure of resources, then it may not be possible for a struggling organization to accomplish. As an example, it is not uncommon for a product development group tasked with creating disruptive innovation to require multiple full time people over multiple years to produce results. This group is dedicated to jumping S-curves—some pan out, others take a long time to move out of the lower portion of the 'S,' and others never do. Even when these resources are allocated, there is no guarantee of results.

Reward structure

Whether an organization's reward structure is explicitly stated or not, personnel quickly learn what is counted as "success." Disruptive innovation often requires significant time, in a direction not recognizable as contributing to incremental progress (during which there may be multiple failed attempts). If incremental progress is what is rewarded, those working on disruptive innovation can quickly lose motivation. On the other hand, if methods are employed to reward creativity, risk-taking and the "potential" of a concept, then motivation for activities that can produce disruptive innovation is enhanced. In addition, research shows that individual rewards as opposed to traditional rewards tend to foster innovation. Examples of traditional rewards are base salary or promotions. Most traditional rewards are given from the top level of the organization. Sometimes traditional rewards are seen as being a function of the number of years in the organization, as opposed to reward based on actual performance. Individual rewards are personal recognition (especially from your peers or supervisor) or recognition that comes from satisfied customers.

Risk tolerance

An organization's attitude toward risk is critical for creating a culture that facilitates disruptive innovation. Many organizations verbalize the positive aspects of taking risks and the tolerance for failure, but their unwritten rules, codified in quarterly reports and midlevel management's assessment of those that report to them, loudly nullify any tendency to take risks. Mantras such as "fail early, fail often" must be backed up by supervisors and top-level management in ways that are clearly seen by those asked to take the risks [5]. Note that organizations that are successful at achieving disruptive innovation both embrace risk and, at the same time, have risk mitigation strategies [6]. As an example of these strategies, "high risk—high payoff" activities are tolerated more easily when they are not in the critical development path of a key product. Other risk mitigation strategies will be discussed in the section of creativity methods.

Physical environment

Physical environment should mimic, or even display, the organization's commitment to seek disruptive innovation. Physical environment often tells a story regarding the organization's leadership structure and communication protocols. For example, if top leadership offices are located far from the majority of the other personnel and if those offices are seen as "off limits" to the majority of the people, then the flow of creative ideas is limited. Physical environment can also create an atmosphere that leans either toward creative thought, or toward a culture that emphasizes simple repetition of both activities and thought. Research shows that creativity occurs when a familiar idea, process or technology is either applied in a different manner or is combined with another idea in a way that is new [7]. This different application or new combination is facilitated by environments that provide freedom and promote a wide variety of thought processes. One additional aspect of physical environment is that in order to create solutions to problems, the organization needs access to the people that will be affected by the solutions (customers or end users). There are numerous techniques for implementing changes to physical environments to enhance the creativity process. The company IDEO has done some great work in this area [5].

Communication culture

Some of the communication issues related to disruptive innovation have been addressed earlier. The keys to providing a communication culture that enhances innovation take two forms. First, the research shows that open communication facilitates disruptive innovation simply because great ideas can come from a variety of sources [8]. In particular, it appears that the combining of ideas is most likely to create innovation when those ideas come from people with very different perspectives (sometimes called the "Medici Effect" [9]). This difference in perspectives could be due to training, experience, age (or numerous other aspects). Therefore, communication protocols that facilitate different types of people addressing the same problem improve the novelty of the solutions. The second key to providing a communication culture that enhances innovation is to provide an environment for people to internally process and refine ideas. This obviously crosses over to the category of physical environments. For example, if an organization only provides physical space where open communication is the norm, then refinement of ideas will be hampered. This contrast of communication cultures is representative of the dichotomy of introverted and extroverted personality types as well. Recent research shows that both are critical in an effective organization and both have different communication strengths and styles [10–12].

Characteristics of key personnel

Research shows that, while innate intellectual capacity (measured by IQ scores) has a significant inherited (genetic) component to it, creative abilities do not [3,13]. Simply put, creativity can be learned. However, in any organization there are those that find the creative process enjoyable and those that do not. Even if other aspects of the organization's culture are aligned to facilitate innovation, if an individual is not motivated to engage in creative processes, their productivity in that environment will be reduced. People can obviously change their desire to engage in innovation-oriented tasks over a period of time, but an initial evaluation of a potential team member's desire to engage in creative processes may help formulate an "innovation team" with greater chance for success. Numerous tests, including MBTI and 6-Hats, have components which can be helpful in identifying potential team members for innovation oriented teams [12].

Creativity/innovation training

A recent large survey of CEOs indicated that creativity is the #1 "leadership competency" of the future [14]. Because research shows that creativity is a learned skill, techniques that develop that skill are critical for development of an organization's innovation capability. These techniques are most effective when they are first learned, then practiced and finally incorporated into the culture and common practices of an organization. Numerous techniques are described in the following.

Methods for enhancing creativity and the resulting innovation

As mentioned previously, creativity and the ability to innovate can be enhanced through training and practice. In this light, a set of concept generation or ideation methods which facilitate innovation are described in the following. Research shows that the quantity, quality and novelty of concepts can be enhanced by use of the following methods [15]. Ideation is the process of developing this solution space.

We divide the innovation training into what we label "ideation methods" and what we call "box-busters." Ideation methods are multi-step processes that produce increased quantity, quality and novelty of the potential solutions to a problem. "Box-busters" are a set of techniques that can be used in the context of different ideation methods to improve that method's ability to expand the solution space.

Ideation methods

The following ideation methods form a suite of options for an organization looking to generate a large quantity of novel and quality concepts. These methods can be applied as a full or partial set depending on the organization's goals and resources. In the following are ten major categories of ideation methods.

1. Customer interaction
2. Background research
3. Functional description
4. Morphological brainstorming
5. C-Sketch rotational drawing exercises (also known as "6-3-5")
6. Analogies
7. Mind maps
8. Physical or multi-sense engagement
9. Group engagement
10. Embracing risk

Each of these major ideation methods are described in the following with sufficient detail to attempt the methods in one's organization.

Customer interaction

Customer needs analysis almost always involves interviewing and surveying customers. However, more extensive customer interaction can be used as an ideation technique. Watching the customer use the product or process, suggesting to the customer the use of new technologies to enhance the product's capability and engaging with high-value groups of customers (called lead users) that are pushing the envelope on how the product is used can all lead to expansion of the design space. Some further interaction methods are described in the following.

- *Interview the customer:* In-person or survey methods are useful. Care should be used to ask general questions first and avoid leading questions [16]. Several web-based survey sites enable broad sharing of a survey to distributed user groups.
- *Observe the customer:* Use of anthropological methods to study the customer's current process and limitations. This may result in broad, possibly non-materiel, solutions. It can also highlight the manner in which the problem must fit within the customer's abilities, culture, and integrated resources.
- *Become the customer:* Experiential observations tend to stick with and motivate the designer more directly than do interview/observations alone. This method is especially useful with the designer is addressing a need for a customer base far removed from their own life experience (e.g., users with disabilities, different ages, different locations/environments) [17].
- *Tech push:* Rapid adoption and experimentation of lower technical readiness level technology into domains not originally intended for. This adoption of existing tech to solve unassociated domain problems can serve as the basis of novel inventions.
- *Crowd sourcing:* Through effective use of surveys, competitions, and observing large quantities of ideas, the designer can put the creativity of the customer's themselves to use. Related to the concept of "co-designing," the user is valued for their own ideas which can be used to enhance or spawn further novel ideas [17].
- *Quality Function Deployment (QFD):* This method once dominated the Japanese manufacturing industry and was then adopted by many American industries to trace the voice of the customer throughout the entire designed product. It is still a valuable method to derive measurable design requirements from original customer needs through use of QFD tools such as the House of Quality [16].
- *Lead user engagement:* Designers interview and observe the lead users in the problem domain. These users may already approach the problem in a novel, though informal, way. These approaches can be observed for advanced ideas. Lead users can also be a valuable source of feedback on early prototypes [18].
- *Contextual analysis:* As the designer engages with customers, a prioritization of customer needs is necessary. A contextual analysis of the customer needs is a useful way to understand the full context (who, where, when, how, etc.) [19].

Background research

Background research is an important part of an overall ideation strategy. It helps to frame the problem and, when accomplished properly, keeps the group from "reinventing the wheel." It can assist in identifying where on the development S-curve the problem is currently located. This, in turn, helps identify where the opportunities lie and whether they are more likely incremental or disruptive innovation in nature. Some of our recent research involves the use of patents and trade publications in the ideation process. Intellectual property considerations should also be an important part of the background research process.

- *Patent research:* the study of existing patents can serve two purposes. First, it can expand the solution space through understanding and combination of existing ideas. This type of study happened to be the basis of the TIPS/TRIZ ideation method [16]. Secondly, the study of patents will help shape intellectual property protection efforts that are critical for tech transfer success [20].
- *Use of S-Curves to identify design innovation opportunities* (*ex: Honda* [21]): For well-understood markets, the innovation S-curve model can be used to identify when incremental or disruptive innovation is appropriate. Background research will identify whether underlying technology readiness supports rapid tech transfer, or in some cases, when market saturation demands a disruptive innovation approach.
- *Reverse the customer flow:* Classic design processes move the designer from customer needs to a solution. Through background research, it may be found that an organization has certain technology, expertise, or resources that a customer needs without realizing it. A designer may simply need to understand and then connect those areas to the appropriate customer base.
- *Image searches:* Through use of existing search engine algorithms, the initial results, may help highlight other considerations in design of products, process, or features. Fringe images are also highly useful in forcing the designer to determine how they may be related in creative ways. This remote link can serve to leap-frog the designer to another idea altogether.
- *Google Scholar:* Similar to patent research, it is important to understand the basis for existing technology. Results from this area of research will form the academic or scientific basis of understanding which is important for further development scoping.
- *Tech journals:* Tech journals will help identify technology of various stages of development and usually focuses on the potential for various applications. This method of background research is especially useful if a tech push approach to ideation is desired.

Functional descriptions

The idea of functional decomposition is to break a problem into descriptions of "what the solution to the problem needs to accomplish." This helps maximize the quantity of ideas in the solution space. To do so, the descriptions of the problem must be related to WHAT the solution needs to do, not HOW it will do it [16].

As a brief example, suppose that a customer orders a product online and as a result the customer needs to receive that product. If the question is specified as "Should we use FedEx or DHS or UPS to deliver the product?" then the solution space is limited. If the relevant question is phrased as "How do we move a product from point A to point B?" then we've opened up the design space to include a far larger variety of possible solutions (maybe a drone is used to do the delivery). However, even this last question is limiting. If the question is rephrased as "How can the customer end up with the product?" then additional solutions where the product is not moved, but somehow the product still ends up with the customer, are added to the solution space. For example, maybe rapid prototyping machines are supplied to all customers and the software code is delivered to build the product—the customer builds their product at their location. The overall functional description earlier can be broken into incremental functional descriptions. For example we may need to: identify customer, communicate with customer, identify purchased product, build product (or locate product), deliver product, confirm delivery. Some of these incremental subfunctions may occur in different sequences depending on the actual solution details. For example, will the product be delivered or will the information to build the product at the customer's site be delivered?

Morphological brainstorming

Morphological brainstorming is most often used in conjunction with the functional description method described earlier. It is a formal approach to combinatorial design. For each of the subfunctions of the problem, a number of different solutions can be imagined. The different combinations of the solutions to the subfunctions constitute the set of possible overall solutions to the problem. The beauty of this process is that, if there are "S" subfunctions and each subfunction has "I" imagined solutions, then the number of unique combinations of different subfunction solutions is "I^S." This quickly populates the overall solution space with a large number of alternative combinations of solutions for each subfunction. Not all combinations of solutions may be possible across subfunctions, but it is a rapid way of developing many system-level designs through novel combinations. These system-level concept variants are then able to be placed in decision models to compare their overall utility in solving the original problem.

C-Sketch rotational drawing

Traditional brainstorming is ineffective for developing a set of solutions (solution space) to an ill-defined problem. Specifically, the research that substantiates this follows this pattern [15]:

- An ill-defined problem is given to two groups.
- Each group has N individuals (with similar background and training)

Group 1	Group 2
• In group 1, each of the N individuals develops a solution space (without communicating with the others in their group). • The solutions spaces from each of the individuals in group 1 are combined and any redundancy is removed.	• Group 2 develops a solution space using a classic group brainstorming process [22]

- Group 1's combined solution space is compared with the solution space produced by group 2.
- Group 2's solution space will not have greater quantity, quality or novelty than the combined solutions space from group 1.

This research finding leads to the conclusion that organizations need methods that enhance the effectiveness of group ideation. One of these methods is "C-Sketch rotational drawing exercises," also known as "6-3-5" [15]. In this method, members of a group (optimally six people) each create three solutions to an ill-defined problem in five minutes (hence the 6-3-5 label). The three solutions are developed by each group member by creating three drawings on a single, large piece of paper. The drawings can also have written descriptions. This is done in a location where all six of the group members are present, but they are not allowed to communicate during the five minutes when they are creating their three solutions. Once the initial five minutes has passed, each member rotates their paper to the colleague next to them and that person has five minutes to augment the drawings by adding to, commenting or recreating, the ideas they received. After five minutes the drawings are rotated again. Rotations occur every five minutes until each group member has worked on every other group member's drawings. Once the rotations have completed, the group can then discuss the different drawings and corresponding solution options. With a nominal group size of six members, this method can easily produce 108 novel and variant ideas. This process has been documented to increase quantity, quality and novelty of the solution space when compared with either individual work or classical group brainstorming ideation [15].

Analogies

Some cognitive psychologists believe that all "new" ideas come from analogy; meaning that the ideas are not really "new" in the strict, literal sense, but come from applying an existing idea in a new manner, or combining two or more existing ideas. In order to use the idea of analogies in design, techniques are needed to help uncover effective analogies. Two categories of analogies have been used to develop designs. First, grammatical techniques leverage the power and versatility of human language. Typically, a seed word (possibly a key part of the functional description developed earlier) is used to produce a list of similar words. A variety of computer programs can be used to develop the analogues words. Second, biological analogies are also often used to produce ideas. Again, a number of different resources (computer programs, web sites, and written material) are used to facilitate this process.

- *Grammatical analogies:*
  - *Wordnet (wordnet.princeton.edu/):* a lexical database that groups words by cognitive synonyms
  - *Visual thesaurus (visualthesaurus.com/):* a thesaurus that presents its results in a rotational hierarchy (i.e., spider diagram) format to enhance understanding of word linkages.
- *Biological analogies:*
  - *The Biomimicry Institute (AskNature.org):* nature inspired strategies, ideas, and resources
  - *Encyclopedia of Life (eol.org):* information about biological solutions
  - *Tree of Life Web Project (tolweb.org/tree/):* information about biological solutions
  - *International Journal of Design & Nature & Ecodynamics:* covers general area of how nature relates to modern scientific thought and design
  - *Materials found in nature (Nature.com/nmat):* provides a forum for materials found in nature
  - *Bio-Inspired Engineering,* Chris Jenkins, Momentum Press, 2012. Text on bioinspiration. Proponent of the idea that design by analogy (DBA) is really backward compared to normal design in that you study natural system and then see if they relate to any current engineering problem as opposed to knowing the problem and looking for a solution.

Mind maps

Mind mapping can be used in combination with many of the other methods presented here. It is a way to organize and display information [23]. This rotational hierarchical format, sometimes also called spider diagrams, is a graphical method for information organization helps to create categories and highlight relationships between the different pieces of information. Mind mapping has been shown to enhance ideation in many ways. Most often it helps increase the quantity of potential solutions to an ill-defined problem by creating categories of solution possibilities where additional ideas can be imagined. The developed solution categories can be used as a starting point for further ideation. This, in turn fills in more of the mind map, which can then lead to further grouping and categories.

Physical or multi-sense engagement

Providing the problem-solver with first-hand experience with the problem often expands the design space. Most often, this first-hand experience involves having the problem-solver use the existing product if one exists. Using the product in different contexts and with different methods or constraints can increase quantity, quality, and novelty of the solution space. If no product/process is available to use, then using whatever currently best meets the core customer need(s) can be helpful.

- *Prototyping strategies:* A deliberate approach to prototyping feasible solutions will include use by the designers and/or user groups. This use of iterative prototypes can highlight areas of the solution of experience that can be enhances through further prototyping [24].
- *Activity diagrams:* Sometimes described as functional flow diagrams, use of activity diagrams can be used to focus on the user's activities. This can help identify parts of an existing process that are high value wasteful [16].
- *Role play:* Role-playing the various users and stakeholders can help to identify interface issues, information needs, and other constraints in advance of live user engagement. Capturing this role-playing can be accomplished through "story-boarding" a scenario of product usage so that the experience can be easily communicated to other non-participants [25].

Group engagement

Although classical group brainstorming is not effective for ideation, there are other group techniques that are effective. One such method is the 6-3-5 exercise. In addition, there are many other techniques used effectively by many organizations. Often these are integrated into the organizations' culture and involve feedback groups like Pixar's Braintrust [26] and 3M's Innovation presentations [8]. They may also include structures for group resources (time and funding) [27] devoted to innovation or may involve techniques to facilitate the Medici Effect (3M rotates their employees from group to group even outside of their area of expertise).

Embracing risk

An organization's risk culture is critical to its ability to innovate. Although many organizations pay lip service to the idea that they tolerate risk, it is tremendously difficult to create a structure that addresses rick in a manner that actually facilitates creativity and fosters innovation. In addition to explicit statements that detail the organization's risk policies, there needs to be definitive evidence, in the form of positive performance reports and rewards, that these policies are real. Also, organizations that successfully navigate the risk issue often employ strategies that allow risk to be embraced, without the full ramifications of the failures that are necessary on the road to achieving disruptive innovation. Some of these strategies involve taking parallel paths where one is "high risk—high payoff" and the other path is less risky but more feasible. Another strategy involves taking risks that are not in the critical path of the product/process development. Organizational leadership can use the following questions to discover opportunities for enhancing a healthy risk culture.

- What percentage of projects fail?—aggressive innovation requires a large %
- How is failure defined?
- How do the different aspects of failure impact performance evaluations?
- Can a high risk solution or aspect of a solution, that is perpendicular to core performance, be developed in parallel to the low-risk solution?

Box Busters: Techniques to expand or reinvigorate the solution space with new ideas

The following is a list of techniques that can be selected to help a design team increase a solution space. The list overlaps with ideation methods, culture aspects, and organizational innovation methods presented earlier in the chapter, but are presented here to provoke thought within a design team.

1. *Do the opposite:* Explore the contrary. Suspend disbelief of a direction originally thought to be impossible, undesirable, or otherwise irrational. Sometimes this approach can yield ideas that are second or third-order removed from the initial "opposite."
2. *Suspend constraints:* Explore limitations. Sometimes the constraints are found to not actually exist due to customer misunderstanding. Other times, temporality removing the constraints allows for new ideas that can be made to fit back inside the constraints later.
3. *Violate physics (or norms, or company policy, or organizational culture, or values, or priorities):* Explore the "impossible." Sometimes referred to as "turning off physics" we try to revert to a completely clean sheet approach to what is physically possible. This unlearned approach can yield a wild idea that can serve as the central concept that could fit back into physics. Sometimes the identification of which part of physics needs to be turned off results in a better understanding of the phenomenon and other functional approaches to solving the problem.
4. *Analogies based in basic function:* Ask "How does nature do this or how do other systems do this?"
5. *Dedicate time to innovate:* Incorporate innovation methods and activities into the standard schedule similar to Google and 3M [27]. Tailor time, resources, and freedom based on an organizations culture and personnel competencies.
6. *Load—relax—capture:* Use when faced with a challenging problem where sustained effort is needed. "Load" or immerse in the problem, technical details, or functions required. Periodically, "relax" into a completely different setting, topic, or activity to allow the mind to restructure, combine, and connect salient information needed to solve the problem. Go have lunch, work out, or engage in a separate meeting or discussion. Then revisit the problem and "capture" the key elements that emerge to help the mind communicate the new connections that have developed [28].
7. *Using new tech in ways it was not intended for:* Explore the ways in which new technology can be used that the inventor may not have intended. Give it to unlearned or other user groups to observe their use of the tech.
8. *Medici effect:* Engaging outsiders, especially those from a different field. This increases the solution space in a combinatorial and even multiplicative manner [9].
9. *S-Curve analysis:* Identify if you need incremental or disruptive innovation.
10. *Avoid typical innovation traps:* Avoid group think, design fixation, structural pressure from power hierarchy.
11. *Function before form:* Focus on understanding WHAT needs to be accomplished, before ideating on HOW it can be done. Resist the urge to jump directly to possible physical forms too early. This eliminates the possibility of many other ways to accomplish an original function.
12. *Build, learn, and iterate:* Use physical mock-ups and prototypes. Use them early in the design, even as quick and low fidelity mock-ups to rapidly increase design feature knowledge [29].

13. *Role play:* Story board the experience from key users' points of view. Script the entire mission of the product or process. Try being the device.

14. *Study the customer:* Ask, and then observe customer in their use of current or proto-typed devices. Consider the use of anthropological methods.

15. *Tech push versus tech pull:* Design takes a balance of top-down methods, and bottom-up understanding of technology. When focused on one direction for too long, consider coming at the program from the other direction.

16. *Engage the unencumbered:* Work with user groups that are without limitations for their ideas. One example is children. They are especially good at implementing the "Suspend Constraints" and "Violate Physics" techniques.

17. *Explore the perfect solution:* Give in to the desire to chase the wants vs. just needs. Then identify impossibilities and explore possibilities [15].

18. *Explore transformation:* Transformation design principles can highlight opportunities to use either mono-or multi-form solutions. Comparing the features and functions of a design can yield efficient use of space, energy, volume, and time [30].

19. *Brain Trust:* Small innovation group with commitment to complete honesty. Remove barriers and expectations of hierarchical structure or repercussions of disagreements or wild ideas. Keys for a brain trust approach [26]:
    1. Deep understanding of the problem.
    2. No power in a brain trust group. Owner still has all the authority.
    3. Honesty is combined with trust.
    4. No competition within brain trust, instead a shared success.
    5. Knowledge that the creator is very likely to become married to their idea in a way that blinds them from thinking an idea needs iteration. So the creator must really want to hear the truth.

20. *Patents or trade publications:* Continue to explore other approaches to related problems to find overlapping ideas or ways that existing inventions can be adapted outside their original domains.

21. *Highlight innovative culture legacy:* Tell stories of past success to set vision and expectations but note that culture comes from shared experience not stories.

22. *Kill innovation barriers:* Ask what parts of an organization's rewards, logistics, or culture inhibits innovation? Seek to actively destroy the barriers with leadership and external organizations that may drive them.

23. *Celebrate "good failure:"* Ask when the last significant failure was. If it has been awhile, it is possible that the organization may be too risk adverse.

24. *Orthogonal performance:* Try to create risk experiences that are orthogonal to primary function. Once the primary functions and solutions are established, consider allowing further exploration in a direction not critical to the success of the primary function. These high risks may pay out huge, but do not put at risk the primary function.

25. *Prototype in parallel:* Create high risk high payoff prototypes in parallel with lower risk ones.

26. *Get true leadership buy-in to innovate:* Get buy-in from upper-level management for innovation and risk—then remove them from the process. Insure they understand the unorthodox methods and the mid-term goals (to include possible failure), and then trade oversight for rapid progress. This takes a great deal of trust in an organization.

27. *Cut out mid-management:* Keep middle management completely out of the picture (e.g., Skunk Works®). Rapid progress in the solution space requires a focus on the primary goal of innovation. Bureaucracy and tertiary processes of an organization will

kill the primary goal quickly if an organization does not value the need to innovate and the methods it requires.

28. *Treat every problem as a new problem:* Repeating or repackaging past successes will not work for new challenges. There must be legitimately new creative ideas and even processes. It's hard because sometimes process is tied to culture or even values in an organization...so where is it ok to change?

29. *Identify conflicts and separate solutions chronologically:* Methods such as the Transformational Design Methodology can be used to separate out the process steps needed to go from one subsolution to the next.

30. *Identify conflicts and separate solutions physically:* Explore physical solutions that may exist in other organizations. Get another entity to do one of the conflicting activities.

31. *Platforms or families of solutions:* Rather than trying to please all user groups with one solution, consider solutions with core structure of functionality that can then be modified or enhances to provide a suite of solutions meeting many customer needs (e.g., smart phones with different RAM or screen sizes).

32. *Quick customization for individual preference:* All design relates to a human user or interface for some function of the design. Focus on the user through deliberate discussion of their interface. Allow for feature customization to enhance the positive user experience (e.g., Keurig coffee makers, drill with different bits).

33. *Default to most desired state or configuration:* Change the default configuration or motivation of the system for the preferred status (e.g., minimum energy configuration or stable state).

34. *Frame the problem in competitive terms:* if the ill-defined problem is framed in a competition environment or warfare context, then several new ideation directions exist:
    1. Flank
    2. Asymmetric warfare
    3. Shock & Awe
    4. Confuse competition
    5. Demoralize competition
    6. "Art of war"—Sun Su
    7. Rules for radicals [31]

In conclusion, this chapter is meant to serve as an innovation guide for organizations. The first two sections described ways to enhance an organization's innovation and the major aspects of an innovative organization's culture. An organization's senior leadership may find these sections useful as they seek to recognize and develop an innovation culture. The final section showed direct methods to apply to a design challenge. These methods were grouped into ideation methods, and box-buster tips. Anyone tasked with developing solutions to problems should find many of these methods useful. Creativity and innovation, similar to many other life-long skills, CAN be developed through training and practice.

References

1. R. N. Foster, Working the S-curve: Assessing technological threats, *Research Management*, 29(4), 17–20, 1986.
2. C. M. Christensen, *The Innovator's Dilemma: When New Technologies Cause Great Firms to Fail*, Harvard Business Review Press, Cambridge, MA, 2013.
3. J. Dyer, H. Gregersen and C. M. Christensen, *The Innovator's DNA: Mastering the Five Skills of Disruptive Innovators*, Harvard Business Press, Boston, MA, 2011.

4. Lockheed Martin Corporation, Skunk Works Origin Story, 2017. (Online). Available: https://www.lockheedmartin.com/us/aeronautics/skunkworks/origin.html. (Accessed 27 Oct 2017).

5. C. Fredman, The IDEO difference, *Hemispheres*, 52–57, 2002.

6. R. Kelty and B. A. Bunten, *Risk-Taking in Higher Education: The Importance of Negotiating Intellectual Challenge in the College Classroom*, Rowan & Littlefield, Lanham, MD, 2017.

7. J. Lehrer, *Imagine, How Creativity Works*, Harcourt Publishers, Boston, MA, 2012.

8. G. C. O'Connor, *Grabbing Lightning: Building a Capability for Breakthrough Innovation*, Wiley, Hoboken, NJ, 2008.

9. F. Johansson, *The Medici Effect, What Elephants and Epidemics Can Teach Us About Innovation*, Havard Business Review Press, Brighton, MA, 2017.

10. M. C. Davis, D. J. Leach and C. W. Clegg, The physical environment of the office: Contemporary and emerging issues, *International Review of Industrial and Organizational Psychology*, 26, 193–237, 2011.

11. D. Goldstein and O. Kroeger, *Creative You: Using Your Personality Type to Thrive*, Simon & Schuster, New York, NY, 2013.

12. S. Cain, *Quiet: The Power of Introverts in a World That Can't Stop Talking*, Random House, New York, 2013.

13. M. Reznikoff, G. Domino, C. Bridges and M. Honeyman, Creative abilities in identical and fraternal twins, *Behavior Genetics*, 3(4), 365–377, 1973.

14. S. Berman and P. Korsten, *Capitalising on Complexity: Insights from the Global Chief Executive Officer (CEO) Study*, IBM Institute for Business Value, Portsmouth, UK, 2010.

15. C. White, K. Wood and D. Jensen, From brainstorming to C-sketch to principles of historical innovators: Ideation techniques to enhance student creativity, *Journal of STEM Education: Innovations and Research*, 13(5), 12, 2012.

16. K. Otto and K. Wood, *Product Design, Techniques in Reverse Engineering and New Product Development*, Prentice Hall, Upper Saddle River, NJ, 2000.

17. T. Kelley, *The Art of Innovation: Lessons in Creativity from IDEO, America's Leading Design Firm*, Doubleday-Random House, New York, 2002.

18. D. G. Johnson, N. Genco, M. N. Saunders, P. Williams, C. Seepersad and K. Holtta-Otto, An experimental investigation of the effectiveness of empathetic experience design for innovative concept generation, *Journal of Mechanical Design*, 136(5), 051009, 2014.

19. M. Green, J. Linsey, C. Seepersad, K. Wood and D. Jensen, Frontier design: A product usage context method, in *International Design Engineering Technical Conference, ASME*, Cleveland, OH, 2006.

20. D. Jensen, J. Wood, P. Knodel, K. Wood, R. Crawford and R. Vincent, Evaluating ideation using the publications popular science, popular mechanics, and make in coordination with a new patent search tool and the 6-3-5 method, in *Proceedings of the American Society of Engineering Education (ASEE) Annual Conference and Exposition*, San Antonio, TX, 2012.

21. J. Rothfeder, *Driving Honda: Inside the World's most Innovative Car Company*, Penguin Group, New York, 2014.

22. A. F. Osborne, *How to Think Up*, McGraw-Hill, New York, 1942.

23. T. Buzan, *The Mind Map Book*, Plumb Books, London, UK, 1993.

24. B. Camburn, B. Dunlap, T. Gurjar, C. Hamon, M. Green, D. Jensen, R. Crawford, K. Otto and K. Wood, A systematic method for design protyping, *Journal of Mechanical Design*, 137, 081102, 2015.

25. K. Crider, L. Cumm, J. Lewis, D. Jensen and J. Wood, Body storming, super heroes and sci-tech publications: Techniques to enhance the ideation process, in *American Society of Engineering Education Annual Conference*, 2011, Vancouver, BC.

26. E. Catmull and A. Wallace. *Creativity, Inc: Overcoming the Unseen Forces that Stand in the Way of True Inspiration*. Random House, New York, 2014.

27. E. Schmidt and J. Rosenberg, *How Google Works*, Grand Central Publishing, New York, 2013.

28. R. G. LeTourneau, *Mover of Men and Mountains, The Autobiography of R. G. LeTourneau*, Moody Publishers, Chicago, IL, 1972.

29. J. Knapp, *Sprint—How to Solve Big Problems and Test New Ideas in Just Five Days*, Simon & Schuster, New York, 2016.

30. A. B. Markman and K. L. Wood, *Tools for Innovation, The Science Behind the Practical Methods that Drive New Ideas*, Oxford University Press, New York, 2009.

31. S. Alinsky, *Rules for Radicals, A Pragmatic Primer for Realistic Radicals*, Random House, New York, NY, 1971.

chapter thirteen

Self-jamming behavior
Joint interoperability, root causes, and thoughts on solutions

Stephen R. Woodall

Contents

Introduction

Since the 1991 Gulf War, the collapse of the Soviet Union, the subsequently celebrated end of the Cold War, and the several, extended, and overlapping wars against terrorism, many changes have occurred which affect the requirements for the design and operation of our military forces. The often-expressed post-Cold War hopes for a more peaceful existence across the family of nations have been tempered by a continuation of armed conflicts and terrorism, across a broad spectrum of intensity, in many spots in the world. In concert with the presumed decrease in the threat to our national existence since our victory in the Cold War, our military forces have been systematically decreased in size in direct proportion to decreases in resources, by nearly 50 percent in some cases. However, reflective of the amount of conflict in the world deemed politically to impact upon our national interests, diplomatic and public demand for the services of the military forces continues steadily to grow, along with their operational tempo.

At the same time, the advance and application of new technologies have allowed our smaller forces to become increasingly more capable, as well as more complex. As a direct result of these new technologies, we are developing and in the process of deploying systems with functionality and capabilities that were unachievable dreams only a decade ago—think of the progress we have made in recent years in the area of deployable ballistic missile defense capabilities. Similarly, we are witnessing revolutions in command and control, battle space visualization, computer data processing, weapons effectiveness and precision, and in intelligence, surveillance, and reconnaissance systems. New long-range, high-precision weapons, combined with dramatic improvements in geographical location and targeting, have significantly enhanced strike and attack capabilities, and added a range of new combat options and tactics. In following the news of the day, we are witnessing the leading edge of the results of years of research and development in our integrated air and missile defense capabilities—today, deployable where our forces go, worldwide. Each military Service—Army, Navy, Marine Corps, and Air—Force is bringing exciting new capabilities to the field.

Where we once thought in terms of platforms and force configurations and dispositions, we now think in terms of systems, and systems of systems. Another way of describing this is the trend in warfighting systems from "platform centricity" to "network centricity." These systems (or networks of systems) are complex, and requiring orders of magnitude increases in data and information, as well as in communications connectivity and bandwidth, to operate and achieve their ideal design functionality. Increasingly, these systems, at the level of a theater of operations, are Joint systems. Common examples of theaters of operations where we are actively involved with our Joint forces today include Northeast Asia (especially the Korean peninsula) and Southwest Asia (especially the Arabian Gulf area).

Joint systems themselves are systems that are designed to be employed by two or more of our military Services operating in concert, in order to improve overall capabilities to coordinate and synchronize operations. The glue that binds these systems together in Joint operations is called "interoperability."

Interoperability

Interoperability is defined as "The ability of systems, units or forces to provide services to and accept services from other systems, units, or forces and to use the services so exchanged to enable them to operate effectively together." At a more technical level, interoperability can be defined as "The condition achieved among communicationselectronics systems or items of communications-electronics equipment when information or services can be exchanged directly and satisfactorily between them and/or their users."

As the complexity of our systems, and systems of systems, in terms of the numbers of lines of computer code, the numbers of components, and data and information processing requirements, continues to grow exponentially, the need for and importance of interoperability between and among the Services, at the Joint force level, has steadily increased. The problem is made even more complex by our frequent requirement to extend the demanding requirements for force interoperability to our Allied and Coalition partners in a range of operations overseas. For example, even as early as the year 2000, we faced such interoperability challenges with our Allies on a daily basis in NATO operations in Bosnia and Kosovo, and in various other United Nations–sponsored peacekeeping operations worldwide. The problems are even more complex today, in anti-terror operations in Afghanistan and Iraq.

An example of a typical system of systems where full Joint interoperability is an elusive, top-level objective is what is known as Command, Control, Communication, Computers (C4). C4 represents a complex integrated system of systems, including Joint and

Service doctrine, Joint tactics, techniques and procedures (JTTP), organizational structures, personnel, equipment, fixed and portable facilities, and communications capabilities—all designed to support a military commander's exercise of command and control across a range of military operations. Another, related, Joint system of systems is Intelligence, Surveillance, and Reconnaissance (ISR)—representing a system combining the collection, processing, integration, analysis, evaluation, and interpretation of information and data concerning enemies, foreign countries, other forces and areas of interest. Increasingly today, C4 and ISR are combined into a single, even more complex system of systems, commonly referred to as C4ISR.

Another system of systems currently under intense development, and growing in visibility and importance as the missile threat from emerging peer competitors and a number of Third World nations increases, is Integrated Air and Missile Defense (IAMD). As a complex system of systems, IAMD includes its own evolving C4ISR operational, systems, and technical architectures, distributed and netted sensors and communications, and a range of afloat (AEGIS warships) and land-based (AEGIS Ashore) ballistic missile defense (BMD) missiles, large, ground-based interceptors (GBIs) guarding against ICBMs, and other anti-air and anti-missile defensive weapon systems (THAAD, PATRIOT) from across the Services.

A pressing issue we face today is that, as the complexity and inter-relationships of all systems grow, the cultural and technical difficulty of achieving true Joint interoperability has exploded markedly. This is true at the level of individual Services, across the four principal Services, and extends to training exercises and actual military and humanitarian operations with our Allies and Coalition partners.

Self-jamming behavior

As a result of these systems approaches to command and control, combined with new, high-data-rate communications, advanced computer processing, and development and deployment of a host of new ISR capabilities, our military decision makers are floating in a sea of data and information, which may or may not be provided in a processed, ready to use form—even if some measure of Joint interoperability exists. In a sense, when we try to exchange massive amounts of information and data—to interoperate—without the right approach to the high-level systems engineering of our systems, we can succeed, literally, in jamming ourselves, decreasing our own operational effectiveness even as we modernize. Indeed, modernization of systems without focused systems engineering efforts to integrate new technology and capabilities, including especially the appropriate interoperability functionality, actually contributes to reductions in overall force interoperability.

Symptoms of reduced or non-existent interoperability include data systems which cannot exchange data, in the same message format, at the same rates. They include intelligence systems which cannot pass critical information to the warfighter when needed, in a format required for timely and effective use. They include combat direction systems and sensors which ideally will share a common tactical picture, but which cannot communicate data with each other, thus creating the need to struggle to attain and maintain a clear, common tactical picture of the battle space. These symptoms slow, rather than speed up, the pace of decisions made by a commander. They also serve to reduce the confidence the commander has in the quality of information provided by his systems—confidence vital to making timely, accurate planning and tactical decisions.

In such situations, our decision makers can be inundated with masses of unprocessed or only partially processed data, where the only possible synthesis and processing of complex data into useable information, knowledge, and understanding is that accomplished in the minds of the decision makers themselves. Perhaps even worse, they may be deprived, due to systems interoperability shortcomings, of the timely availability of much or all of the strategic and tactical information they need to make the best decisions. I refer to this phenomenon, where we have failed to achieve true Joint interoperability, for whatever reason, as "self-jamming behavior."

As a consequence of self-jamming behavior, our forces are increasingly in situations where the incompatibility of our warfighting systems serves to decrease our overall warfighting effectiveness. As discussed earlier, this problem is particularly critical when we operate with other nations. Apt examples include operations with our NATO allies, or combined operations with other partners or allies—from conflicts such as the Gulf War, to other sorts of operations included in the broad scope of Military Operations Other Than War (MOOTW), such as humanitarian operations, or noncombatant evacuations. In the 1999 air war against the Serbs, our Allied and Coalition forces achieved "interoperability" only after reducing all military communications to uncovered (non-encrypted) circuits— thus signaling, free of charge, most Allied intentions and force movements to the enemy forces. Such were the innate incompatibilities of our collective Allied C4ISR systems of systems, even after fifty years of NATO existence and efforts to achieve at least a basic level of Allied force interoperability! Although improvements have been made in recent operations, NATO and combined operations performed as elements of the war on terror, much remains to be done to achieve full interoperability. Even today, we're not very good at "sharing" our warfighting data, and access to our battle networks.

Obstacles to joint interoperability

For another recent, relevant example, such incompatibility, both in information and command and control systems, and in the linking and integration of combat direction systems across a theater-wide area, we need only look back to the 1991 Gulf War. Here, the underlying interoperability problems were resolved by sub-division of the battle space into land and maritime zones. This permitted the naval forces in the Arabian Gulf and the Red Sea to inter-operate their sea-based systems to cover the sea surface areas and the airspace above, while the Air Force and the Army systems covered the land and the airspace above the land. While this physical deconfliction resolved some of the technical interoperability difficulties, it also served to reduce the overall effectiveness of the force, by significantly reducing the area coverage and reach of many of our most modern systems, including the AEGIS weapon systems in our warships, the Army PATRIOT batteries ashore, and the Air Force AWACS surveillance aircraft covering the airspace above the land.

For example, even though the many AEGIS guided missile Cruisers assigned to Gulf War forces in theater were capable of eliminating enemy air threats deep over the land into Iraq and over Kuwait, they were prohibited from engaging any targets which were over land—even if they were the firing platform and air defense system best suited and best positioned to do so. Analogous restrictions were placed on Air Force and Army systems with respect to threats over water. The result was a significant reduction in the overall effectiveness of the force, denying to the Commander-in-Chief and the Joint Force Commander the ability to employ his forces' total capabilities to their maximum effectiveness.

Today, significant recognition is given to the interoperability problem, and the need for achieving resolution. As noted as early as in 2000 by the Joint Staff in their Joint Vision 2020, "The Joint force has made significant progress toward achieving an optimum level of interoperability, but there must be a concerted effort toward continued improvement."

However, we are, in many ways, in much the same situation today we were in during the Gulf War. This means that we are less ready than the achievable ideal regarding Joint interoperability to fight a theater-wide engagement. What are the obstacles to achieving interoperability of our forces, and reducing or eliminating our selfjamming behavior?

Root causes

To understand the difficulties we face in achieving Joint interoperability, and thus reducing our self-jamming behavior, we need to identify and understand the root causes of our current state of marginal interoperability. These root causes can be divided into three key areas: Service cultures, organization for acquisition and systems design, and doctrine and training.

Root cause—service cultures

The Services differ significantly in their individual cultures, and their approaches to command relationships, and thus their approaches to requirements for the ideal C4ISR functionality. Taking the Services in turn:

- The Army is hierarchical, stove-piped by functionality and specialty, and centralized in planning, while operations at the lower levels can be decentralized.
- The Marine Corps approach is similar to the Army's, although the organization is much more horizontal, and generally has more freedom for taking initiative in execution of operations.
- The Navy, which historically has operated on the oceans and seas with minimal oversight, is comfortable with the delegation of many elements of operational command and authority to subordinate warfare area commanders, with all commanders operating in a highly data-intensive and collaborative environment. For the Navy, the authority of commanders to decide and act on their own initiative is perhaps the highest of all the Services. As the Navy's role in ballistic missile defense worldwide continues to grow, the need for and importance of Joint Interoperability becomes more vital by the day.
- Finally, the Air Force approach to command is reflective of their aerospace power tenet of centralized command and planning, and decentralized execution—where operational initiative is often subordinated to zeal in execution of the plan.

These very real cultural differences impact on every aspect of achieving Joint interoperability, from approaches to doctrine, tactics, and training, to views on the ideal systems approaches to the functional design of command and control, information management, and warfighting capabilities. Related to these differences in how the ideal way of planning and commanding operations is perceived, achieving consensus between and among the Services on any point related to Joint interoperability has proven and continues to be exceptionally difficult.

Root cause—organization for acquisition and systems design

The area of organization, especially for acquisition and systems design, presents some of the greatest obstacles to achievement of Joint interoperability. Some of the more striking points include these:

- Within our Department of Defense, there is no single organization or Service responsible for Joint interoperability; thus, there is no one accountable for failure to achieve interoperability. No single organization develops the requirements. No single organization controls resources to be applied to achievement of Joint interoperability. No single organization is responsible for development of the standards for, or the operational testing and evaluation of, achievement of Joint interoperability capabilities. That is, still, to this day—there is nobody "in charge."
- Across the four principal military Services, there is no consensus regarding what the ideal operational, systems, or technical architectures for Joint interoperability ought to be, and how they should be system engineered. In other words, the questions "How much interoperability is enough?" and "What constitutes the architecture for the ideal level of Joint interoperability functionality?" are not yet answered, or even fully answerable.
- We currently design systems, and systems-of systems, the way our acquisition organizations and major program offices are organized. Service by Service, we are organized to design and engineer new systems in stove-piped, highly structured, intensely bureaucratic organizations—where concerns about inter-Service and Joint interoperability are often not considered until a system design is so far along that the best that can be done regarding interoperability is to design awkward and expensive "interfaces" with other, relevant systems. In other words, we still design many new systems without sufficient attention being paid, early in the requirements definition and system development process, to engineering in Joint interoperability from a top-level Joint systems perspective.
- The issue, in some ways, is less technical than programmatic. We must continue to deal for decades with some of our legacy systems, where interoperability will, as a matter of design, remain limited. And, simply enforcing standards for design as an interoperability fix can be carried only so far. For example, in the fielding of our most modem Joint tactical data link, Link 16 (originally known as TADIL J), in recent years, all Services followed a common, clearly established military data link standard. However, since each Service implemented the standard in different ways (from within their own Service's acquisition "stove-pipe"), there remain today design and performance differences between the individual Service implementations of Link 16, which continue today to produce "self-jamming behavior" in Joint operations. Systematically engineering our way out of these self-inflicted Joint interoperability problems will be costly, and will continue for years ahead.

Root cause—doctrine, tactics, and training

Our current joint doctrine is too general and it is particularly weak on issues of requirements, standards, and criteria for Joint interoperability. As root causes:

- There is no shared vision for attaining and maintaining a minimum requirement for Joint interoperability, nor is there a consensus on a doctrinal basis for such a vision.
- Service parochialism and cultural differences routinely delay consensus building in the development of new Joint doctrine.
- Beyond the Services' differences, there is no consensus across theaters of operation with respect to Joint tactics, techniques, and procedures (JTTP); additional differences are often generated due to disagreements with Allies and Coalition partners. There is

neither an overall consensus regarding to how to address this issue, directly related to Joint interoperability, nor is there anyone in charge of resolving the conflicts.

- Without a consensus on the ideal form for doctrine and JTTP for Joint interoperability, there can be no consensus for the ideal approaches regarding training for Service and Joint forces regarding attaining and maintaining interoperability-related functionality—both technical and procedural.

What can be done? Thoughts on solutions to self-jamming behavior

Removal of the obstacles to achievement of Joint interoperability, and thus eliminating the self-jamming behavior of our armed forces, will require significant, deliberate efforts. Thoughts on solutions must include addressing, at a minimum, the underlying root causes of the difficulties in achieving an agreed approach toward systems engineering the ideal level of Joint interoperability, beginning with a consensus on overall, top-level requirements and architectures.

Solutions related to service cultures

Directed movement toward building a consensus across Service cultures regarding requirements for Joint interoperability began with the establishment of a new organization within the Joint Staff, called the Joint Theater Air and Missile Defense Organization (JTAMDO), in March of 1997. Today, that organization is called the Joint Integrated Air and Missile Defense Organization (JIAMDO). Although its declared mission is "to produce Joint operational concepts, requirements, and architectures to guide the development of the Family of Systems that provides the Joint Force Commander dominant air and missile defense capability," related objectives clearly seek common Joint operational, systems, and technical architectures and Joint interoperability as the approach to achieve evolving Joint integrated air and missile defense (IAMD) requirements. However, as a fairly new and small organization, its practical influence over the major stakeholders—the Services, the Ballistic Missile Defense Office (BMD today, the Missile Defense Agency (MDA)—and the now defunct, disestablished Joint Forces Command (JFCOM)—remains limited.

Solutions related to organization for acquisition and systems design

Regarding required changes to organization for the Joint acquisition and systems design of systems related to Joint interoperability, under the initiatives begun following JTAMDO's reorganization as JIAMDO, much groundwork has been accomplished in creating collaborative bodies in the form of Integrated Product Teams (IPTs). These teams, involving the direct participation of the Office of the Secretary of Defense, representative of the Unified Joint Commanders in Chief, the Joint Staff, and the Services themselves, are working toward a consensus regarding the ideal way ahead for Joint operational and other architectures, including agreed upon top-level requirements and technical standards, for Joint interoperability. Unfortunately, the regular, working members of the teams tend to be too junior to have the authority make or influence the hard decisions for their organizations.

Related efforts of consensus building where Joint interoperability will be a key element include Joint collaborative planning and engagement capabilities, combat identification capabilities, automated command decision aids, integrated weapons fire control systems, and common, integrated capabilities for intelligence, surveillance, and reconnaissance. All of these efforts, as they evolve into Joint warfighting capabilities, will provide our armed forces the ability to plan, collaborate, and fight seamlessly across an entire theater of operations, from attack and strike operations to active defense, in a single "Joint Engagement Zone," rather than fighting in self-imposed maritime and land-based stove-pipes, as was done in the Gulf War.

Although the Services' stove-piped organizations for system design and acquisition still exist, the requirements related to Joint interoperability are beginning to be taken into account in their systems engineering approaches to system design far earlier in the design process than in the past. This trend offers hope that, beyond the lives of many of our current, less interoperable legacy systems, true Joint interoperability will become as important to the Services in system design considerations as Service-specific operational requirements.

Solutions related to doctrine, tactics, and training

Although somewhat dated today, Joint Vision 2020 noted that "Although technical interoperability is essential, it is not sufficient to ensure effective operations. There must be suitable focus on procedural and organizational elements, and decision makers at all levels must understand each other's capabilities and constraints. Training and education, experience and exercises, cooperative planning, and skilled liaison at all levels of the Joint force will not only overcome the barriers of organizational culture and differing priorities, but also will teach members of the Joint team to appreciate the full range of Service capabilities available to them."

However, today there is still no single Joint doctrinal publication that addresses the key issues and elements of Joint interoperability, nor has there been developed a complete consensus across the Joint community regarding top-level requirements for interoperability. There remains much work to be done to conceive, to draft and to build consensus on an ideal set of Joint interoperability requirements, including the complementary Joint doctrine and Joint tactics, techniques, and procedures (JTTP) to complete Joint force implementation. Finally, with agreed upon doctrine and JTTP, development of broad procedural and technical training and education in the elements of Joint interoperability, applicable across the Joint forces, can begin.

Summary

As our Services struggle, within limited resources and with growing operational commitments, to field new platforms, systems, and systems of systems, to upgrade and modernize current platforms and capabilities, and to maintain operational readiness, much can be done to improve overall force operational capability through steady, deliberate progress toward a state of true Joint interoperability. To the extent that our forces are not interoperable and our systems incompatible, we will continue to exhibit self-jamming behavior.

To achieve the ideal state of Joint interoperability will require an extraordinary, historically unprecedented level of cooperation between and among the Unified Commanders in Chief, the Department of Defense, the Joint Staff, and the individual Services. Residual obstacles will always remain in differing Service cultures, different approaches to organization for system design and acquisition, and differing views on doctrine and tactics. However, the effectiveness of our military forces for the remainder of this new century will depend directly upon their success, as a Joint team, in overcoming these obstacles—and eliminating all self-jamming behavior.

Bibliography

Feasibility of Third World Advanced Ballistic and Cruise Missile Threat, Volume I: Long Range Ballistic Missile Threat (National Defense Industrial Association: Systems Assessment Group, National Defense Industrial Association Strike, Land Attack, and Air Defense Committee, Arlington, Virginia, October 1998).

Joint Publication 1-02, Department of Defense Dictionary of Military and Associated Terms (United States Government: Government Printing Office, Washington, DC, 2000).

Joint Theater Air and Missile Defense (JTAMDO) Strategic Planning Offsite: Final Report (Joint Theater Air and Missile Defense Organization, The Joint Staff, Arlington, Virginia, March 2000.

Joint Vision 2020 (Director for Strategic Plans and Policy, J-5, The Joint Staff: Government Printing Office, Washington, DC, June 2000).

chapter fourteen

4D Weather Cubes and defense applications

Jaclyn E. Schmidt, Jarred L. Burley, Brannon J. Elmore, Steven T. Fiorino, Kevin J. Keefer, and Noah R. Van Zandt

Contents

Introduction

Innovative technologies and integrated system concepts to address air, land, maritime, and space capability gaps need innovative field test and modeling and simulation (M&S) tools. These are essential to transitioning the suitable capability in a timely and cost-effective manner. Modeling and simulation provides a low-cost option for assessing these innovations to understand whether operational requirements will be met and identify areas needing further attention so as to ensure capability gaps are comprehensively addressed under various conditions. This chapter introduces an atmospheric characterization and radiative effects M&S tool vital to the development and operational impact of Command and Control, Intelligence, Surveillance, and Reconnaissance (C4ISR), as well as integrated kinetic and non-kinetic (i.e., high energy laser or HEL) capabilities. The tool is anchored on a verified and validated first-principles, atmospheric characterization model developed by the Air Force Institute of Technology's Center for Directed Energy (AFIT CDE).

AFIT CDE developed a robust simulation and analysis tool, known as 4D Weather Cubes, to provide physically realistic visible-spectrum images and data tables that accurately translate propagation and atmospheric effects measured or assessed at one part of the spectrum to any other spectral region. Although its initial development was to enhance visualization tools, Weather Cubes are utilized as data analytics to simplify machine learning (e.g., fire control and target recognition systems) dependent on human processing of spatially and temporally dynamic data whether very heterogeneous, conventional meteorological parameters, or hyper-spectral atmospheric propagation spanning ultra-violet through radio frequency information. Anchored on inputs from the Laser Environmental Effects Definition and Reference (LEEDR) and the National Oceanic and Atmospheric Administration's (NOAA's) Global Forecast System (GFS) numerical weather prediction (NWP) model, each Weather Cube is tagged with a universal time reference and area of interest whether city, region, or world-wide. These expansive data arrays of meteorological and environmental data demonstrate how clouds, precipitation, and aerosol particles affect ultraviolet through radio frequency radiative propagation. They are configured for both desktop as well as high performance and cloud computing and generally encompass large volumes reaching from the surface to the top of earth's atmosphere that blend first principles atmospheric processes and constituents (e.g., water droplets, aerosols) with archival and predictive NWP data to characterize atmospheric and radiative effects. Integrating Weather Cubes into military, US Department of Defense (DoD), and other nonmilitary data analytics and visualization tools provide a computationally inexpensive way to account for realistic atmospheric effects for many applications including kinetic and non-kinetic capability assessment, mission planning and human in the loop fire control, machine-learning forensics of massive imagery intelligence databases, and climatological impact studies.

Weather Cubes

Background on input models

AFIT CDE developed perhaps the first directed energy weapon (DEW) simulation package, called the High Energy Laser End-to-End Operational Simulation (HELEEOS), which included a correlated, probabilistic climatological database to assess HEL system performance for realistic atmospheres.[1] The capability to incorporate weather impacts into other models though correlated, vertical profiles of atmospheric effects generated by HELEEOS gained traction far beyond the DEW community. Therefore, a separate model, called the Laser Environmental Effects Definition and Reference (LEEDR), was developed that allows the export of the first principles atmospheric characterizations for other DEW simulation codes, military or DoD mission planners, or even nonmilitary scientific research such as climate change impact studies.[2]

LEEDR is a validated and verified, fast-calculating, first principles, worldwide, surface-to-100 km, atmospheric characterization and radiative transfer code with spectral range considerations between ultraviolet and radio frequency wavelengths (200 nm to 8.6 m). The two primary goals of LEEDR are (1) create correlated physically realizable vertical profiles of meteorological data and environmental effects, such as gaseous and particle extinction, optical turbulence, and cloud-free line-of-sight (CFLOS), and (2) to allow graphical access to and export of the probabilistic data from the Extreme and Percentile Environmental Reference Tables (ExPERT) database. Vertical profiles generally consist of temperature, water vapor content, pressure, optical turbulence, and extinction effects due to atmospheric particulates and hydrometers at various wavelengths. LEEDR contains detailed databases and characterization algorithms for worldwide climatological data, temporally and spatially varying boundary layer definitions, particulate and molecular data, and allows users to input observed surface conditions. Internally consistent line-by-line and correlated-k distribution radiative transfer algorithms capable of assessing path transmittance, path radiance, and celestial contributions to an observed signal level in true three-dimensional geometry on a spherical earth with any relationship between the target and observer have been incorporated into LEEDR calculations.[3] Recent code upgrades include the implementation of a fast-calculating, two-stream-like multiple scattering algorithm that captures azimuthal and elevation variations and fully solves for molecular, aerosol, cloud, and precipitation single-scatter layer effects with a Mie algorithm at every atmospheric layer. Validation studies that compared LEEDR multiple scattering radiance outputs to published sky radiance observations for clear and cloudy conditions, as well as radiance data from the Moderate Resolution Imaging Spectroradiometer (MODIS) on board the Terra and Aqua satellites, were performed and showed promising results.[3]

One unique feature of LEEDR is its boundary layer characterization. A well-mixed boundary layer that varies with height based on location, season, and time of day is assumed, meaning that potential temperature and water vapor mixing ratios remain approximately the same throughout the layer. Adiabatic parameterizations are applied to boundary layer calculations, since temperature and dew point temperature values decrease at known rates with height at dry and moist adiabatic relationships, allowing for relative humidity to increase with height and reach a maximum value (approximately 100 percent) near the top of the boundary layer. Hygroscopic aerosols swell in size due to water uptake when relative humidity values are greater than 70 percent, increasing aerosol scattering and thus creating a spike in atmospheric extinction. Experiments performed at AFIT validated the modeled extinction spike utilizing LIDAR data, and thus this feature is important when modeling atmospheric effects. The inclusion of surface observations via a surface data entry feature has also been shown to more accurately characterize key meteorological effects and parameters within the atmospheric boundary layer through fast-calculating dry and moist adiabatic relationships.[2,4]

In addition to internal databases consisting of global, probabilistic climatological information for meteorological and aerosol optical properties, NWP data has been incorporated to supply gridded observations or real-time, correlated weather forecasts or past re-analysis grids into LEEDR.[5] Therefore, assimilating these data sources into LEEDR enables post-event, nowcast, and forecast analyses of atmospheric and radiative effects for real-world scenarios and significantly improves its capabilities to perform predictive and analytical investigations for a variety of defense applications. The GFS model is the primary model of choice for incorporating NWP into LEEDR. GFS data, publically available for download through NOAA's Operational Model Archive Distribution System (NOMADS), is a global, hydrostatic, operational weather forecast model that is generated four times per day at a half-degree horizontal resolution. Meteorological data is available at 64 atmospheric levels, extending from the surface to approximately 30 km in altitude. Other NWP models, including military and DoD weather models, are in the process of being fully integrated into LEEDR to expand the current NWP consideration.

Weather Cube description

The development of Weather Cubes came from a collaboration with the Air Force Life Cycle Management Center's (AFLCMC) Simulation and Analysis Facility (SIMAF) division to improve Extensible Architecture for Analysis and Generation of LinkEd Simulation's (EAAGLES) Optical Kill Chain simulations by rendering full atmospheric effects on weapons and sensors for enemy threat engagements. Owned and managed by the US government and maintained by the SIMAF division at Wright-Patterson Air Force Base in Dayton, Ohio, EAAGLES is a designed software framework that supports the acquisition, test, and training communities. War-gaming capabilities at the time were limited to visible spectrum imagery only, missing the mark for simulating atmospheric effects when considering spectral bands outside of visible light, such as the infrared, millimeter-wave, or radio frequencies. AFIT CDE used LEEDR to provide better, physically-based atmospheric characterizations, effects, and rendering to improve OKC simulations, which find, fix, track, engage, and assess the process of addressing enemy threats. This innovative tool, consisting of numerous LEEDR profiles that can define a large horizontally and vertically varying volume of atmospheric parameters and effects, bridged the gap by providing visually stunning and realistic-looking visible-spectrum images and graphics—like those in computer games—that accurately translate to propagation and atmospheric effects outside of the visible spectrum.

Generated from the coupling of LEEDR and NWP data, Weather Cubes are 4-dimensional data arrays that define meteorological and multi-spectral atmospheric effects for a volume of interest (latitude, longitude, and altitude) at a specific date and time. Note that a new set of Weather Cubes is created with an NWP input date and time change. These volumes capture realistic multi-directional variations of atmospheric parameters based on weather conditions that may have or actually did occur. LEEDR code has been optimized to generate in batch mode vertical profiles at varying latitudes and longitudes within an expansive area. A series of mesh files, containing output variables indexed to a location and altitude, are made available at various wavelengths for each set of Weather Cubes. Although computationally taxing under normal circumstances, Weather Cube data processing scripts have been migrated onto High Power Computing (HPC) systems to exponentially reduce runtime from days to mere seconds, depending on the size of the area of interest. HPCs offer the capability to generate volumetric data for the entire globe, potentially in an operational manner to aid the mission planners and warfighters in daily tasks.

Microphysical and optical properties characterizations for clouds, rain, and aerosols are defined by LEEDR. Currently, half-degree GFS data is typically used to define the atmosphere from the surface to approximately 30 km. Above this altitude, regional standard atmospheres are used to characterize the upper portions of the atmosphere. As previously mentioned, the benefits of using GFS data is that it enables post-event, nowcast, and forecast analyses of radiative effects for real-world scenarios. Another GFS benefit concerns operational security; incorporation of GFS output involves ingesting a global data file into the Weather Cube processing and maintains the integrity of regional analyses without the concern of revealing an operational location or area of interest. Only GFS grid points relevant to a given set of latitudes and longitudes are considered in calculations, and outputs can be indexed to provide an additional level of security. The same indexing practice can be applied to other sensitive HEL system information (e.g., wavelengths, sensor bands). Therefore, LEEDR and thus Weather Cubes ensures that sensitive information about areas of interest remain undisclosed.

Weather Cube variable outputs include but are not limited to temperature, pressure, relative humidity, dew point, atmospheric density, wind vectors, vertical velocity, cloud and precipitation types and locations, visible refractive index gradients (C_n^2), extinction due to molecular, particulate, and hydrometeor absorption and scattering, single scattering albedo, and phase functions from the surface to 100 km with a vertical resolution of 100 meters and for any number of user-defined wavelengths. Outputs are stored in a matrix indexed by latitude, longitude, and altitude; and each matrix is saved to a mesh file in binary format. The volumetric mesh files can also be viewed as a look-up table and therefore ingested into other software codes for further analysis.

Recent enhancements

Initial Weather Cube productions required pertinent cloud and rain layers to be custom inserted into data processing code since GFS data does not directly output this type of information. Although the weather was realizable for the region, the process lacked efficiency and resulted in uniform cloud layers. Recent enhancements have been made to the data processing by inferring cloud and rain fields from GFS atmospheric variables to create a more realistic set of sky conditions with varying cloud base and height altitudes, as well as associated rain layers with varying rain rates. A simple, yet physically-based, sky characterization algorithm was developed at AFIT CDE to generate cloud and rain fields based on GFS relative humidity, vertical velocity, and 3-hour surface precipitation totals. The NWP-generated cloud and rain fields are incorporated directly into the Weather Cube data processing. The resulting Weather Cubes provide a realistic, physically-based representation of sky conditions.

The cloud field algorithm determines the presence of cloud based on relative humidity and vertical velocity thresholds for two different atmospheric layers, within and above the boundary layer. Within the atmospheric boundary layer, which is simulated to temporally vary from 500 to 1524 m AGL, relative humidity values must be 100 percent. Above the boundary layer, the relative humidity must be equal to or greater than 70 percent in order for cloud to be considered. Cumulus and stratus cloud types are determined by vertical velocity. Note that vertical velocity (ω) is the Lagrangian rate of change of pressure with time in units of Pascals per second. Figure 14.1 outlines the algorithm requirements and cloud type considerations.

A rain field is determined if both of the following requirements are met. First, a cloud field must be present for a given grid point. Secondly, a 3-hour precipitation value must be reported for the same location. If the aforesaid conditions are valid, a rain shaft is designated from the middle of the cloud to the surface of the Earth. An hourly rain rate (mm/hr) is then determined by simply dividing the 3-hour total precipitation total by a factor of three. Microphysical and optical properties from LEEDR are assigned to this

	Relative Humidity	Vertical Velocity (ω)	Wind Speed	Cloud Base Height	Cloud Type
Within the Boundary Layer	$\geq 100\%$	Upper Limit: 0 Pa/s Lower Limit: -0.12 Pa/s	< 2.5 m/s (<5 kts)	0 m AGL	Fog
	$\geq 90\%$	Upper Limit: -0.12 Pa/s Lower Limit: -5.99 Pa/s	N/A	N/A	Stratus
	$\geq 90\%$	Upper Limit: -6.0 Pa/s Lower Limit: - Infinity Pa/s	N/A	N/A	Cumulus Continental Clean
Above the Boundary Layer	$\geq 70\%$	Upper Limit: -0.12 Pa/s Lower Limit: -5.99 Pa/s	N/A	N/A	Stratus
	$\geq 70\%$	Upper Limit: -6.0 Pa/s Lower Limit: - Infinity Pa/s	N/A	N/A	Cumulus Continental Clean

Figure 14.1 Cloud field determination thresholds.

GFS Hourly Rain Rate (mm/hr)	Weather Cube Rain Rate
0 < Rain Rate ≤ 3.5	Very Light Rain (2 mm/hr)
3.5 < Rain Rate ≤ 8.75	Light Rain (5 mm/hr)
8.75 < Rain Rate ≤ 18.75	Moderate Rain (12.5 mm/hr)
18.75 < Rain Rate ≤ 50	Heavy Rain (25 mm/hr)
50 < Rain Rate ≤ ∞	Extreme Rain (75 mm/hr)

Figure 14.2 Rain field determination thresholds.

converted rain rate value based on its placement within one of five predetermined rain rates categories, ranging from very light rain to extreme rain, as seen in Figure 14.2.

A study was conducted to validate the cloud and rain field algorithms using data from the mid-Atlantic region of the US east coast. NASA's MODIS data were used to verify the cloud coverage, cloud top heights, and optical depths, and rain field placement and rain rates were compared to the National Weather Service (NWS) Next Generation Radar (NEXRAD) data from the Wilmington, North Carolina (KTLX) site. In order to fully evaluate both algorithms, two case studies were analyzed: (1) Hurricane Arthur, a well-organized meteorological phenomena that occurred in July 2014, and (2) a trough moving through the Carolinas region that generated scattered showers and thunderstorms on August 18, 2016 at 1800 UTC. A total of 12 wavelengths were processed for each set of Weather Cubes, but only four of those wavelengths are shown. Figure 14.3 displays a set of Weather Cubes for the Hurricane Arthur case on July 3, 2014 at 1800 UTC. Cloud fields are clearly evident in the 1.06, 3.0, and 11.0 μm propagation; note the low-level water vapor effects near the surface at 11.0 μm. Rain fields are clearly visible in the 2.5 cm propagation characterization.

NWP-generated cloud fields compared very well with MODIS data for both cases, as seen in Figure 14.4. Discrepancies between the two datasets can be attributed to the lower resolution GFS input data for Weather Cube processing, but cloud coverage patterns and heights in the Weather Cubes are consistent with that observed by MODIS. When considering a hurricane, the towering cumulus clouds that primarily compose the storm's structure are covered by a dense, cirrus cloud shield. Optical depth analyses were done by assuming cirrus cloud cover due to the fact that MODIS's algorithms determine optical depth based on the first designated cloud that it intersects, a cirrus cloud. The same approach of assuming a cirrus cloud optical depth was considered for the August 18, 2016 case.

A first-order validation was performed on rain fields against base reflectivity from the lowest elevation angle or 0.5-degree tilt of NEXRAD radar. Reflectivity values (Z) were converted to rain rates (mm/hr) through Marshall-Palmer distribution relationships. Due to rain droplet sizes being on the order of microns, radar reflectivity is measured in decibels of reflectivity (dBZ), a logarithmic method that differentiates between precipitation sizes (i.e., drizzle, hail). NEXRAD rain rates were averaged per half-degree grid point for a one-to-one comparison with Weather Cube rain placement and rates. Due to the limitations of operating radar systems, rain fields were analyzed only for areas of NEXRAD coverage, as seen in Figure 14.4. The overall locations of rain fields within the Weather Cubes for both case studies were comparable to that of NEXRAD coverage, but the rain rates differed greatly from the values reported by the KTLX radar.

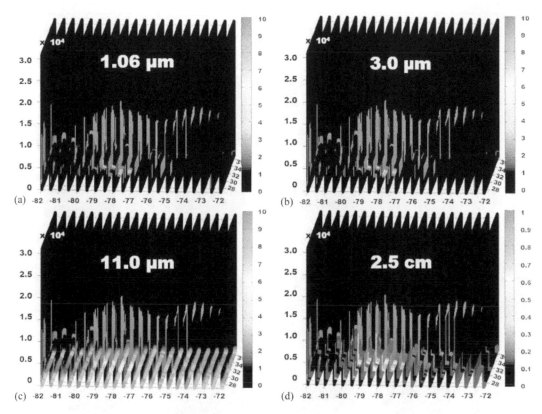

Figure 14.3 Weather Cubes depicting atmospheric effects at (a) 1.06 μm, (b) 3.0 μm, (c) 11.0 μm, and (d) 2.5 cm propagation for Hurricane Arthur on 3 July 2014 1800 UTC, just prior to making landfall near Wilmington, N.C. The deep gray color indicates cloud extinction at all 4 wavelengths (a-d). Low-level water vapor effects near the surface at 11 μm (c). Note the scale change and rain fields located near the surface for 2.5 cm (d).

Differences in observed and modeled rain rates can be partially attributed to the fact that only the KLTX 0.5-degree tilt base reflectivity was considered. KTLX rain rates might have varied if all tilts had been evaluated to account for potential precipitation at higher altitudes. Another discrepancy is due to the curvature of the Earth and elevation differences play a role in what portion of the atmosphere is observed at various distances from the radar. The maximum range of the base reflectivity product is 124 nautical miles (about 143 miles) from the radar location; and volume coverage patterns (VCP) with varying numbers of elevation tilts during precipitation events are considered to observe the vertical structure of a storm. Figure 14.5 illustrates the full suite of NEXRAD elevation angles that are used to create the VCPs. As the elevation tilt and distance from the radar increases, the radar is observing higher portions of the atmosphere at maximum range compared to areas close to the radar due to the elevation tilt. Higher reflectivity values on the maximum range of the radar coverage were observed by the KLTX site, and thus higher rain rates were assigned. Although it appears that higher reflectivity values at higher altitudes result in greater surface precipitation totals, this is not always the case. Estimated rainfall amounts have the potential to be lower due to evaporative processes of precipitation falling through dry layers. A more detailed study is needed to fully evaluate the rain field algorithm.

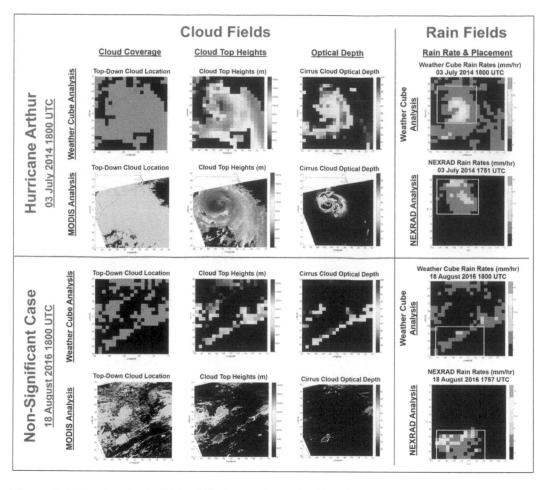

Figure 14.4 Cloud and rain field validation study results. Top plots are Weather Cube comparisons to MODIS (satellite) and NEXRAD (radar) for a well-developed weather event (hurricane). Bottom plots are Weather Cube comparisons to MODIS (satellite) and NEXRAD (radar) for a less-developed weather event (weak cold front and thunderstorms).

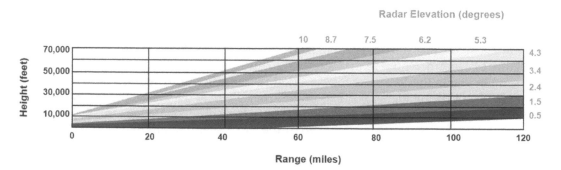

Figure 14.5 NEXRAD Volume Coverage Pattern.

Defense applications

The original motivation for developing Weather Cubes was to enhance military gaming simulations to better render atmospheric effects when assessing enemy threats. But this innovative tool goes far beyond advancing war-gaming simulations, opening doors to furthering remote sensing, directed energy performance assessments, and climate research. Weather Cubes offer novel contributions to the advancement of developing and deploying defense technologies. Recent applications include enhancements to various modeling and simulation codes, remote sensing capabilities, and HEL system performance evaluations. Additionally, the methodology to directly correlate weather occurrences to HEL performance has been accomplished by binning weather conditions, including precipitation and fog, based on statistics from thousands of resulting Weather Cubes for a region. The ability to predict that a line of sight is free of clouds, evaluate system performance based on a given set of weather conditions, and assess vertical profiles of optical turbulence are only a few examples of how Weather Cubes can aid the warfighter in making crucial decisions.

Cloud-free line-of-sight

The presence of clouds continues to be a critically important parameter for DEW operational system performance analyses and remote sensing applications. Many existing CFLOS climatologies are obtained from the 14th Weather Squadron (14WS), formerly known as Air Force Combat Climatology Center (AFCCC), and provide ground-to-space probabilities but do not account for elevation and azimuthal variations. This ground-based, climatological database is available for 415 sites considering two seasons (winter and summer, based on January and July data respectively) and for view angles of 0° to 80° zenith. The assumption that the probability of a CFLOS (PCFLOS) for a data-void region is similar to the climatology of a nearby CFLOS site can lead to very inaccurate representations of weather impacts, as sky conditions can vary drastically over short distances.

As previously discussed, recent enhancements to Weather Cubes which include the implementation of NWP-inferred cloud fields led to the development of a cloud model that consequently can be utilized for CFLOS studies. Monte Carlo simulations considering variations of platform and target altitudes, as well as slant ranges, can be applied to thousands of different cloudy sky realizations by using years of Weather Cubes. The result produced azimuthally-dependent PCFLOS statistics for any location world-wide at any time of day, any altitude of interest, and for any view angle. Furthermore, spectrally-dependent PCFLOS can be developed due to the wide range of spectral considerations of LEEDR and thus Weather Cubes. In other words, clouds affect spectral bands very differently, and CFLOS climatologies can be developed for each spectral band. For instance, molecular extinction of a cloud is detrimental for HEL systems operating at NIR spectrum wavelengths, but electromagnetic propagation is minimally affected at radio frequencies. Characterizing whether a line of sight is cloud-free for any location worldwide is a vast improvement over current data available and would greatly enhance operational analyses and defense readiness.

A preliminary study was conducted to demonstrate this capability and validate results against the current CFLOS climatology. Weather Cubes were generated for January and July 2015, at four times a day, centered on Joint Base Langley-Eustis, VA. Monte Carlo simulations were used to analyze the 248 Weather Cubes to determine if lines of sight were cloud-free including eight platform heights and a two-degree azimuthal resolution. Resulting PCFLOS

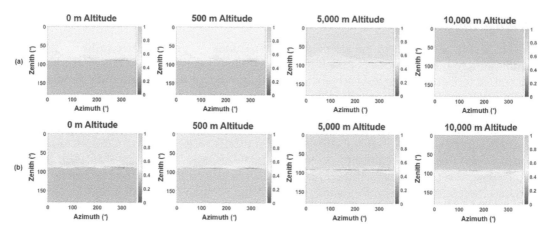

Figure 14.6 Daytime, seasonal CFLOS with platform altitude variations at Joint Base Langley-Eustis. Plots are for winter (a) and summer (b) daylight hours at various altitudes (10 m, 500 m, 5000 m, and 10,000 m) and display all zenith angles (0 up to 180 down) and all azimuthal angles (0 to 360).

values provided seasonal and temporal variations and were compared to the 14WS CFLOS climatology data at the same location as a first-order validation. January and July data represented winter and summer seasons, respectively. Daytime hours consisted of 1200 and 1800 UTC, and 0000 and 0600 UTC were considered nighttime hours. Although the sample set was small, Figure 14.6 displays the results at four platform heights for two seasons. Note that these plots show the advantages of this tool over the current database available.

Data from the 557th Weather Wing's World-Wide Merged Cloud Analysis (WWMCA) was also leveraged in validating the preliminary PCFLOS study. WWMCA utilizes analysis of data from multiple environmental satellites, conventional surface observations, and other supporting databases. These include the NOAA Polar Orbiting Environmental Satellites (POES), Defense Meteorological Satellite Program (DMSP) satellites, the geostationary orbiting satellites (GOES) and the Japanese meteorological satellite (MTSAT), the European Space Agency's METEOSATs.[6] Cloud information is available for up to four cloud layers and 38 cloud parameters per file at a quarter-degree resolution dating back to approximately 40 years. WWMCA cloud top height, cloud base height, and total cloud cover were used for CFLOS analyses. Figure 14.7 shows PCFLOS generated from Weather Cubes and WWMCA for a platform altitude of 10 m at Joint Base Langley-Eustis AFB. Differences in PCFLOS values at 20° zenith and 302° azimuth were only 5 percent for this small sample set. Future research will be focused on expanding the validation study to 10+ full years of Weather Cube analyses.

Optical turbulence modeling

Optical turbulence is an important phenomena that can greatly influence the propagation of light, causing a degradation in transmission, beam quality, and imaging capabilities. This effect is a product of fluctuations in the index of refraction of the air due to temperature, humidity, and pressure gradients, with generally the strongest effects being within the surface and atmospheric boundary layers. Much effort over the years has been directed toward the development of standard turbulence profile models, and a significant number of these models have been incorporated into LEEDR (e.g., Hufnagel-Valley, Clear1). But these standard models do not always characterize realistic values for locations outside

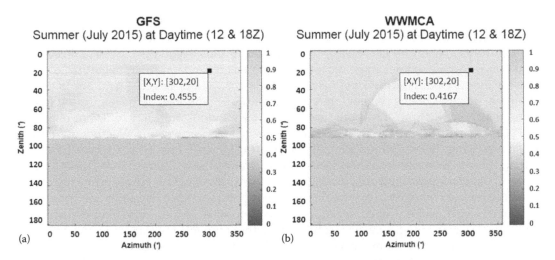

Figure 14.7 PCFLOS generated from (a) GFS and (b) WWMCA data for a view angle of 20° zenith and 302° azimuth and a 10 m AGL platform at Joint Base Langley-Eustis, Virginia.

of the region in which the data collection occurred for model development. Although optical turbulence values can be derived from satellite data, gaps in data coverage pose a potential issue with quantifying temporal variations. Accurately characterizing spatial and temporal variations in the refractive index structure parameter (C_n^2) is necessary to understanding how optical turbulence plays a role in DEW system and sensing capabilities and how to overcome these limitations.

Weather Cubes offer the opportunity to research horizontal and vertical profiles of C_n^2. Refractive indices can be derived from NWP using the Tatarskii approach.[7] The same methodology used in the development of CFLOS climatologies can be applied to optical turbulence for any location worldwide. Compiling C_n^2 statistics from 10+ years of Weather Cubes presents the variability in optical turbulence strength, as well as provide C_n^2 estimates for any time of day and the associated weather conditions (e.g., cloud cover). Figure 14.8 displays spatially varying C_n^2 for two locations on 18 August 2016 1800 UTC.

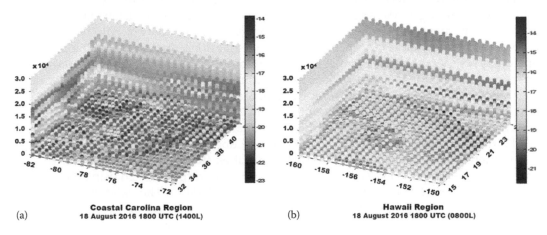

Figure 14.8 Optical turbulence (C_n^2) cubes displaying spatial variations for two regions encompassing (a) the North and South Carolina coast and (b) Hawaii on 18 August 2016 1800 UTC.

The left cube is representative of the coastal portions of North and South Carolina at 1400L, and the right cube shows C_n^2 values for a region encompassing Hawaii at 0800L.

An additional advantage of using NWP models is the availability of data at three hour intervals, or even one hour intervals for some models recently put into operation, assures that for operational requirements below altitudes of 30 km, forecast data are available within 90 minutes of any time of interest.[8] For reasons mentioned at the beginning of this section, forecasting optical turbulence can have tremendous impact on efficient planning and effective execution of civilian and military assets and operations, whether for active/passive sensing, laser communications, or creating desired effects at a target based on sufficient beam intensity. It has been noted that C_n^2 values using Tatarskii- NWP calculations below 100 meters trends towards unrealistically high values of C_n^2. This is primarily due to the NWP model's inability to resolve fine-scale temporal and spatial gradients in governing parameters such as local wind speed and direction, relative humidity, temperature, and pressure. Though not specifically motivated by this tendency, researchers have pursued alternative semi-empirical models to adequately address near-surface optical turbulence. These have sought to leverage various combinations of observed standard meteorological and topographical parameters including temperature, winds, relative humidity, surface albedo, heat flux, and time of day. Based on first-principles such as Mohnin-Obukhov Similarity theory and indeed Tatarskii theory as well, they tend to extend their derived surface optical turbulence results to 100 m through use of power law relationships, which generally are validated through observation.[9] More research continues to advance a hybrid-Tatarskii model which puts further emphasis on macro-meteorological parameters such as atmospheric pressure gradients and less on topography and local vegetation types parameters to address surface-to-100 m C_n^2 values.

Characterizations of plumes

Weather Cubes can be leveraged to exploit remote sensing capabilities to detect and track numerous types of plumes (i.e., volcanic eruptions, industrial explosions, dust storms). Atmospheric absorption and scattering due to plume effects will differ based on the optical properties of the particulates and the operating wavelength of the remote sensing device. The visual aspect of multi-spectral Weather Cubes demonstrates how plume particulates differently affect each spectral band, ranging from UV-to-RF. For instance, the NEXRAD operating at 10 cm has the capability to detect plumes of various kinds, as well as potentially diagnose the motion. Reflectivity values will differ not only based on the type of plume (i.e., volcanic eruptions, industrial explosions, dust storms) but also by the operating wavelength of the radar. As prior mentioned, NWP models provide predictions on many atmospheric variables such as temperature, pressure, and wind. Weather Cubes, generated from LEEDR and NWP forecast data, contain wind direction and speeds at different atmospheric levels that can be utilized in forecasting plume movement, offering emergency management offices the resources to make quick decisions.

Figure 14.9 displays an example set of Weather Cubes displaying a volcanic plume from a hypothetical eruption from the coordinates of Lihue (Kauai Island), Hawaii at various spectral bands. These $360 \times 360 \times 30$ km volumes of data were generated using half-degree resolution GFS data from August 19, 2016 0000 UTC (1400 L). The validated cloud and rain algorithms defined sky characterizations. Microphysical and optical properties for climatological aerosols were characterized via the Global Aerosol Data Set, a 5×5 degree, worldwide database comprised of various aerosols species and concentrations specific to each region. An extreme, summer volcanic plume as defined in the 1985 US Air Force

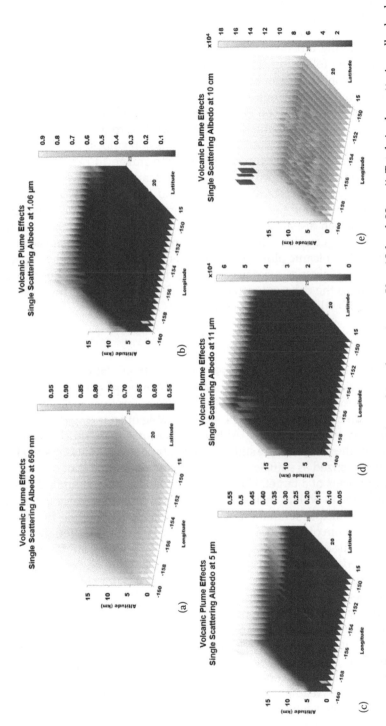

Figure 14.9 Spectrally-dependent Weather Cubes depicting a hypothetical eruption on Kauai Island, Hawaii. Total single scattering albedo shows how clouds, rain, and aerosols affect propagation at (a) 650 nm, (b) 1.06 μm, (c) 5.0 μm, (d) 11.0 μm, and (e) 10 cm.

Geophysics Handbook was inserted for six grid points, including Kauai Island and areas east and south of the island.[10] The volcanic aerosol generator in LEEDR places volcanic aerosols at altitudes of 9–50 km with a peak in extinction near 30 km.

Spectrally-dependent volcanic plume effects can be seen in Figure 14.9, displaying total single scattering albedo profiles for electromagnetic propagation at particular wavelengths within the following spectral bands: Visible, SWIR, MWIR, LWIR, and RF (microwave). Note that it is clearly evident how clouds, rain layers, and aerosols affect each spectral band differently. Total single scattering albedo is defined as the total scattering divided by total extinction, and a transparency scheme based on transmission values was applied to the images. High transmission values correlate to a transparent atmosphere, depicted as white. The opposite is true of low transmission values.

Changes were made to the 11 μm Weather Cube image to better showcase plume effects. For the 11 μm volcanic plume only, total single scattering albedo was replaced with the total volcanic plume single scattering albedo and multiplied by a factor of 1×10^{-4}. This allowed the 11 μm plume to be visible among the molecular, non-volcanic aerosol, and cloud albedos, which are all very low at 11 μm; these changes are reflected on the "stretched" scale. The 10 cm plot is highly transparent due to all transmission values are near one; to ensure the volcanic plume could be seen, it was designated as opaque. These three different plotting schemes were necessary to display the same plume in each spectral band demonstrates the different radiative transfer physics occurring in each spectral band that is captured by the LEEDR-driven Weather Cube technique.

Performance binning of high energy lasers

As HEL technologies advance so as to draw efficiently on their platform's power source (i.e., electricity), they provide future commanders concepts of employment creating unlimited magazines. A prime example is the US Navy's Laser Weapons System (LaWS) mounted on the USS Ponce–declared operational in 2014—and capable of engaging drone aircraft, small boats, and potentially ballistic missiles. Although there has been much interest in recent years, the majority of these systems are still in the prototype stages. A successful HEL engagement might be characterized as sufficient energy focused on a small spot size to heat up and damage vital components of a target. Comprehensive field testing is limited due to limited resources, both funding as well as inclusive, operationally relevant range environments. So supplementary, first-principle simulation models offer a reliable method for assessing HEL performance for a plethora of regions and engagement scenarios. A thorough understanding of the limitations specific to the HEL system itself and its operational environment, including weather effects, is needed to secure a properly justified capability assessment.

Certain environmental parameters, such as high water vapor content, cloud cover, and strong optical turbulence, can greatly hinder the efficiency of the HEL system. Meteorological conditions can greatly vary depending on the location, season, and time of day. Therefore, knowing the variability of weather for an engagement will aid developers and war fighters in mitigating atmospheric effects in operational settings. Weather Cubes are able to solve this complex problem by quantifying typical (mode, median) and extremes in weather conditions based on realistic atmospheres over a statistically viable period of time (i.e., 10+ years). Most importantly, Weather Cubes always create atmospheric parameters that could all occur together no matter how extreme the conditions because they are based on actual observations or state-of-the-art scientific predictions of those atmospheric conditions. This is not the case with "standard" atmospheres or composite or averaged atmospheres. In their simplest form, Weather Cubes are expansive lookup tables (LUT) of meteorological and environmental data specific to a universal time reference and locations of interest that can be fully integrated into other simulation models to assess HEL performance, such as HELEEOS.

HELEEOS is a fast-running, scaling law code that addresses atmospheric effects and beam propagation. It is comprised of the three following codes that provide robust, time efficient solutions compared to computationally expensive, first-principles wave optics models: (1) LEEDR defines the atmospheric characterization and radiative transfer effects; (2) directed energy propagation metrics and irradiance capabilities of HELEEOS leverage the Scaling for HEL and Relay Engagement (SHaRE) code developed by MZA Associates Corporation; and (3) the Adaptive Optics Compensation of Thermal Blooming (AOTB) micro wave optics model developed by SAIC—Nutronics enables HELEEOS to very realistically capture thermal blooming, non-linear interactions with turbulence, and instabilities associated with employing one's beam control/adaptive optics sub-system. The infusion of such realistic atmospheric effects into the simulations allows HELEEOS to better assess variability and uncertainty in HEL system performance arising from spatial, spectral and temporal variations in operating conditions. As a result, HELEEOS-derived analyses are making important direct contributions to the joint warfighting community by helping to establish clear and fully integrated future program requirements.[2]

One advantage of utilizing Weather Cubes with HELEEOS is that it combines the atmospheric characterization and radiative transfer of LEEDR with HEL propagation modeling, drastically reducing analysis runtime. Additionally, the cubes provide greater atmospheric fidelity in all three spatial dimensions, which is especially important for longer-range HEL engagement scenarios. Users can input engagement specifications, such as platform and targets heights, laser firing times, and target velocity, to simulate realistic engagements. In addition to Weather Cubes integration into HELEEOS, AFIT CDE developed a metric to simplify highly complex and variable phenomena while directly correlating those effects with system design complexity and performance. The weather effects metric reduces complex environmental effects to a single quantity to assist system performance assessment and quantifies the benefits of using adaptive optics. The metric is comprised of three Strehl ratios that tie system performance to: (1) "higher-order" wave-front effects due to turbulence, (2) beam distortion due to heating of air along laser's propagation path, which is thermal blooming, and (3) extinction of the beam along the propagation path due to weather effects. The weather effects metric generally defines performance results in power-in-the-bucket (PIB) or peak irradiance as a function of range. Figure 14.10 is an example of the metric and shows a knee-in-the-curve, indicative of where in this case the benefit of a system's adaptive optics sophistication has greatest compensatory benefit in light of ambient weather conditions.

A characteristic performance study on a surface-to-air engagement (surface to 100 m over a 3.7 km path) was conducted for a low power HEL system targeting an unmanned aircraft system (UAS) at a short range. Weather Cubes were generated using 2015–2016 (four times a day—00 UTC, 06 UTC, 12 UTC, 18 UTC) GFS data, resulting in approximately 2,700 realistic atmospheres and consequently 2,700 performance results. Centered on the National Capital Region, these cubes encompass a 90 km × 90 km × 100 km region and provide temporal and spatial variations in atmospheric conditions. Turbulence effects were characterized employing a hybrid Tatarskii surface turbulence model, much in line with turbulence modelling discussed earlier. The surface turbulence model, developed by MZA Associates Corporation, was based on a continuous, 30-day field measurement campaign collected near the NCR. Surface values were blended with the 100 m value of the Tatarskii model, and the Tatarskii model defined index of refraction values from 100 to 100 km. Weather Cubes were integrated into HELEEOS for higher-fidelity simulation

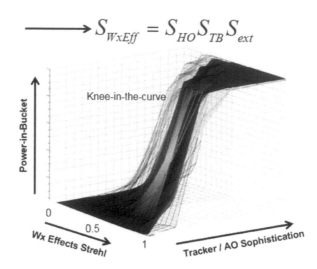

$$S_{WxEff} = S_{HO}S_{TB}S_{ext}$$

Figure 14.10 Weather Effects Metric reduces complex environmental effects by simplifying high-order optical turbulence (S_{HO}), thermal blooming (S_{TB}), and extinction (S_{ext}) effects into one value.

of atmospheric variability along the engagement path, and peak irradiance values were used to quantify performance values over the two year time period.

Since clouds can greatly hinder HEL performance, results were analyzed for engagements considering both cloud and cloud-free lines-of-sight for each engagement scenario providing an estimate of the percentage of time that clouds could affect the HEL engagement. This did not disregard the presence of clouds and adverse conditions but offered a more realistic analysis of HEL performance as the warfighter would not attempt to engage in such visual obstructions in real-world scenarios. The presence of clouds or other adverse weather (i.e., rain, fog) conditions that were intersected along the engagement path were indicated by high extinction coefficients stored within the Weather Cube data structure for these types of weather.

The results for this representative analysis were catalogued into 30 bins of performance values (e.g., 1–2, 2–3, 3–4 kW/m², etc.). Thirty percentile bins were chosen to satisfy the minimum sample size for a statistically relevant parametric assessment, ensuring reasonable confidence in a normal distribution. Both histograms and probability distribution functions (PDF) showcased the frequency and/or the percentage of instances the HEL system engaged the target at various performance thresholds when operated in ambient atmospheric conditions defined by the 2700 Weather Cubes. In addition to the 30 bins, performance percentiles were derived from the resulting statistical analysis. This provides further insight into the limitations and threshold specifications of the HEL system for the given region. Figure 14.11 displays the results for low power CUAS engagements for atmospheres where no clouds were encountered along the path. The grey bars display the percentage of time a particular performance threshold occurred within the resulting data set; performance values (i.e., peak irradiance) increase along the x-axis with the lowest performance values occurring nearly 26 percent of the time for this depicted analysis. The bins were specifically defined to ensure equal numbers of instances of performance outcomes, each outcome in turn tied to a specific Weather Cube.

The vertical lines delineate specific performance percentiles deduced from the entire set of performance results and overlays these according to the performance bins. Note that performance percentiles are labeled in an inverse manner, with the 0th-percentile being

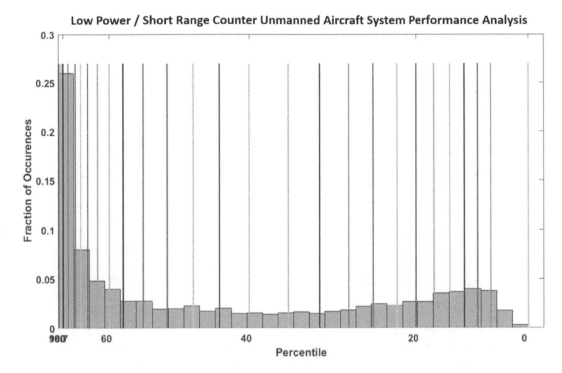

Figure 14.11 Histogram of low power/short range counter unmanned aircraft system performances over the 2015–2016 calendar years. The grey bars are representative of peak irradiance values (values not shown), and performance percentiles are represented by vertical lines. Engagement geometry as surface to 100 m over a 3.7 km path.

representative of maximum HEL performance and 100th-percentile representative of minimal performance. Each resulting performance value within a percentile bin is directly correlated to a specific set of atmospheric conditions (i.e., a Weather Cube) but is simultaneously representative of the entire performance percentile bin. This means that despite variations in key meteorological conditions, such as temperature and relative humidity, all Weather Cubes associated with each bin resulted in approximately the same HEL performance for this surface-to-air engagement.

One can extend the utility of this type evaluation by exploiting the underlying linkage of a specific performance outcome with a particular Weather Cube having a unique date-time-location identity. As such, it is possible to designate the most representative set of atmospheric conditions that one in turn could use to adequately characterize high and low system performance and values in between for any number of alternative HEL system designs—even though every Weather Cube volume in a percentile bin produced nearly identical HEL performance. Such designated representative "reference" atmospheres are simply Weather Cubes at a particular date, time, and location that are considered to best represent the conditions associated with a specific performance threshold and hence percentile. For the representative assessment here, reference atmospheres were determined for each of the 30 performance bins based on the following surface weather parameters: temperature, relative humidity, and optical turbulence. The distribution of all three meteorological conditions was evaluated via cluster analysis: the Weather Cube with the least amount of Cartesian distance from the centroid data point was selected as the reference atmosphere for the performance percentile bin. Figure 14.12 shows the distribution of

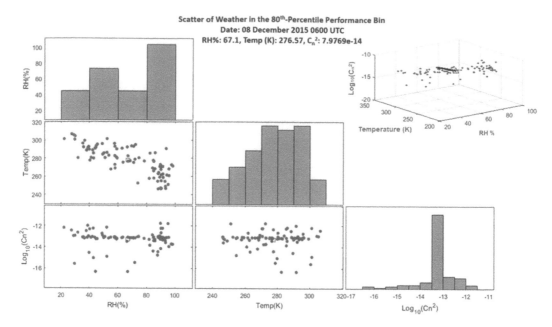

Figure 14.12 Distribution of surface temperature, relative humidity (RH), and optical turbulence (C_n^2) values for the 80th-percentile performance bin. Cluster analysis (top-right scatter plot) was used to determine that 08 December 2015 at 0600 UTC was the best representative or reference atmosphere for the 80th-percentiles performance.

surface temperatures, relative humidity values, and optical turbulence strengths (C_n^2) associated with the set of Weather Cubes that fell into and resulted in performance defining the 80th-percentile bin. The scatter plot on the top right of the Figure 14.12 provide a quick overview of the how these surface meteorological parameters vary for the set of Weather Cubes. Note the nearest point selection according to the centroid of the collection of all data. By definition that selected point is tied to a unique Weather Cube and is now designated the reference Weather Cube for the 80th percentile performance. The bar graphs and scatter plots by meteorological parameter provides even deeper insight.

A collection of reference Weather Cubes provide developers and warfighters a tool to assess whether alternative HEL system concepts will perform and achieve the desired mission effects. Additionally, the reference Weather Cubes inherently have all of the vertical profiles of meteorological data to assess how these alternative HEL concepts would perform using multiple concepts of employment. If this HEL system is not generating the necessary power-in-the-bucket when operating at the surface, would it be more beneficial to elevate the HEL? Would the HEL perform the same if the target were at a different altitude or range? Weather Cubes provides the user a plethora of opportunities to simulate these engagements and assess the feasibility of the system in other realistic scenarios. A deeper dive into the weather conditions, not only at the surface but aloft, offers a snapshot into the proposed questions. Figure 14.13 displays vertical profiles of (a) temperature and dew point temperature, (b) relative humidity, (c) optical turbulence, and (d) atmospheric attenuation effects for a reference atmosphere. Presuming the primary weather parameter causing the HEL degradation is the strong optical turbulence at the surface, these effects could be minimized if the HEL platform were raised to an elevation of 250 meters above the surface, for example.

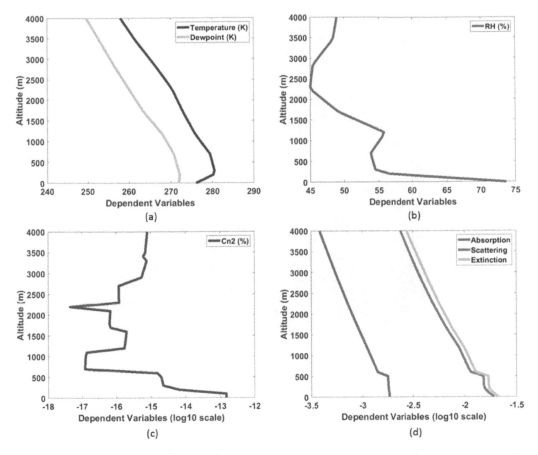

Figure 14.13 Vertical profiles of (a) temperature and dew point temperature, (b) relative humidity, (c) optical turbulence (C_n^2), and (d) atmospheric attenuation effects for the 80th-percentile reference atmosphere.

Imaging software enhancements

Weather cubes also have a lot to offer for long-range imaging applications, such as target tracking, remote sensing, and LADAR. In such applications, the quality of the images depends strongly on atmospheric conditions. Factors such as optical transmission, clouds, turbulence, path radiance, and backscatter often change rather significantly with both altitude and location. Unfortunately, numerical simulations of imaging systems typically assume that the atmosphere changes only with altitude, an assumption which poorly represents many real-world scenarios. However, including regional variations via Weather Cubes provides additional realism to the simulations, yielding more reasonable estimates for system performance.

This section covers several of the aspects of imaging simulations which are enhanced by Weather Cubes, namely path radiance, backscatter, and turbulence. All of the results were generated using the computer model PITBUL (Physics-based Imaging and Tracking of Ballistics, UAVs, and LEO satellites), which was developed at AFIT.[11] This model is used to evaluate the performance of imaging and tracking systems, often for high energy laser applications.

When light passes through the air, some of it scatters off of molecules and aerosols, thus creating path radiance. Path radiance is often caused by the scattering of sunlight, however it may also come from earth shine, sky glow, or even the moon or stars. Because

Figure 14.14 A synthetic, false-color image of a Boeing 747 flying within the atmospheric boundary layer. The large horizontal and vertical variations in path radiance are predominately caused by the sun's position and changes in atmospheric conditions, respectively.

the light scatters in all directions, some of it travels into the imaging system, where it is generally considered a source of noise. Properly modeling this path radiance is key to defining imaging system performance, and the horizontal variations captured by Weather Cubes become important over long paths.

Figure 14.14 shows a Boeing 747 flying within the atmospheric boundary layer. It is a false-color image captured in the visible spectrum. The sun is just off to the right, which creates large horizontal variations in path radiance due to changes in the scattering coefficient with angle. The image also shows large vertical variations. The dark region at the bottom is the earth's surface (deciduous broadleaf forest). It is much darker than the path radiance in this case. Above the horizon, most of the vertical variations are caused by changes in aerosol size and concentration within and just above the atmospheric boundary layer. All of this highlights the importance of capturing both vertical and horizontal changes in atmospheric conditions.

Next, we turn our attention to backscatter of active (laser) illumination. LADAR, active target tracking, and certain remote sensing systems use a laser to illuminate the target. Similar to path radiance, some of the laser light scatters in the atmosphere, sending photons back towards the imaging system and creating noise known as backscatter. When the path between the imaging system and the target is long, atmospheric conditions can change drastically, thus impacting the amount of backscatter. So, the 3D information of Weather Cubes can be very important.

As an example of backscatter, Figure 14.15 shows both (a) the illumination pattern on target and (b) the return light. It is again a false-color image. The illumination pattern in (a) is broken up into bright and dark patches due to atmospheric turbulence, as

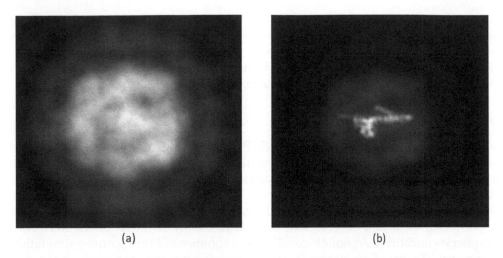

(a) (b)

Figure 14.15 Active (laser) illumination of a generic UAV. The plots show both (a) the scintillated laser illumination on target and (b) the return light. The return includes both direct reflection off the target's hard body as well as light backscattered off of molecules and aerosols.

we discuss shortly. The resulting image of the target (b) includes both backscatter and direct reflection off the target's hard body. The direct reflection is very bright. Around that direct reflection, we see the backscatter of the laser illumination, which significantly reduces target contrast in this case.

The third effect is atmospheric turbulence. Turbulence can blur and distort images of the target. Further, when the imaging system uses active illumination, turbulence can break up the illumination beam into bright and dark patches via a process known as scintillation. Figure 14.15a shows a prime example of such scintillation, which causes variations in illumination intensity.

Both scintillation and image distortion can severely degrade system performance. Also, both effects are impacted by changes in turbulence strength over the path, and Weather Cubes add such information to the simulations. Figure 14.16 shows a synthetic image of

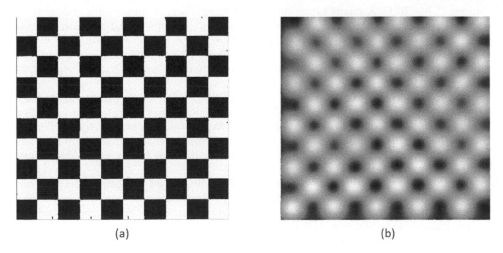

(a) (b)

Figure 14.16 A checkerboard target both (a) before and (b) after turbulence distortions. The turbulence effects vary over space (and time), as one can see in (b).

a checkboard target both (a) before and (b) after addition of turbulence. The turbulence causes the checkers to shift and blur. As one can see in (b), these distortions vary in space. Also, their strength depends on the turbulence along the whole path between platform and target, information which is provided by Weather Cubes. So, we have seen that Weather Cubes enhance the modeling of path radiance, backscatter, and turbulence.

Summary

Weather Cubes, anchored to LEEDR and NWP data, provide expansive data arrays of spatial and temporal variations of meteorological and environmental data that demonstrate how clouds, precipitation, and aerosol haze affect ultraviolet through radio frequency radiative (light) propagation. Furthermore, Weather Cubes always create correlated, atmospheric parameters that have actually occurred or might occur together because they are based on actual observations or state-of-the-art scientific predictions of those atmospheric conditions. Although developed to enhance military gaming simulations to render physically-based atmospheric effects, this robust model and analysis tool has far exceeded its initial purpose and can be used to further remote sensing and imaging capabilities, directed energy performance assessments, and climate research. Detailed applications of this innovative tool have been discussed, and the results have been shown.

References

1. Fiorino, S. T., R. J. Bartell, G. P. Perram, D. W. Bunch, L. E. Gravley, C. A. Rice, Z. P. Manning, and M. J. Krizo, 2006. The HELEEOS atmospheric effects package: A probabilistic method for evaluating uncertainty in low-altitude high energy laser effectiveness. *J. Dir. Energy*, 1, 347–360.

2. Fiorino, S. T., R. J. Bartell, M. J. Krizo, G. L. Caylor, K. P. Moore, and S. J. Cusumano, 2008b. Validation of a worldwide physics-based, high spectral resolution atmospheric characterization and propagation package for UV to RF wavelengths. *Proceedings of SPIE*, Vol. 7090 of, San Diego, CA, SPIE Optics and Photonics, 70900I.

3. Burley, J., S. Fiorino, B. Elmore, and J. Schmidt, 2017. A fast two-stream-like multiple scattering method for atmospheric characterization and radiative transfer. *J. Appl. Meteorol. Climatol*, 56(11), 3049–3063.

4. Shirey, S. M. (2016). A relative humidity based comparison of numerically modeled aerosol extinction to LIDAR and adiabatic parameterizations (Master's thesis). Air Force Institute of Technology, Wright-Patterson Air Force Base, OH, 65p.

5. Fiorino, S. T., R. M. Randall, M. F. Via, and J. L. Burley, 2014. Validation of a UV-to-RF high-spectral-resolution atmospheric boundary layer characterization tool. *J. Appl. Meteorol. Climatol.*, 53, 136–156.

6. Burley, J., S. Fiorino, B. Elmore, and J. Schmidt (2018). A Remote Sensing and Atmospheric Compensation Tool for Assessing Multi-Spectral Radiative Transfer through Realistic Atmospheres and Clouds. Manuscript submitted to *J. Atmos. Oceanic Tech.* for publication.

7. Burchett, L. R. (2016). Methods for passive remote turbulence characterization in the planetary boundary layer (Doctoral dissertation). Air Force Institute of Technology, Wright-Patterson Air Force Base, OH, pp. 211.

8. Meier, D. C. (2015). Operational exploitation of satellite-based sounding data and numerical weather prediction models for directed energy Applications (Doctoral dissertation), Air Force Institute of Technology, Wright-Patterson Air Force Base, OH. Retrieved from http://www.dtic.mil/docs/citations/AD1003080.

9. Bendersky, S., N. S. Kopeika, and N. Blaunstein, 2004. Atmospheric optical turbulence over land in middle east coastal environments, prediction modeling and measurements, *Appl. Opt.*, 43, 4070–4079.

10. Jursa, A. S., and US Air Force Geophysics Laboratory, (1985). Handbook of geophysics and the space environment. *Hanscom Air Force Base, Mass.: Air Force Geophysics Laboratory, Air Force Systems Command*, US Air Force, Springfield, VA.
11. Van Zandt, N. R., J. E. McCrae, and S. T. Fiorino. PITBUL: A physics-based modeling package for imaging and tracking of airborne targets for HEL applications including active illumination, *Proceedings of the SPIE 8732: 87320H*, SPIE International, Bellingham, WA, 2013.

chapter fifteen

Innovative approach to infrastructure resilience
A case study of evaluating Department of Defense sites for small modular reactors

Olufemi A. Omitaomu, Bandana Kar, Randy J. Belles,
Michael P. Poore, Gary T. Mays, and Budhendra L. Bhaduri

Contents

Introduction

The Department of Defense (DOD) spends about $4 billion annually on energy required to power its installations. Facility managers at DOD have objectives and constraints that are very different from their private sector counterparts, since most of DOD's energy consumption goals are often mandated by Congress through legislation or by President of the United States through Executive Orders [1]. There are approximately 700 DOD military bases in the United States; and because the power requirements for these sites include supporting base missions and base infrastructures that often include tens of thousands of on-site workers and site residents, the power requirements could be considerable. Due to their national security mission, many of these sites are good candidates for reliable, dependable, compact, and secure on-site power generation capabilities [2]. A Geographic Information Systems (GIS) based site suitability approach is extensively used to evaluate and rank available land areas for facility locations [3]. This approach has been used for identifying commercial buildings, waste disposal sites, and nuclear waste sites [3,4]. However, previous applications of this approach are for site-specific or regional studies. In 2008, the Department of Energy's Oak Ridge National Laboratory (ORNL) started an effort to develop a high-resolution computational framework to identify suitable areas for different energy sources at the national scale. In 2011, the idea was formalized into a decision support tool called the **O**ak **R**idge **S**iting **A**nalysis for Power **G**eneration **E**xpansion (OR-SAGE). The tool uses industry-accepted approaches and criteria, an array of geospatial data sources to screen sites and identify candidate locations to site different power generation technology applications. The basic premise of the tool requires development of exclusionary, avoidance, and suitability criteria to evaluate sites for a given energy source, such as small modular reactors (SMRs). For specific applications of the tool, it is necessary to identify site selection criteria (SSC) that encompass key benchmarks and environmental characteristics for that application. These criteria might include population density, seismic activity, proximity to water sources, proximity to hazardous facilities, avoidance of protected lands and floodplains, susceptibility to landslide hazards, and others.

For evaluation purpose, the OR-SAGE tool divides the contiguous United States into 100 m by 100 m cells, and applies successive SMR-appropriate SSC to each of the 700 million cells representing the contiguous United States [5]. If a cell meets each SMR criterion, then the cell is included as a candidate to be integrated in the possible siting of an SMR. While some SSCs parameters preclude siting a facility because of an environmental, regulatory, or land-use constraint, other SSCs help identify less favorable areas, such as proximity to hazardous operations. More than forty datasets represent the SSCs, which when combined help identify areas with challenges and advantages for energy sources of interest.

The tool has been applied to various applications [2,4–6]. This chapter presents an application of the OR-SAGE tool for screening a sample set of DOD's military base sites for possible powering with SMRs. The purpose of this siting evaluation is to demonstrate the capability of an innovative DOE tool for DOD and Emergency Management applications.

Approach and methodology

The key to the approach implemented in this study was to use industry-accepted practices in screening sites and then to employ the proper array of data sources to identify candidate areas using computational capabilities and geospatial technologies available at ORNL. The focus of the ORNL electrical generation source siting study is to identify candidate areas from which potential SMR sites might be selected, stopping short of performing any detailed site evaluations or comparisons. This approach is designed to quickly screen for potential sites based on different environmental characteristics. Although this is a top-down approach to SMR siting at a national and regional scale, the fundamentals of this approach could be applied for bottom-up analysis of specific sites that may be suitable for SMR deployment as demonstrated in this chapter.

It is desirable to have the capability to compare areas that meet all the designated SMR site selection criteria, because some areas may prove to be more desirable than other areas. Thus, undertaking a cross-comparison of areas based on SMR SSCs could better inform a decision about where to exert efforts to engineer around a siting issue. Using this tool, cross-comparison, and scoring of the DOD sites has also been demonstrated [6].

Multi-criteria evaluation

Multi-Criteria Evaluation (MCE) is a decision-making technique that is used for tasks that may have a diverse set of possible outcomes with conflicting objectives and may be influenced by numerous criteria [7–10]. Although the technique emerged in early 1970s, and its origin lies in mathematics and operations research, MCE is widely used for various environmental decision-making tasks involving social/economic and environmental criteria, such as site selection for waste disposal, site suitability for shelter location, and resource management and evaluation analysis [8,10,11]. The MCE technique, like other approaches and theories, such as Multi-Criteria Decision Analysis, Multi-Objective Decision Making, Multi-Attribute Decision Making, Multi-Attribute Utility Theory, public choice theory, allows users to determine an option based on certain criteria [8–11].

Weighted Linear Combination (WLC) is a type of MCE that allows stakeholders to rank certain criteria based on their usability and effectiveness for specific decision-making tasks [3,10]. Each criterion is classified based on certain threshold, and then each classified criterion is multiplied with its corresponding rank/weight, and finally, all multiplied outputs are added together to determine a ranked spatial distribution of all criteria [3,10,12]. In this study, although a similar WLC approach was implemented, all criteria used to determine suitable sites for SMR locations were assigned an equal weight given that each criterion is crucial to siting a SMR. The equation used in the deployment of OR-SAGE tool can be written as [3]:

$$ss_i = \left(\sum_{j=1}^{n} FR_j * w_j \right)$$

where:
    ss_i is the summary suitability score for location i
    FR_j is the rating for criterion j
    n is the total number of criteria included in the tool
    w_j is weight assigned to each criterion j

Evaluation of selected department of defense sites

There are approximately 700 DOD military bases in the United States (Figure 15.1). It was assumed that only larger DOD sites would be candidates for hosting an SMR facility; therefore, only bases with area greater than 1000 acres (since SMR requires only about 50-acre area), and those that offer greater flexibility considering their mission objectives were considered. Based on the analyses, about 170 military bases passed all the siting criteria; however, the siting details for two Air Force bases, two Military bases and one Naval base are discussed in this chapter. For additional information, interested readers should refer to Reference [2].

Although the current and/or future load profiles for the sites were not considered, the initial screening characterized all land in the contiguous United States regarding the potential for hosting a near-term SMR design. This analysis did not specifically consider proximity to load requirements or national interests (e.g., critical loads) brought about by the missions, mission support, and residents of large military bases. Instead, it assumes load requirements exist at such a site, and potential benefits of replacing or augmenting the power provided by off-site electric power plant also exist.

Figure 15.1 Polygons (in blue) depict the US military base locations relative to SMR aggregate map (in green).

Near-term SMRs are based on light-water reactor (LWR) technology with compact design features that are expected to offer a host of safety, siting, construction, and economic benefits [2]. These smaller plants are ideally suited for small electric grids and for locations that cannot support large reactors, thus providing utilities with the flexibility to scale power production as demand changes by adding modules or reactors in phases for additional power. The near-term SMR designs are based on existing pressurized-water reactor (PWR) technology. They are characterized as "integral" PWRs (iPWRs) since these plants have major equipments such as pumps, steam generators, and pressurizers all located within the pressure vessel in an integrated, compact design. Individual reactor units in these designs are typically in the 25- to 250-megawatt-electric (MW(e)) power range. Modular installations of iPWRs can range up to 540 MWe based on proposed vendor configurations. Note that other longer-term advanced SMR designs—such as high-temperature gas reactors, liquid metal reactors, and molten-salt reactors—were not analyzed in this study; however, the screening parameters selected for the near-term iPWR reactors are expected to also encompass these advanced SMRs, except for cooling water.

Review of small modular reactor site selection and evaluation criteria

Based on preliminary design information and expert judgment, it is assumed that an SMR iPWR base design package (single unit or multi-module) from each vendor can easily be accommodated on a 50-acre footprint. For locating a SMR on a 50-acre land, a binary approach was used to exclude the following land criteria. A more detailed discussion of each individual SSC selected for SMR siting is available in [5]. The following criteria are tracked on a cell-by-cell basis for the entire contiguous United States.

1. Land with a population density greater than 500 people per square mile (including a 10-mile buffer) is excluded.
2. Wetlands and open water are excluded.
3. Protected lands (e.g., national parks, historic areas, wildlife refuges) are excluded.
4. Land with a moderate or high landslide hazard susceptibility is excluded.
5. Land that lies within a 100-year floodplain is excluded.
6. Land with a slope of greater than 18 percent (~10°) is excluded.
7. Land areas that are more than 20 miles from cooling water makeup sources with flow of at least 65,000 gallons per minute (gpm), based on a 540 MWe modular iPWR installation, are excluded for nominal SMR plant applications.
8. Land too close to identified fault lines is excluded (the length of the fault line determines the standoff distance).
9. Land located in proximity to hazardous facilities (airports and oil refineries) is avoided.
10. Land with safe shutdown earthquake (SSE) peak ground acceleration (2 percent chance in a 50-year return period) greater than 0.5 g is excluded.

The OR-SAGE tool tracks the SSC parameters for each 100 by 100 m cell. As a result, not only can the cells that are clear of all the SSC layer exclusions be displayed visually, but also cells that are tripped by one, two, and three or more exclusions can be tracked and displayed. This is known as the "SMR composite map" (Figure 15.2a), which is a powerful aspect of the OR-SAGE tool, because it allows areas with a limited number of siting challenges to be identified. These sites with certain challenges could be made suitable for SMR siting with engineering solutions.

Results from the analyses of these DOD sites demonstrate that OR-SAGE provides useful insights for evaluating options and challenges related to powering these sites with an SMR. The sites are typically quite large—a criterion for their initial selection—and have considerable land areas that satisfy many of (or all) the siting criteria examined. Note that site-specific hazards such as training ranges, ordnance handling, storage areas, etc., were not considered. Some on-site hazards such as airfields were qualitatively considered.

Beyond designating areas as suitable (i.e., meeting all the siting criteria at a specific set of threshold values), specifically, the green space in the SMR composite map shown in Figure 15.2a, it is desirable to have the capability to cross-compare areas that meet all the

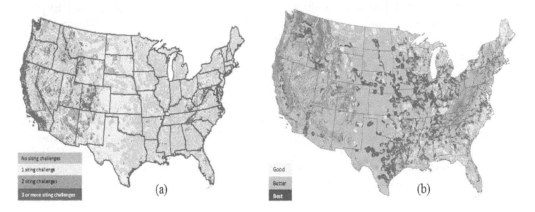

Figure 15.2 Characterization of land areas in the contiguous United States based on selected input values. (a) Sites that satisfied one, two, and three or more siting criteria; (b) Sites ranked based on their suitability based on all the siting criteria.

designated SMR SSC. This allows targeting areas that may ultimately prove to be more desirable than other areas. This scoring technique was demonstrated for suitable areas for SMR siting and shown in Figure 15.2b. In this figure, the highly-suitable areas are shown in dark green, moderately-suitable areas are shown in medium green, and areas with low-suitability are shown in light green. All three green hues meet all the SMR siting criteria at the select set of values as shown in Figure 15.2a. The high-suitability areas comprise of 21.6 percent of the contiguous United States, which includes land near major cities such as Chicago, Nashville, Atlanta, Dallas, and Houston. The moderately-suitable areas represent 15.5 percent of the contiguous United States. This includes suitable land with tremendous water resources but less power demand such as seen in Montana. Additional scoring elements could add greater differentiation. Such a cell by cell comparison methodology could assist in the evaluation of SMR site selections.

Site evaluation results

Each DOD site was also evaluated visually using Google Earth and similar Internet mapping resources to identify proximity to nearby towns, structures or facilities representing potential hazards. We present detailed results for five DOD sites with at least 1000 acres that were evaluated for SMR siting using the OR-SAGE tool and all the SMR SSC.

Eglin air force base

Eglin Air Force Base, located in the western Florida panhandle (Figure 15.3), covers approximately 420,000 acres (about 650 square miles) in Santa Rosa, Okaloosa, and Walton Counties of Florida. Nearby towns include Niceville a few miles south of the center of the base; Destin, approximately 5 miles south; Pensacola, approximately 20 miles west; Crestview, approximately 5 miles north; and Freeport, approximately 2 miles southeast of the base's perimeter. This base hosts the 33rd Fighter Wing; 53rd Wing; Air Force Research Laboratory Munitions Directorate; 6th Ranger Training Battalion; Joint Deployable Analysis Team; Defense Threat Reduction Agency Research and Development Counter Weapons of Mass Destruction, Weapons and Capabilities Division; 96th Test Wing; Armament Directorate; 919th Special Operations Wing; 20th Space Squadron; an Air Force Operational Test and Evaluation Center detachment; Naval School Explosive Ordnance Disposal command; 728th Air Control Squadron; 7th Special Forces Group; and others. Descriptions of the military units and base history are readily available on the Internet [13,14].

Approximately 50 percent of the air force base area that covers approximately 650 square miles is forested. The permanent population surrounding the base within 1 mile of the perimeter is approximately 337,000. The base has about 15,000 workers and about 8,000 people reside on the base in about 2,300 households [13]. Because of Base Realignment and Closure (BRAC) Commission actions in 2005 affecting other bases, overall base population at Eglin AFB may grow to 38,000 in the next few years, and Eglin may see military construction totaling $732 million over this time-period.

Eglin Air Force Base is a large, federally controlled site with a well-trained, well-armed security force. Many of the activities and missions carried out on this base are state-of-the-art, high technology endeavors. Personnel living or working on the base tend to have considerable familiarity with technology- and security-related operations associated with

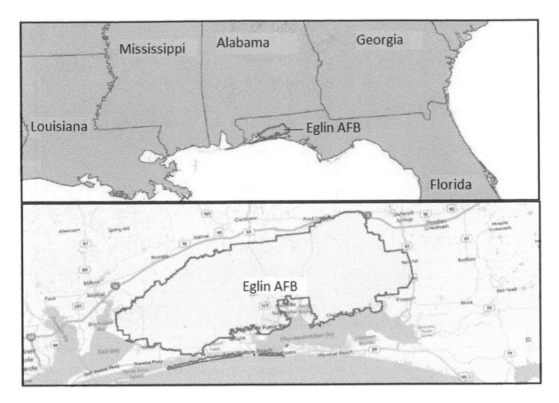

Figure 15.3 Eglin Air Force Base.

nuclear power plant operations as well as with necessary construction-related activities. Power demands on the base associated with military missions, and local infrastructure for current residents and workers tend to be considerable and need to be feasible for a site-located SMR. Growth in electrical demand and energy could occur at this base due to missions, military staffs, residents, employees, and constructions due to BRAC-related consolidations and closures of other bases.

Figure 15.4 depicts the results of OR-SAGE tool for Eglin Air Force Base, and Figure 15.5 shows the composite map for the air force base. According to these figures, the base has partial site issues, and approximately 71 percent of the 420,000-acre site meets multiple conventional standards for SMR siting on the base. Although the OR-SAGE tool excludes an area up to 5-mile radius buffer surrounding a commercial airport, the airfields on site were not automatically removed from consideration due to special circumstances of this OR-SAGE application. Because military facilities are already an SMR SSC exclusion factor for commercial SMR siting, the airfields on the base must be considered separately. Excluding land covered by a 5-mile radius from the center of an airfield essentially excludes an additional 40 percent of Eglin Air Force Base for SMR siting consideration. Further exclusion of runways reduces the available land area to approximately 30 percent of the total Eglin Air Force Base land area. Because the site meets current Nuclear Regulatory Commission (NRC) Red-Guide (RG) 4.7 recommendations for population density, no other requirements were considered for SMR population siting, and the site was classified as suitable for siting an SMR.

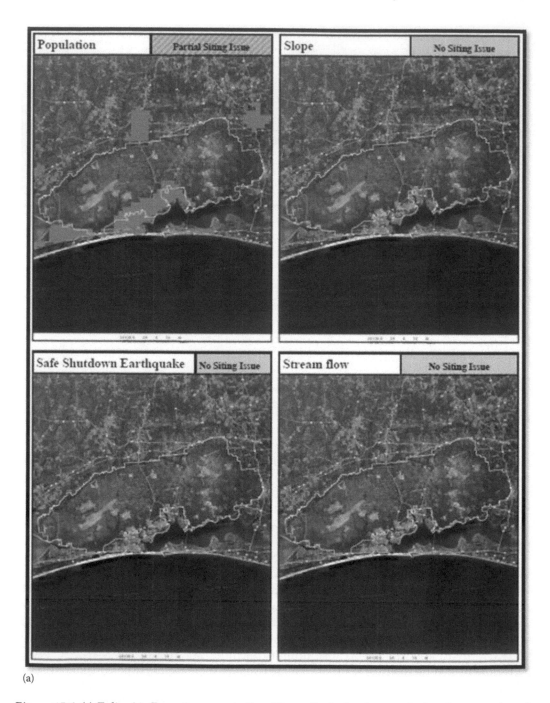

(a)

Figure 15.4 (a) Eglin Air Force Base meets the siting criteria for slope, safe shutdown earthquake, and streamflow but partially meets the criterion for population.   (*Continued*)

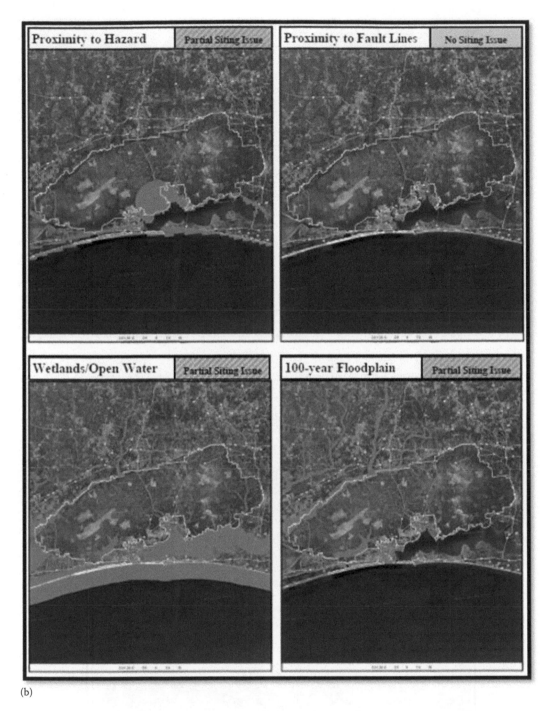

(b)

Figure 15.4 **(Continued)** (b) Eglin Air Force Base meets the siting criterion for fault lines but partially meets the criteria for proximity to hazards, wetlands/open water, and 100-year floodplain.

(Continued)

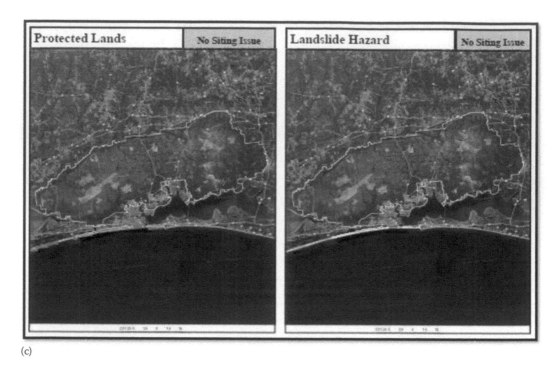

(c)

Figure 15.4 **(Continued)** (c) Eglin Air Force Base meets the siting criteria for protected lands and landslide hazard.

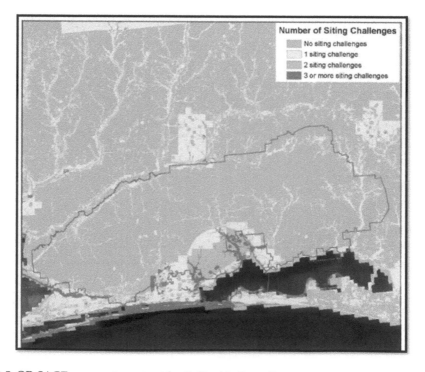

Figure 15.5 OR-SAGE composite output for Eglin Air Force Base.

Beale air force base

As shown in Figure 15.6, Beale Air Force Base is located on about 21,000 acres (about 33 square miles) in the central area of northern California, about 10 miles east of the city of Marysville. Beale Air Force Base is home to the 9th Reconnaissance Wing. Interstate 80 is about 20 miles east of the base, and Interstate 5 is about 35 miles west. Approximately 4,000 military personnel are on-site at any given time [15], and about 1,300 people live on the base. Descriptions of the installation, missions, and base history could be obtained from ref. [15,16].

Figure 15.7 depicts the results of using OR-SAGE for Eglin Air Force Base; the composite map for the base is shown in Figure 15.8. The results indicate that approximately 75 percent of the 21,000-acre Beale site meets multiple conventional standards for siting an SMR on the base facility. The airfield on site was also not automatically removed from consideration. Excluding land from consideration within a 5-mile radius buffer surrounding the airfield essentially excludes all but a few hundred acres of Beale Air Force Base suitable for siting an SMR.

The airport runway runs almost north and south on the northwestern edge of the base. If an off-axis relaxation in the exclusion distance requirement to the airfield is permitted from 5 to 2 miles, then approximately 50 percent of the site would be suitable for siting an SMR. If a 3-mile exclusion distance is applied, then approximately 25 percent of the site would be suitable for siting an SMR. Note that the airport buffer criterion is an

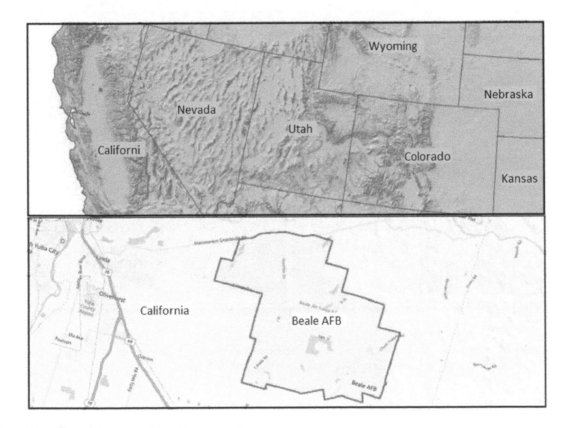

Figure 15.6 Beale Air Force Base.

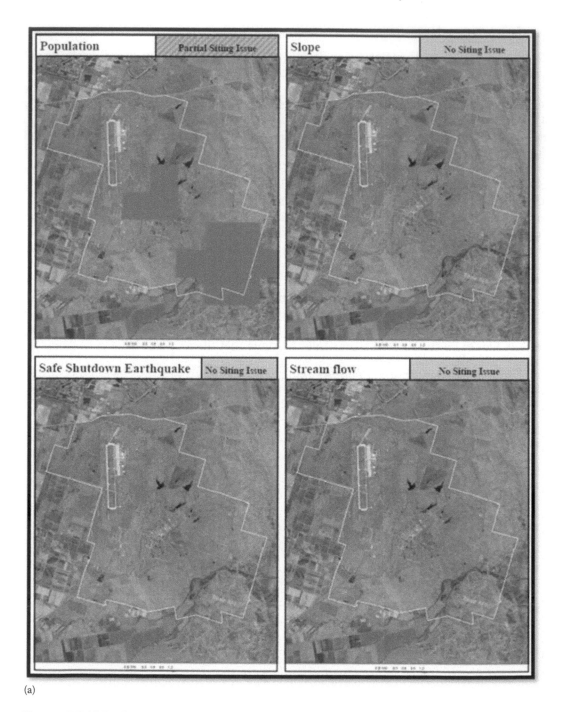

(a)

Figure 15.7 (a) Beale Air Force Base meets the siting criteria for slope, safe shutdown earthquake, and streamflow but partially meets the criterion for population. *(Continued)*

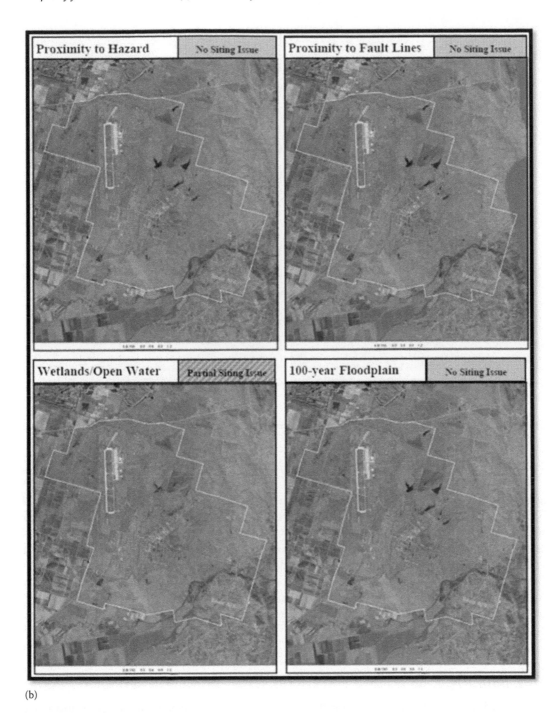

(b)

Figure 15.7 **(Continued)** (b) Beale Air Force Base meets the siting criteria for proximity to hazards, fault lines, and 100-year floodplain but partially meets the criterion for wetlands/open water.

(Continued)

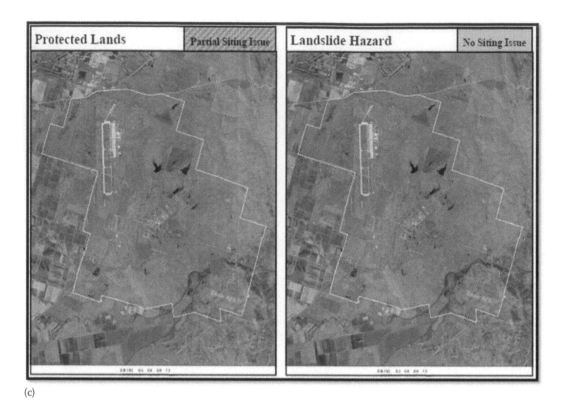

(c)

Figure 15.7 **(Continued)** (c) Eglin Air Force Base meets the siting criterion for landslide hazard but partially meets the criterion for protected lands.

avoidance recommendation. This along with the underground construction of a typical SMR may support a relaxation of the airport avoidance buffer distance. Nuclear power restrictions in place in California were not considered for this study since the siting tool is based on site characteristics and does not incorporate state or local policy, land use, or zoning issues. Like the Eglin site, this site meets the current NRC RG 4.7 recommendations for population density without additional consideration for relaxed SMR population siting requirements. Unless a relaxation in the avoidance area associated with the on-site airfield is permitted, this site is not a likely candidate for consideration of siting an SMR.

Fort Benning

Fort Benning is located on about 165,000 acres (~260 square miles) on the Chattahoochee River border between Georgia and Alabama, approximately the north-south center of each state, although over 90 percent of the base area is in Georgia (Figure 15.9) [17]. Fort Benning is home to the US Army Armor School, US Army Infantry School, Western Hemisphere Institute for Security Cooperation, 75th Ranger Regiment, 3rd Brigade–3rd Infantry Division, and many other corps, units, institutes, and agencies. The base is accessible via Interstate 185 that enters Columbus, Georgia, and the base from the north, and via

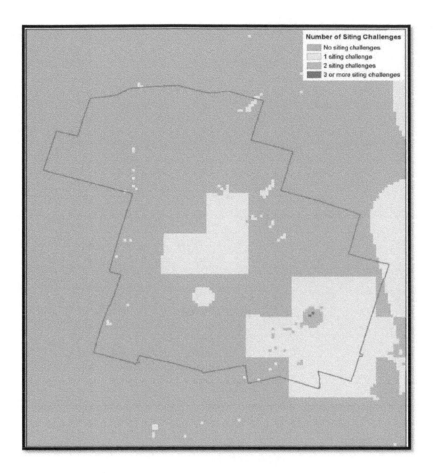

Figure 15.8 OR-SAGE composite output for Beale Air Force Base.

highway 27 that bisects the base from northwest to southeast. The base also has numerous rivers and streams.

Approximately, 115,000 soldiers train at Fort Benning each year, and about 130,000 soldiers, employees, families, and contractors are on-site at any given time [18]. Services and resources are available on the base for military staff, families, employees, and service contractors like a small town, including lodging and housing, schools, hospital, various shopping facilities, restaurants, library, cultural and recreational amenities, and other goods and services [17,18].

Given that Fort Benning is a well-armed security base, power demands on the base associated with military missions and local infrastructures for current residents and workers would also be considerable, thereby requiring the installation of an on-site SMR. Figure 15.10 depicts the suitability of Fort Benning based on the SSCs using OR-SAGE tool. As evident from this figure, the Fort Benning site has partial site issues with population and landslide hazard [18]. This essentially divides the base into three distinct areas that meet all SMR SSC. Approximately 50 percent of the 165,000-acre site meets multiple conventional standards for consideration of siting an SMR on the base facility, and three large areas are suitable for SMR siting.

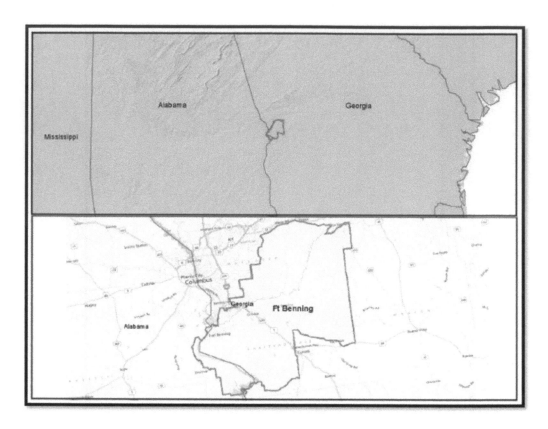

Figure 15.9 Fort Benning.

Figure 15.11 depicts the composite map derived from using the criteria discussed earlier using the OR-SAGE tool. The siting challenges in Fort Benning for locating SMR are predominantly in the northwestern part of the base. Because this site like other air force bases meets current NRC RG 4.7 recommendations for population density, additional considerations for relaxed SMR population siting requirements based on reduced source term were not used. However, the composite map does not reflect specific hazards associated with the site, such as ordnance storage areas, weapons ranges, etc., that could render some areas of significant size unsuitable for siting a reactor. Nonetheless, this site should be considered as favorable for siting an SMR.

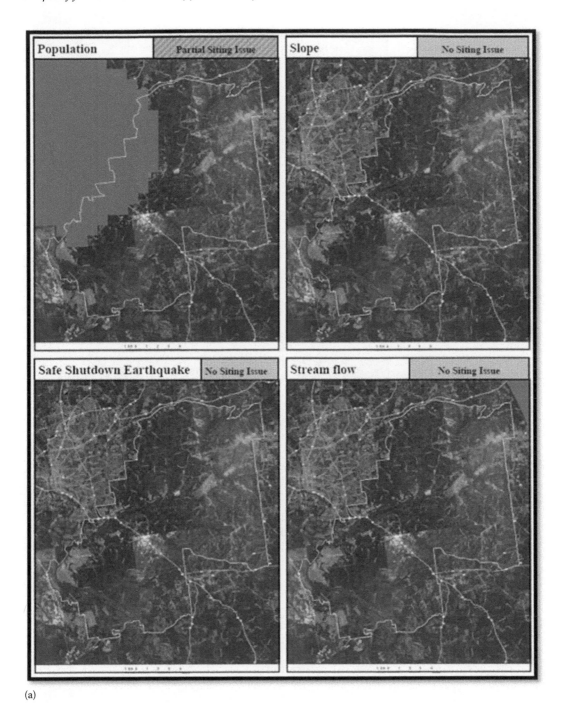

(a)

Figure 15.10 (a) Fort Benning meets the siting criteria for slope, safe shutdown earthquake, and streamflow but partially meets the criterion for population. *(Continued)*

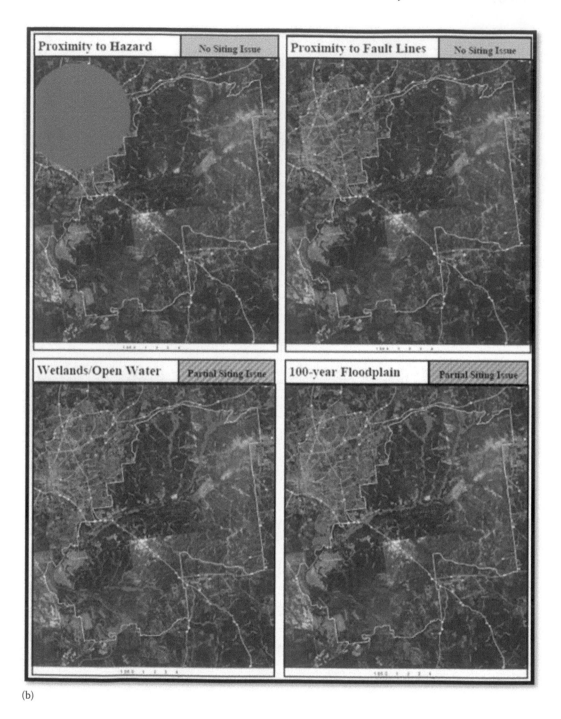

(b)

Figure 15.10 **(Continued)** (b) Fort Benning meets the siting criteria for proximity to hazards, fault lines, but partially meets the criteria for 100-year floodplain and for wetlands/open water.

(Continued)

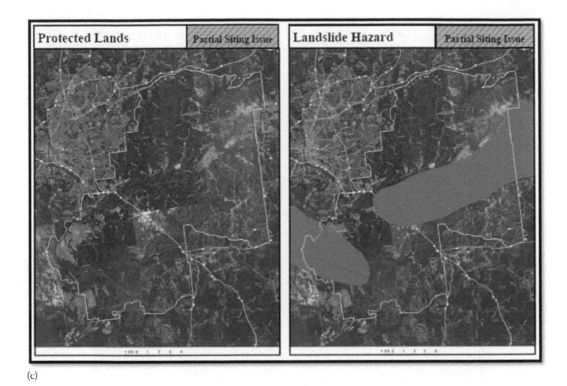

(c)

Figure 15.10 **(Continued)** (c) Fort Benning partially meets the siting criteria for protected lands and landslide hazard.

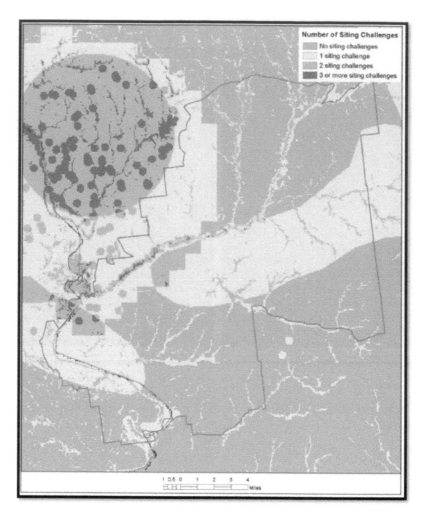

Figure 15.11 OR-SAGE composite output for Fort Benning.

Fort Campbell

Fort Campbell is located on about 93,000 acres (about 145 square miles) on the western border of Tennessee and Kentucky between the towns of Hopkinsville, Kentucky, and Clarksville, Tennessee, about 60 miles northwest of Nashville, Tennessee (Figure 15.12) [19]. Fort Campbell is home to the 101st Airborne Division, 5th Special Forces Group (Airborne), 160th Special Operations Aviation Regiment, 52nd Ordnance Group, US Army Medical Command, Installation Management Command, Network Enterprise Technology Command, US Air Force 19th Air Support Operation Squadron (ASOS) and Detachment 418 Weather Squadron, and other tenant groups, corps, units, institutes, or agencies. Interstate 24 is east of the base.

The population of Fort Campbell is about 30,000 active duty soldiers and over 50,000 family members [19,20]. Services and resources are available on the base for military staff, families, employees, and service contractors like a small town. The nearest major

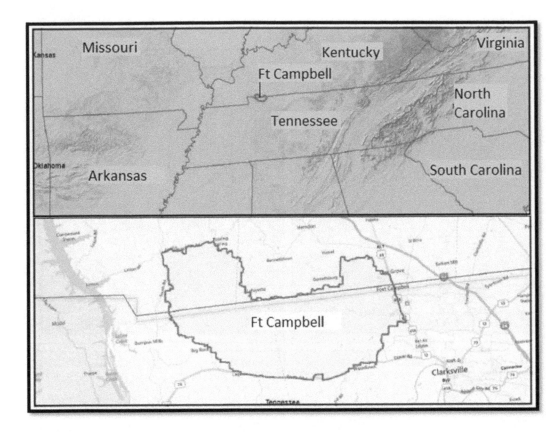

Figure 15.12 Fort Campbell.

fault line based on USGS data is noted to be 570 miles west in Oklahoma. The maximum safe shutdown earthquake for the site is below 0.3 g peak ground acceleration, and the maximum slope on the site is about 16 percent. Cooling water for reactors is available from the Cumberland River, and the base has access to major highways, water transport, and rail transport.

Figure 15.13 displays the suitability of the site based on individual criterion used in the OR-SAGE tool. Evidently, the site meets four criteria partially–population density, location of wetlands, 100-year flood plain, and protected lands; however, overall the site meets all the criteria discussed earlier for siting a SMR. According to the suitability determined by the tool, Fort Campbell has a partial site issues with population in the eastern portion of the base [20]. Because the base meets current NRC RG 4.7 recommendations for population density without additional consideration for relaxed SMR population siting requirements based on reduced source term, the site should be classified as favorable for siting an SMR.

Based on the composite map generated by the tool (Figure 15.14), approximately 80 percent of the 93,000-acre site meets multiple conventional standards for consideration of siting an SMR on the base facility. However, the composite map does not reflect specific hazards associated with the site, such as ordnance storage areas, weapons ranges, etc., that could render some areas of significant size unsuitable for siting a reactor.

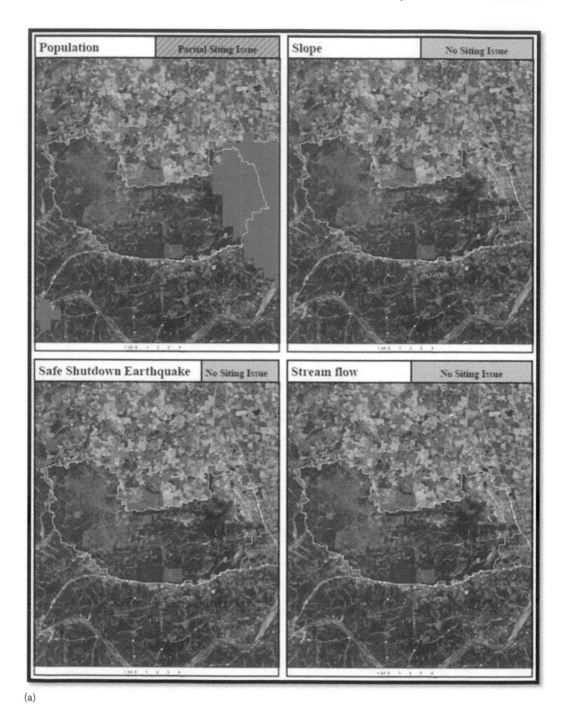

(a)

Figure 15.13 (a) Fort Campbell meets the siting criteria for slope, safe shutdown earthquake, and streamflow, but partially meets the criterion for population. *(Continued)*

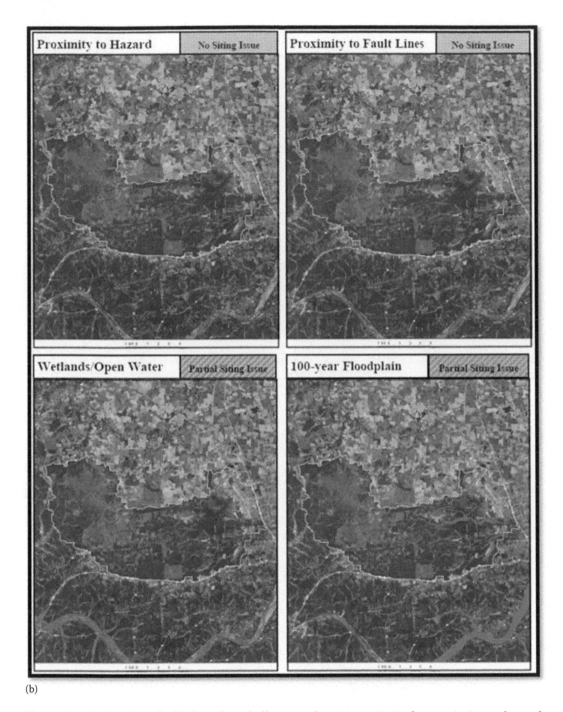

(b)

Figure 15.13 **(Continued)** (b) Fort Campbell meets the siting criteria for proximity to hazards, fault lines, but partially meets the criteria for 100-year floodplain and for wetlands/open water.

(Continued)

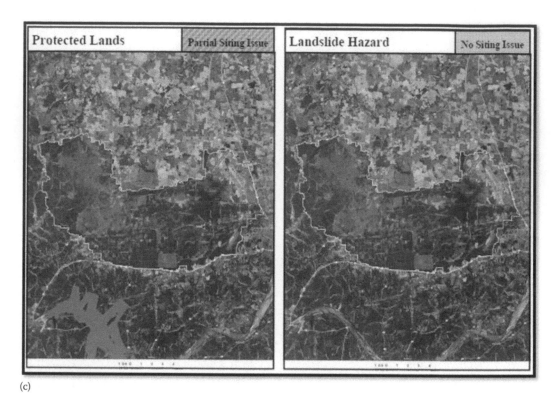

(c)

Figure 15.13 **(Continued)** (c) Fort Campbell meets the siting criterion for landslide hazard, but partially meets the siting criterion for protected lands.

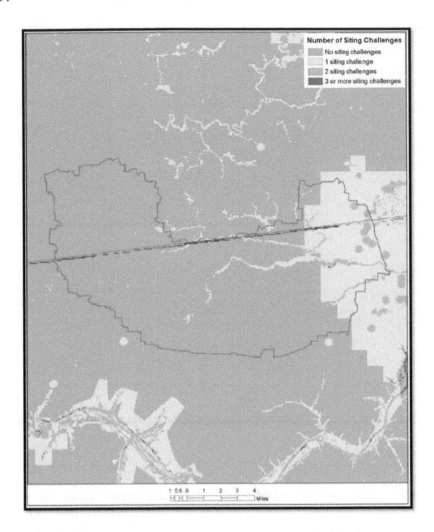

Figure 15.14 OR-SAGE tool generated composite map for Fort Campbell.

Naval surface warfare center-crane division

As shown in Figure 15.15, the Naval Surface Warfare Center-Crane Division (NSWC-Crane) is located on about 55,000 acres (about 85 square miles) in south-central Indiana, near the town of Crane [21]. The NSWC-Crane provides numerous military development and support operations, including expeditionary warfare systems, fleet maintenance and modernization, radar, power systems, strategic systems, small arms, surface and airborne electronic warfare, night vision systems, undersea warfare systems, and systems development for the DD(X) destroyer and the littoral combat ship [21,22]. Interstate 69 is approximately 5 miles from the northwest corner of this site and highway 231 is on the western edge of the site.

Approximately 5,000 soldiers, employees, contractors, and families are on-site at any given time [22]. The nearest major fault line based on USGS data is noted to be 650 miles southwest in Oklahoma, and the maximum safe shutdown earthquake for the site is below 0.3 g peak ground acceleration. The maximum slope of the site is about 22 percent. Cooling water for a reactor could be obtained from the White River just southeast of the site.

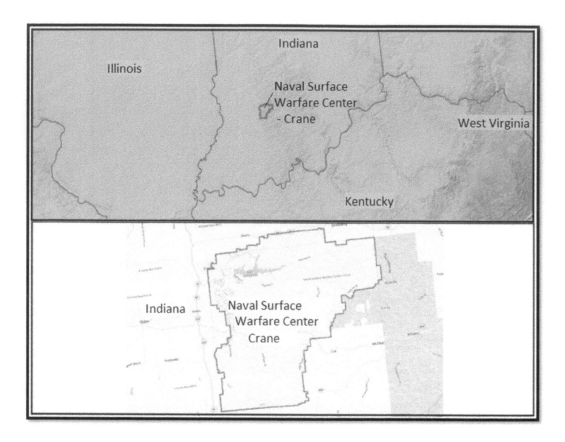

Figure 15.15 NSWC-Crane.

The NSWC-Crane site has limited partial site issues (Figure 15.16). As evident from the composite map (Figure 15.17), these concerns primarily affect the periphery of the site. Approximately, 95 percent of the 55,000-acre site meets multiple conventional standards for siting an SMR on the base facility. Like other sites, the composite map does not reflect specific hazards associated with the site, such as ordnance storage areas, weapons ranges, etc., that could reduce suitability of some areas of significant size for siting a reactor. Furthermore, the site meets the current NRC RG 4.7 recommendations for population density without additional consideration for relaxed SMR population siting requirements based on reduced source term. Therefore, this site should be also classified as favorable for siting an SMR.

Conclusion

Evidently, the OR-SAGE tool is effective in identifying potential sites for locating SMRs in DOD facilities based on a number of criteria pertaining to physical and social characteristics. Although there are 170 such facilities spread across the United States, for this study, the site suitability analysis was conducted for a handful of facilities, the results for which are presented here. However, the tool did not account for existing policies, such as those related to airfield sites on air force bases, which may increase or decrease site suitability. The tool also generalized the ranking of each criteria by assigning equal weight to

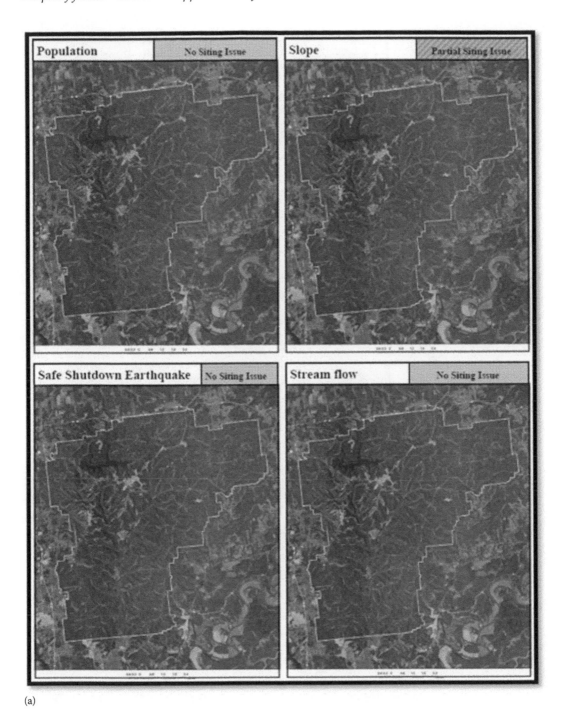

(a)

Figure 15.16 (a) NWSC-Crane site meets the siting criteria for population, safe shutdown earthquake, and streamflow, but partially meets the criterion for slope.   (*Continued*)

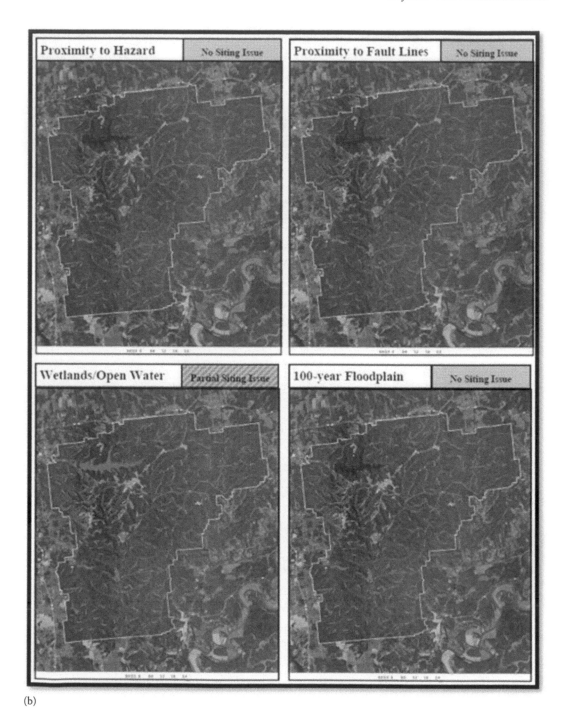

(b)

Figure 15.16 **(Continued)** (b) NWSC-Crane site meets the siting criteria for proximity to hazard, proximity to fault lines and 100-yr flood plain, but partially meets the criterion for presence of wetlands/open water. (*Continued*)

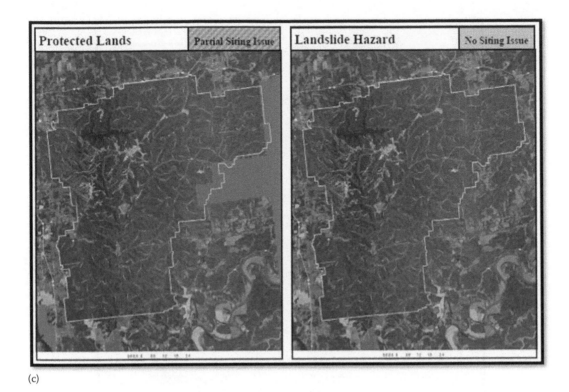

(c)

Figure 15.16 **(Continued)** (c) NWSC-Crane site meets the siting criteria for landslide hazard, but partially meets the criterion for presence of protected lands.

each criterion, but future studies could analyze the implications of weighted criteria with respect to policy guidelines. For instance, all sites discussed here met the current NRC RG 4.7 recommendations for population density; hence, these sites were considered suitable. While physical criteria are crucial for SMR siting, population density is crucial while dealing with siting critical facilities (i.e., hospitals) during emergency situations in addition to physical criteria. Therefore, incorporating a ranking scheme for site suitability assessment in the tool would provide additional insights.

In the current version of the tool, the factors have been classified using a binary approach since this is appropriate for licensing requirements. However, it would be useful to classify siting criteria based on certain intervals and thresholds. For instance, land that was present within a 100-year flood plain was eliminated. However, given the recent flooding following hurricane Harvey in 2017, it might be useful to classify land areas based on their presence in 100, 500, and 1000-year flood-plains, and rate each class to increase accuracy. Land areas not within a 100-year flood-plain may be suitable for SMR siting but may still require additional design consideration in case of, say, a 1000-year flood event.

The tool developed in this study, and the methodology implemented for development of this tool and by Kar and Hodgson (2008) [3] could be used for other decision-making tasks. For instance, such a tool could be implemented to assess suitability of existing infrastructures, such as evacuation routes, emergency evacuation shelters,

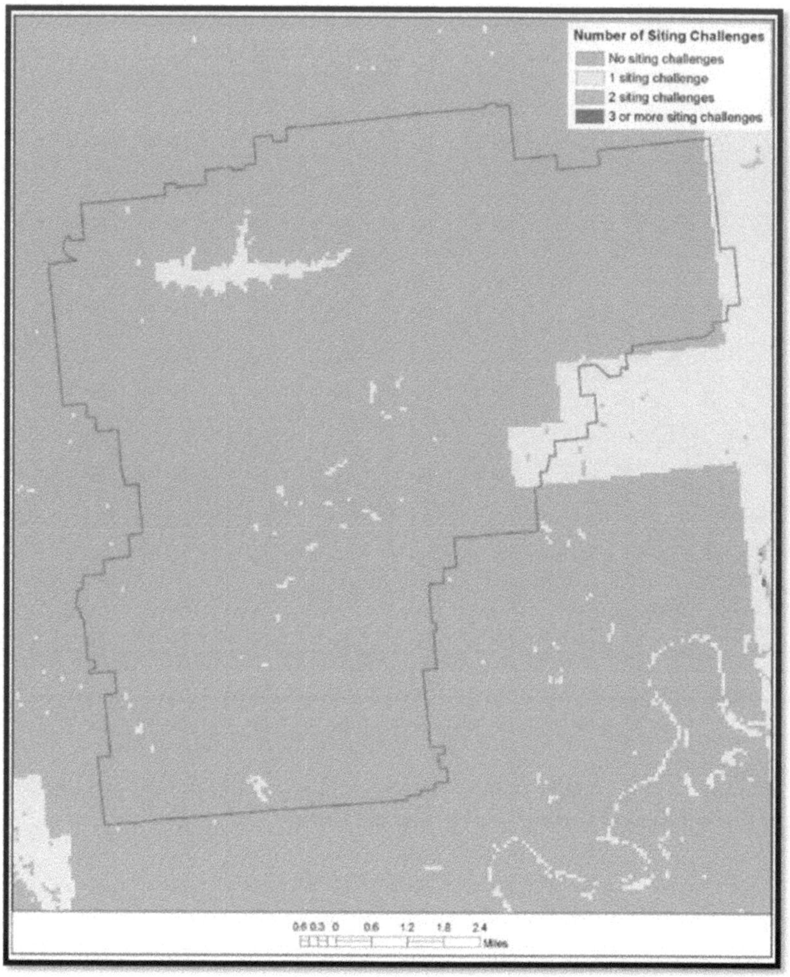

Figure 15.17 Composite Map of NWSC-crane derived by OR-SAGE tool.

critical assets/facilities (e.g., temporary housing, human services, and health care facilities identified under Emergency Support Functions #6 and 8) to help with disaster risk reduction following a hurricane like Harvey and Irma in 2017. Real-time data and information obtained via crowdsourcing and drones could also be integrated into this tool to help emergency managers and first responders undertake emergency management operations in real-time. Also, the tool could be used to train and prepare stakeholders for future hazard events. The output of a site suitability tool could also be combined with location-allocation models for siting critical assets and facilities in areas that are at high-risk to hazard events for DOE, DOD, and civilian purposes.

Acknowledgment

This material is based upon work supported by the US. Department of Energy, Office of Science, Office of Nuclear Energy. This manuscript has been authored by UT-Battelle, LLC under Contract No. DE-AC05-00OR22725 with the US. Department of Energy. The

References

1. Lawless, S., Maples, A. and Purcell, C. (2017). Energy security technology transfer. *The Military Engineer*, 109(708), 58–59.
2. Poore, W.P., Belles, R.J., Mays, G.T., and Omitaomu, O.A. (2013). *Evaluation of Suitability of Selected Set of Department of Defense Military Bases and Department of Energy Facilities for Siting a Small Modular Reactor*, ORNL/TM-2013/118, March.
3. Kar, B. and Hodgson, M.E. (2008). A GIS-based model to determine site suitability of emergency evacuation shelters. *Transactions in GIS*, 12(2): 227–248.
4. Omitaomu, O.A., Blevins, B.R., Jochem, W.C., Mays, G.T., Belles, R., Hadley, S.W., Harrison, T.J., Bhaduri, B.L., Neish, B.S., and Rose, A.N. (2012). Adapting a GIS-based multicriteria decision analysis approach for evaluating new power generating sites, *Applied Energy*, 96: 292–301.
5. Belles, R.J., Mays, G.T., Omitaomu, O.A., and Poore, W.P. (2012). *Updated Application of Spatial Data Modeling and Geographical Information Systems (GIS) for Identification of Potential Siting Options for Small Modular Reactors*, ORNL/TM-2012/403, September, Oak Ridge, TN: Oak Ridge National Laboratory.
6. Belles, R.J., Copinger, D.A., Mays, G.T., Omitaomu, O.A., and Poore, W.P. (2013). *Geographic Information Systems Evaluation of Sample Coal Plant Sites to be Repowered with Small Modular Reactors*, ORNL/TM-2013/109, March.
7. Carver, S.J. (1991). Integrating multi-criteria evaluation with geographical information systems. *International Journal of Geographical Information Systems*, 5(3): 321–339, doi:10.1080/02693799108927858.
8. Proctor, W., and Drechsler, M. (2003). Deliberative multi-criteria evaluation: A case study of recreation and tourism options in Victoria Australia, *European Society for Ecological Economics, Frontiers 2 Conference*, Tenerife, Canary Islands.
9. Department for Communities and Local Government (DCLG). (2009). *Multi-Criteria Analysis: A Manual*, Department of Communities and Local Government [online]. Available from: www.communities.gov.uk (Accessed September 24, 2017).
10. Drobne, S. and Lisec, A. (2009). Combination and ordered weighted averaging, *Informatica*, 33: 459–474.
11. Steele, K., Carmel, Y., Cross, J., and Wilcox, C. (2009). Uses and misuses of multicriteria decision analysis (MCDA) in environmental decision making. *Risk analysis*, 29(1), 26–33.
12. Malczewski, J. (2000). On the use of weighted linear combination method in GIS: Common and best practice approaches, *Transactions in GIS*, 4(1): 5–22.
13. Eglin Air Force Base Official Site—http://www.eglin.af.mil/ (Accessed September 28, 2017).
14. Eglin Air Force Base Guide—http://www.eglinguideonline.com/ (Accessed September 28, 2017).
15. Beale Air Force Base Official Site—http://www.beale.af.mil/ (Accessed September 28, 2017).
16. Beale Air Force Base Guide—http://www.military.com/base-guide/beale-air-force-base (Accessed September 28, 2017).
17. US Army Maneuver Center of Excellence—http://www.benning.army.mil/ (Accessed September 28, 2017).
18. 2017 Fort Benning Tour Guide—http://virtual.mybaseguide.com/publications/g30/fort-benning/ (Accessed September 28, 2017).
19. Official Site of Fort Campbell and the 101st Airborne Division (Air Assault)—http://www.campbell.army.mil/Pages/Default.aspx (Accessed September 28, 2017).

20. Military One Source—http://www.militaryonesource.mil/ (Accessed September 28, 2017).
21. Naval Sea Systems Command, Warfare Centers—http://www.navsea.navy.mil/Home/Warfare-Centers/NSWC-Crane/ (Accessed September 28, 2017).
22. Naval Support Activity Crane—https://www.cnic.navy.mil/regions/cnrma/installations/nsa_crane.html (Accessed September 28, 2017).

chapter sixteen

Three innovations for defense acquisition reform

Roy L. Wood

Contents

Three big ideas for defense acquisition reform

Typical acquisition reform efforts have focused on making changes in the margins, achieving marginal results. Fundamental structural and process changes need to be made for any significant improvements to be seen. The changes suggested in this chapter are not difficult to make from a policy perspective, but will challenge entrenched roles, perceived entitlements, and a status quo organizational culture.

This chapter offers bold reform ideas in three specific areas: achieving the benefits of competition above the prime contractor level by competing capability requirements among the military Services, reforming the technology development and transition process, and shifting the workforce model toward a majority civilian acquisition workforce. Some of these ideas are not new and have been recommended at various times, but have never been fully embraced or implemented. The evidence of decades of acquisition reform indicates that the marginal reforms typically taken are not making the desired changes the Department says it needs and wants. Implementation of the reforms suggested here could provide outcomes that actually make a difference.

Improving competition: The first challenge addressed is how the DoD can reap the benefits of healthy competition with a shrinking industrial base. Competition is widely recognized as an important way to keep defense acquisitions affordable, yet this is increasingly difficult with a smaller and more specialized set of industries. Workarounds, like dual sourcing, split buys, and leader-follower procurements have propped up the industrial base, but sub-optimized the advantages of real competition (Wydler et al., 2012).

Encouraging competition among subcontractors has also been recommended, but government involvement in ensuring competition at this level has met with only limited success. Some of the previous strategies to this end have been to encourage prime contractors to invoke head-to-head competition among potential subcontractors, create innovative teaming arrangements, or create second sourcing arrangements. The government has also set up separate competitions for specific items and provided them to the prime as government furnished equipment. However, these techniques to increase competition below the prime have been met with limited success and raised concerns over violation of *privity of contract* (Federal Acquisition Regulation [FAR] Part 42) and increased risks of placing the government in the role of lead system integrator (GAO, 2010).

If the shrinking industrial base is creating conditions where real competition is not possible, and government overreach into the subcontracts is not tenable, then the idea of competing at a level *above* the prime contractor should be considered. That is, create a more competitive environment where the military Services and agencies "compete" with each other to provide a given capability. This innovative solution will have many of the same inherent advantages that are seen in prime and subcontractor competition. This idea will be discussed later in more detail.

Improving innovation and technology transition: The second challenge is the well-known difficulty in transitioning new technology into acquisition programs. New technologies are often developed in the laboratory and matured to the point where the concepts can be demonstrated, but not sufficiently ready to integrate into an acquisition program. This gap, the so-called "valley of death," has long existed between science and technology (S&T) and acquisition organizations. The valley of death problem has been extremely resistant to resolution. This chapter will also offer several potential solutions, including leveraging commercial models of technology transition, removing barriers to technology innovation, and instituting a more disciplined management of acquisition system baselines to facilitate transition.

Improving the acquisition workforce: The third challenge is possibly the most controversial, and yet probably the one with the most leverage to improve acquisition outcomes. Multiple studies and initiatives have been undertaken in recent years to improve the performance of the acquisition Program Manager (PM; Ahern, 2009; Fox, 2014). These efforts ranged from extending the tenure of PMs, to improving the quality and quantity of training, to offering incentives and rewards for good performance. Yet PM performance, by many study standards, remains subpar (Francis, 2014).

A large proportion of these Program Managers are military officers. Given the challenge that military officers have in mastering both operational and acquisition facets within a typical 20-year career, the recommendations in this chapter are to shift acquisition leadership to a primarily civilian corps, using Military Officers in more appropriate and sustainable roles to monitor progress toward meeting program requirements. The chapter also suggests ways to improve how civilians are managed that will bolster the experience and leadership of civilians as program managers.

Big idea #1: Improving competition

"Real competition is the single most powerful tool available to the Department to drive productivity" (USD[AT&L] Better Buying Power web portal).

Full and open competition is the holy grail of defense acquisitions. Competition is believed to lower costs to the customers, incentivize productivity and efficiency, and spur innovation among competitors. To win a competitive contract, a defense company must provide a responsive proposal for a product or system at an affordable price that meets the military requirement. To position itself to win a competitive procurement, a company must continually assess its capability to produce technical and innovative solutions to meet government needs, while keeping its cost structures lean and competitive to produce these goods at more attractive prices than its competitors. Again, and again, the government has seen evidence that competition encourages this behavior in the defense industry and has gone to great lengths to sustain a viable industrial base where competition can flourish. In short, competition is good, and more is better.

Yet, since the mid-1990s, the defense industrial base has shrunk and consolidated to an unprecedented level. With fewer businesses in the industry, it has become increasingly difficult for the government to encourage fierce head-to-head competition for many of its products and systems. The remaining industries have tended to become more specialized, oligopolistic providers of particular categories of products. For example, the Navy can choose from only two commercial shipyards for submarine construction. For tactical aircraft, only Lockheed-Martin, Northrop-Grumman, and Boeing compete. Under these conditions, government source selections have had to be as concerned with sustaining a competitive industrial base as with getting the best deal on any particular item. Defense costs continue to rise, in part because of this less competitive industrial base (Harrison, 2012).

Government efforts to create pseudo-competitive solicitations among the prime contractors, and to find ways to encourage competition at the subcontractor level, have met with varying degrees of success. Smaller numbers of new program starts have exacerbated the dilemma and created an environment where losing a single large procurement for ships or aircraft, for example, could force competitors out of the business, leaving the government with a single monopolistic provider in that sector. The Navy, for instance, juggles competition among its new amphibious, auxiliary, and destroyers to maintain multiple, viable shipyards (Cavas, 2015). In doing so, the Navy is avoiding true head-to-head competition and thereby distorting any real economic advantages in favor of spreading workshare.

As competition among primes becomes more challenging, the government has tried to promote and encourage competition at the subcontractor level. Organizationally and contractually, however, this is difficult for government to do directly beyond setting an expectation that the prime contractor will pursue rigorous competition down the supply chain. For the government to more intrusively attempt to manage competition among subcontractors risks violating the privity of contract prerogatives of the prime contractor.

Privity of contract provides that only the parties to a contract can confer rights or impose obligations relative to the contract. Since subcontracts are between the subcontractor and the prime, the government has no legal rights to meddle in the subcontract details. In other cases, the government has decided to compete subcontracts directly to procure subsystems using a "component breakout program" from a vendor. The government then provides the subsystems as government furnished equipment (GFE) to the prime (OUSD[AT&L], 2014). Many government organizations are hesitant to use such a strategy, however, because of inherent risks of removing control from the prime and placing the government in the proxy role of system integrator.

Given this difficulty of achieving real competition at the prime or subcontractor levels, then perhaps the government should seriously consider instituting competition *above* the prime. This would involve creating a much more competitive environment between the Services and Agencies inside the Department of Defense.

Today, when a warfighting capability gap is identified by a Combatant Commander and described in an Initial Capabilities Document, an Analysis of Alternatives (AoA) is conducted to assess which potential solution would best provide this capability. The AoA should assess a range of material solutions, together with operational concepts and costs. Unfortunately, the process of creating the AoA is usually assigned to a single Service where a preconceived, Service-centric solution often emerges. The Government Accountability Office (GAO, 2009) noted that, "while AOAs are supposed to provide a reliable and objective assessment of viable weapon solutions, we found that Service sponsors sometimes identify a preferred solution or a narrow range of solutions early on, before an AOA is conducted."

A more robust and objective process would be to "compete" Initial Capabilities Documents (ICDs) among the Military Services and let each of these "bidders" conduct its own Service-centric Analysis of Alternatives (AoA) to provide the capability. Rather than having only the predictable replacement of an Air Force bomber capability with another bomber, for example, more novel and affordable solutions are also likely to emerge from the Navy or the Army. Competitive AoAs would become more rigorous, with both technical solutions and cost estimates coming under greater cross-Service scrutiny. The best competitive AoA, as judged by the Combatant Command and Joint Requirements Oversight Council (JROC), would then be "awarded" to the winning Service to manage through the conventional acquisition process. Armed with a more thorough and complete AoA, the government would be better equipped to negotiate with industry for a capability the joint forces require and have a much better understanding of the cost of such a system.

Adding this extra layer of competition could help address a number of current shortfalls and issues. First, it would force the Combatant Commanders and JROC to write ICDs that are focused on warfighting capabilities rather than allowing or telegraphing a Service-centric solution. For example, a generically-written capability for destroying targets at long ranges could be accomplished with manned or unmanned bombers; cruise or ballistic missiles launched from aircraft, ships, submarines, or land sites; rocket-assisted shipboard or ground artillery; or potentially other more innovative solutions.

In such a competitive scenario, one can imagine the Navy and Air Force going head-to-head with aircraft and missile alternatives, and the Army and Navy competing on missiles or artillery, and each of the solutions competing on affordability. Likewise, and importantly, each of the Services would be able to consider their own proposed alternatives backed by realistic and supportable concepts of operations, or CONOPS, that would also have unique associated costs and opportunities. For instance, the Army proposal would have to include CONOPS provisions, and associated costs, to transport their proposed artillery to the battlefield, while the Air Force would include logistics and maintenance considerations for an aircraft solution. Each Service would have to justify how its proposed solution would fit into the current inventory and war fighting strategy.

Second, engendering Service competition for real resources would create an environment where the Service Chiefs are incentivized to ask hard questions about solutions the other Services put forward, and be better prepared to answer questions about their own proposals. This would force—and enforce—a cross-Service competitive rigor that does not exist today (Fay, 2015). Today, with little incentive for one Service to call the bluff of another, overestimated claims of performance or underestimated cost estimates go unchallenged until too late in the acquisition process.

From the literature on internal competition, Birkinshaw (2001) points out three advantages to an organization: First, it increases flexibility; second, it challenges the status quo; and third, it motivates greater effort (pp. 21–22). For the DoD, these three factors would hold true as well. Flexibility is critical during this time in history of rapid changes in potential threats, and opportunities presented by new technologies. As militaries are wont to assume that the next war will be like the last one, it is critical to encourage building a more flexible and responsive military that can cross swords effectively with different or more powerful adversaries. Competing at the Service level would prevent the DoD from being stuck with proposals for the usual things from the usual players.

Like the first point, creating competition among Services would help break the current status quo. The Services have become quite comfortable in their mission stovepipes, each continuing to receive about an equal 30 percent of the annual Defense budget. Like the current DoD, Birkinshaw (2001) points out that large firms also become inertia-ridden over the years, victims of their own success. Customers and their needs are taken for granted, and management systems and processes take on a life of their own. Practices and beliefs become ingrained. Such a system is hardly conducive to revolutionary new ideas (p. 22). This sounds very much like the DoD. Despite complicated and lengthy AoAs, amazingly few produce accepted solutions outside the status quo. Most new systems are incremental improvements over previous ones, becoming one-for-one mission replacements of aircraft carriers, bombers, and ground vehicles. In 2004, the Joint Defense Capabilities Study noted that Service planning does not consider the full range of solutions available to meet joint war fighting needs. Alternative ways to provide the equivalent capability are not adequately considered—especially if the alternative solutions are resident in a different Service or Defense Agency (p. iii).

Birkinshaw's (2001) third point is that competition motivates greater effort. Firms—and Services—could be expected to be more aggressive, innovative, and forward-leaning when faced with a direct threat to budgets and resources. One might imagine, for example, that the Navy and Air Force would (finally) engage in a more thorough and lively discussion of the mix of sealift versus airlift capability if the results had the real potential to change the resource and mission mix of each Service. Similarly, each of the Services would be forced to scrutinize the output of their various laboratories and warfare centers if they were forced to compete with each other on superior technology and innovation. This alone could have positive and long-lasting impacts on future war fighting capabilities, as science and technology efforts in each of the Services are ramped up and better integrated into the acquisition process.

Big idea #2: Improving innovation and technology transition

From 2011 to 2015, the GAO repeatedly identified technology immaturity as a major contributor to program problems (GAO, 2011-2015). Unsurprisingly, they consistently found that technology that is not fully mature and ready for transition to acquisition introduces significant cost, schedule, and performance risks. Prominent examples include the F-35, the most costly defense program in history (Thompson, 2013), and the DDG-1000 Zumwalt destroyer class (Hagerty et al., 2008; GAO, 2008), truncated to three ships after substantial cost increases and schedule delays (US Navy Fact File, 2014; GAO, 2008). Both these programs depended on many cutting-edge technologies that were immature at program inception and required substantial concurrent development and maturation as the acquisition program was in *execution*.

As a result, the high costs and prolonged timelines for fielding these systems have long been a frustration for operational commanders. As former Chief of Naval Operations, Admiral Gary Roughead, pointed out, "As a Service Chief, my greatest frustration was to be briefed on an exquisite acquisition timeline that delivered an initial operating capability more than a decade hence when the need was immediate" (US Senate, 2014, p. 148). Getting a partial capability now and full capability later is undoubtedly better, in most cases, than going for a long period of time with no capability. Designing programs with off-the-shelf technologies today, while embarking on a more rational offline technology development strategy for later insertion, appears to be a way to achieve this aim.

Improve the technology insertion baseline process

To reduce system development time and field an improved capability, it is necessary to enforce a proven and rigorous technology insertion strategy in acquisition programs. This is not a new idea, but one that seems to get lost in the euphoria of planning a new program. Many successful programs, like the Navy's AEGIS and submarine programs, have been very disciplined in the use of time-certain baseline upgrades that provided the technology community the opportunity to prove out new technologies and be prepared to enter the acquisition process at certain predefined points in the acquisition life cycle (Holzer & Truver, 2014; Mitchell, 2010).

New program starts should survey the state-of-the-shelf for mature and available technologies to use in the initial baseline, providing most—but perhaps not all—of the capabilities the system may ultimately need. The Service or Agency should then embark on S&T efforts to create and mature new technologies outside the acquisition program that close the capability gap. The acquisition program should identify specific baseline upgrade points in the acquisition program schedule. If any given technology can be matured to meet the desired schedule, then it can be integrated into the next baseline; if not, it is shifted to a future insertion point. This approach requires significant planning and discipline by the acquisition program manager and close coordination with technology developers to establish hard deadlines and performance expectations. Buy-in from the operational and requirements communities is also needed so they understand the capabilities and limitations of the baseline approach. In practice, this strategy will get new capabilities to the field sooner and promote predictable upgrades over time.

Improve the technology transition process

Technology and innovation happen—it's unavoidable, and frankly, it's fun. S&T organizations happily invest decades and millions of dollars in science projects that produce potential and promise, but too many of these projects never transition into acquisition programs. S&T organizations mature the technology to a point where the science is proven, but have no responsibility for productizing the technology. This leaves responsibility for maturation beyond the laboratory or prototype to acquisition programs and their contractors. Accepting immature technology, however, is anathema to acquisition program managers whose role it is to reduce an acquisition program's cost, schedule, and performance risks. This maturity gap between proof of concept and productized technology is the well-known "valley of death."

A straightforward way to close this gap may be found in the way commercial companies transition their technologies to products. In 2006, the GAO conducted a study that contrasted DoD and commercial technology transition practices (GAO, 2006). The GAO found that the best commercial practices involved the S&T community keeping responsibility for maturing technologies well beyond the point that the DoD currently does, and assigning relationship managers to work with both the S&T and production managers to facilitate technology transition. The GAO's specific recommendation for the DoD was to allocate a portion of 6.4 funding (advanced component development and prototyping) to the S&T community specifically for technology maturation. Unfortunately, the DoD rejected this recommendation, and only partially concurred with the idea of a relationship manager. Both these ideas continue to have considerable merit.

Commercial firms, like those in the GAO study, have had successes by holding their S&T organizations responsible for maturing technologies to the point where the risk of productizing them is minimal. This includes responsibility for prototyping and testing the technologies in realistic environments—exactly the process needed by the DoD to allow acquisition PMs to have confidence to accept new technologies for integration with minimal risk for cost and schedule overruns or performance failures. The DoD should reconsider its reluctance to accept the GAO's recommendation and institute more rigorous S&T involvement and responsibility in technology maturation prior to transition.

Similarly, the DoD should also adopt the industry best practice of assigning *relationship managers*. Given that DoD S&T and acquisition managers "speak different languages," have widely different cultures, and use processes that are often not mutually supportive, there is a clear need to bridge these gaps between S&T and acquisition. Relationship managers could serve to better facilitate problem-solving and foster better two-way communications between S&T and acquisition organizations.

Likewise, relationship managers could work with technology professionals to identify promising technologies that are not obvious candidates for existing acquisition programs. Technologists are quick to point out that average users don't know what they don't know. For example, there was no known demand for an iPod or smartphone until the technology was introduced. For most consumers, it would be hard to imagine life before those technologies. Users may not be able to imagine what new technology is within the art of the possible, while some technologists go happily about their business without an appreciation for what the users may actually need.

This technologist–user disconnect, prevalent in the DoD, could easily be bridged by individuals who are "bilingual" and experienced in both S&T and acquisition. One way to accomplish this would be to post a senior S&T professional on the staff of each Program Executive Office (PEO). S&T liaison officers who "speak technology" would work closely with the PEO and its acquisition program offices to help find technology solutions across the PEO's portfolio of programs. This arrangement could create the much- needed linkage between the S&T network and program managers to meet match technology needs with emerging technologies.

Improve innovation by removing barriers

Defense has become more isolated and less innovative, in part by its own actions and those of Congress. The Department has driven industry to consolidate, established substantial barriers to new entrants, and contributed to risk aversion among non-Defense businesses. According to former Principal Deputy Under Secretary of Defense for Acquisition, Technology, and Logistics, the Honorable David Oliver,

> While that defense industrial base was once robust enough to tolerate many failures, that circumstance no longer exists. Our defense industry is no longer based upon the entire vibrant American commercial industry, as it was in World War II. Instead, during the Cold War, the defense industry grew into an isolated one. America is now the only Western nation with an isolated (by regulation and practice) defense industry. The rest of the Western world has adopted different approaches which seek to better access the technologies being developed in the commercial industries and is accelerating ever faster away from the American defense acquisition model. (US Senate, 2014, p. 142)

Congress and the Department have imposed significantly greater oversight, restrictions, and requirements on defense contractors than commercial companies could or would tolerate. Commercial contractors or subcontractors generally work under contractual provisions derived from the Uniform Commercial Code, a 270-page document (USLegal, 2014), while government vendors must labor under the regulatory burden of the FAR, DFAR, and supplements totaling over 4,000 pages. Government contractors must deal with a bid and proposal process far more involved than those needed for most commercial contracts. Defense contractors are required to have certified cost accounting systems to work for the government, and are subject to audits and penalties for violating any of the rules or referenced clauses in those thousands of pages of regulations. Defense contractors are subject to contract termination for cause or simply for the convenience of the government, and the government can unilaterally change the contract without the contractor's consent.

Further, should a contractor be willing and able to navigate the maze of obstacles and barriers and then perform superbly, delivering great value to the government, there is no guarantee of follow-on work. If a contractor wishes to work on classified projects, there are substantial additional requirements for clearing the facility and workforce. Also, many companies find themselves with severe restrictions on releasability of information to foreign governments, companies, or individuals, even talented "green card" holders educated in US universities. They also face the potential for products or components to be designated dual use (Dual Use, 2012) or fall under the International Trafficking in Arms Regulations (ITAR), severely limiting a company's markets and sales (ITAR, 2014).

In total, these are significant barriers that keep many companies from choosing to deal with the government and the Department of Defense. The legal obligations and potential penalties, arcane regulations and restrictions, and bureaucratic hurdles represent a tremendous time and resource investment. It should not be surprising that many of the most innovative companies choose not to work on government contracts, and those companies that do are likely to be, or become, more bureaucratic, cautious, and risk averse.

If the Department truly wants to attack a root cause that will improve innovation by lowering barriers to innovative companies, this is an area ripe for reform. Improvements would require re-engineering of the FAR and DFAR—a task easier said than done. The Department would have to work closely with Congress, and there would undoubtedly be some pushback from current defense industries that benefit from the high barriers to new competition. Nevertheless, this is a big idea worth pursuing.

Big idea #3: Improving the defense acquisition workforce

"Although there is a pressing need for the Defense Department to perform the active manager role, the current approach to program management is fundamentally flawed. After fifty years, we know that an Army or Air Force colonel or Navy captain (0–6) with limited industrial management knowledge and experience is ill prepared to direct and oversee a first-of-a-kind multi- hundred million dollar industrial program with hundreds of complex challenges and dilemmas" (Fox, 2012, p. 200).

Importance of military involvement in acquisitions

Most leaders in the Pentagon would agree that it is important for military operators to be involved in the procurement of military equipment and supplies. In the current acquisition system, military serve in relatively small numbers in virtually all capacities, including as systems engineers, contracting officers, financial managers, logisticians, and others. However, in the key position of Program Manager, military tends to dominate. Overall, military members represent only 10 percent of the total acquisition workforce, but in program management, a disproportionate 42 percent are military (Gates et al., 2013). The military members bring broad leadership, enthusiasm, and operational experience to the business of procuring military equipment and, in many cases, firsthand judgment about the military utility of a system's design. These aspects of having a military involvement in a program are largely beneficial, but there are a number of deep-seated problems.

First and foremost, acquisition is a difficult, high-stakes business. Fox and Miller (2006) described the skills required of a program manager, most of which are not core warfighting skills, and must be gained through training and experience in acquisition: "Managers [of large, complex programs] must augment a strong foundation of conventional management skills in planning, organizing, and controlling, with knowledge of the requirements, resources, and constraints of a specific project as it progresses" (p. 109). Military program managers must be able to negotiate the complicated planning, programming, and budgeting system and have a good understanding of government financial management. The PM must also become knowledgeable in the technology of the program to be able to understand complex engineering issues and make tradeoff decisions. The military PM must manage both a largely civilian workforce of direct reports and a vast web of contractors. Traditional military leadership training may not equip a PM to operate in that environment. Again, Fox points out that "skilled project managers focus more on monitoring and influencing decisions, and less on giving orders" (p. 124).

Experience challenge with dual-track military

It has always been problematic for military personnel planners to effectively allocate time in a typical military member's career to both gain the required operational experience *and* sufficient acquisition experience to manage large, complex procurements. To gain broad operational experience in the field, officers are typically rotated through increasingly challenging positions every 18–36 months. Their promotions in an "up-or-out system" depend on this mobility. Longer tours can be seen as career stagnation by promotion boards. Operational career paths are strictly regimented, with success depending on doing well in command and other must-do tours, including Joint Service tours.

Each of the military Services treats operators differently when they are transitioning into acquisition careers, but most officers start the transition mid-career or later. With only a few, if any, short acquisition tours under their belts, many military PMs are ill-equipped to lead large, complex acquisitions. Independent analyst Katherine Schinasi observed in testimony to Congress, "An operational commander does not make good business decisions. He was not trained to do so nor is he rewarded. Military advancement depends on frequent rotations; sound program management and accountability relies on continuity" (US Senate, 2014, p. 157).

Norm Augustine, former Lockheed CEO and government executive, noted in recent congressional testimony that, "The issue most assuredly is not one of dedication or native ability: the issue is a lack of relevant experience and the freedom to exercise that experience. One hundred managers with one year's experience should never be considered to be the same as five managers each with 20 years' experience" (US Senate, 2014, p. 12).

Few, short acquisition tours preclude deep experience

Short tour lengths carry over to assignments in senior PM positions, and this tenure issue has also been identified and addressed by a mandate that PMs sign agreements to serve at least four years or until the next major milestone of a program (DoD, 2005). Unfortunately, this agreement is largely unenforceable, as many PMs reach retirement eligibility or are promoted out of the position before their tenures are reached.

Shorter tours are not only a systemic problem, but many feel that a military PM may be incentivized to serve in this position for as little time as possible before moving on. Similarly, few military program managers are ever given a second program, since this would be viewed broadly as not moving up and, again, likely to stop a promising career in its tracks. Lessons learned by a military leader on one program are therefore not transitioned to another, losing significant opportunities to create a learning organization and improve future outcomes.

Civilian defense acquisition workforce—with military requirements advisors

As an alternative to the current system, which continues to resist reform efforts because of the deep, systemic challenges, the Department of Defense should consider transitioning to a civilian acquisition corps *with military requirements advisors assigned to large programs or PEOs*. This change would solve many of the problems associated with inexperienced military Program Managers, but would require a more robust civilian acquisition career management scheme to work well.

Longer tenures in a dedicated single career

An all-civilian workforce would not immediately be a panacea for improving acquisition outcomes. There are challenges associated with the way civilian acquisition professionals are assigned and developed, most having a less aggressively managed career than the typical military officer. Yet, with a potential 40-year career to devote exclusively to acquisition, these members have significantly more career capacity to develop the requisite skills and experience to become expert program managers and functional leaders than their military counterparts. There are no ill effects to a civilian who remains in a particular program office or in a leadership role for a decade or more—a time frame that would doom any military member's career.

Compensating for lack of military operational experience

A major drawback to having an all-civilian workforce, of course, is the perceived lack of firsthand military field experience. This is mitigated somewhat by veterans' hiring preference that encourages former military members move into the civil service. A recent RAND study noted that former military members represent an important and growing source of future civilian acquisition workforce leaders (Gates et al., 2013, p. 50).

To more intentionally ensure that military equities and nuances are represented in defense programs, military operators could, and should, be assigned as advisors to the program manager of every large program or to PEOs with many smaller programs. This officer should be collocated with the program or PEO, but have reporting responsibility back to the Service or sponsoring organization that initiated the capability requirement.

In this scenario, the military advisor would be immersed in the day-to-day business of the program office, observing tradeoffs and advising engineers and managers when questions arise about how a particular piece of equipment or feature would be used in the field. This military advisor would also be well positioned to report progress and potential operational issues back to the requirements originator for action or clarification. As an advisor, the military member would not need the in-depth training or experience to actually run the program, nor would they be obliged to accept extended tours of duty in a program office that could hurt their opportunities for promotion.

Fewer incentives for short-term decision-making

Unlike their military counterparts who are sometimes incentivized by short tours to make short term decisions, civilian leaders in programs for long career assignments would be better served to take the long view, knowing that they must live with the decisions they make. Further, civilians who gained experience and were successful in one program could be assigned to larger, more challenging ones, taking with them the knowledge and experience they gained along the way. Since civil servants are not subject to the "up-or-out" policy of the military, pressures and disincentives that would be career ending for a military member would have far less influence on a civilian's career. Civilians would be more apt to make decisions based on the long-term good of the program, rather than the immediate good of their careers. This is especially true, since they would be faced with the downstream likelihood of having to live with the consequences of their decisions.

Summary and recommendations

There has been, and continues to be, a long and continuing saga of defense acquisition reform efforts. Most have either failed, or only succeeded in making improvements at the margins. This chapter presented three big ideas to help move acquisition reform from treating symptoms to addressing root causes. The ideas impact the way material solutions are developed by introducing competition among the Services and Agencies for the privilege of managing programs and their resources; to the way technology is matured and inserted, suggesting a more rational and rigorous approach to the transition process; and in making fundamental changes to the way programs are managed, by moving toward a civilian workforce with military requirements advisors. As with most change efforts, those with the most potential for gain are also the ones most difficult to plan and implement. All three of these innovations are possible, but will require substantial willpower and collaboration inside the Department and between the Department and the Congress. The question remains, is the Department ready for real acquisition reform?

References

Ahern, D. (2009). *OSD Study of Program Manager Training and Experience*. Washington, DC: Office of the Secretary of Defense.

Birkinshaw, J. (2001). Strategies for managing internal competition. *California Management Review, 44*(1), 21–38.

Cavas, C. P. (2015). USN ship strategy focuses on industrial base. Defense News. Retrieved from http://www.defensenews.com/story/defense/naval/ships/2015/03/23/shipbuilding-navy-stackley-ingalls-bath-nassco-national-steel-lpd17-amphibious-ships-lha-lha8-assault-ship-taox-fleet-oiler-ddg1000-ddg51-arleigh-burke-zumwalt-amdr-radar-raytheon-aegis/25083347/

DoD. (2005). *Operation of the Defense Acquisition, Technology, and Logistics Workforce Education, Training, and Career Development Program* (DoD Instruction 5000.66). Retrieved from http://www.dtic.mil/whs/directives/corres/pdf/500066p.pdf

Dual Use. (2012). *Dual Use Export Licenses*. Washington, DC: US Department of Commerce. Retrieved from http://www.export.gov/regulation/eg_main_018229.asp

Fay, M. (2015). Pentagon should service competition, not customer service. Retrieved from http://www.realcleardefense.com/articles/2015/02/24/pentagon_should_service_competition_not_customer_service_107656.html

Fox, J. R. (2012). *Defense Acquisition Reform 1960–2009: An Elusive Goal*. Washington, DC: US Army Center of Military History.

Fox, J. R. (2014). *Report to Congress on Department of Defense 2014 Study of Program Manager Training and Experience*. Washington, DC: Office of the Secretary of Defense.

Fox, J. R., & Miller, D. B. (2006). *Challenges in Managing Large Projects*. Fort Belvoir, VA: Defense Acquisition University Press.

Francis, P. L. (2014). *Where Should Reform Aim Next?* (GAO-14-145T). Washington, DC: GAO.

GAO. (2006). *Stronger Practices Needed to Improve DoD Technology Transition Processes* (GAO-06-883). Washington, DC: Author.

GAO. (2008). *Defense Acquisitions: Assessments of Major Weapon Programs* (GAO-08- 467S). Washington, DC: Author.

GAO. (2009). *Defense Acquisitions: Many Analyses of Alternatives Have not Provided a Robust Assessment of Weapon Systems Options* (GAO-09-665). Washington, DC: Author.

GAO. (2010). *Defense Acquisitions: Additional Guidance Needed to Improve Visibility into the Structure and Management of Major Weapon System Subcontracts* (GAO-11-61R). Washington, DC: Author.

GAO. (2011). *Defense Acquisitions: Assessments of Selected Weapon Programs* (GAO-11-233SP). Washington, DC: Author.

GAO. (2012). *Defense Acquisitions: Assessments of Selected Weapon Programs* (GAO-12-400SP). Washington, DC: Author.

GAO. (2013). *Defense Acquisitions: Assessments of Selected Weapon Programs* (GAO-13-294SP). Washington, DC: Author.

GAO. (2014). *Defense Acquisitions: Assessments of Selected Weapon Programs* (GAO-14-340SP). Washington, DC: Author.

GAO. (2015). *Defense Acquisitions: Assessments of Selected Weapon Programs* (GAO-15-342SP). Washington, DC: Author.

Gates, S. M., Roth, E., Srinivasan, S., & Daugherty, L. (2013). *Department of Defense Acquisition Workforce: Update to Methods and Results Through FY 2011*. Washington, DC: RAND Corporation.

Hagerty, J. C., Stevens, P. D., & Wolfe, B. T. (2008). *DDG 1000 versus DDG 51: An Analysis of US Navy Destroyer Procurement*. Monterey, CA: Naval Postgraduate School. Retrieved from http://www. dtic.mil/cgi-bin/GetTRDoc?AD=ADA494009& Location=U2& doc=GetTRDoc.pdf

Harrison, T. (2012). *The Effects of Competition on Defense Acquisition*. Washington, DC: Center for Strategic and Budgetary Assessments.

Holzer, R., & Truver, S. C. (2014). Not your "Father's AEGIS." *Center for International Maritime Security*. Retrieved from http://cimsec.org/not-fathers-aegis/13697

International Trafficking in Arms Regulations (ITAR). (2014). *International Traffic in Arms Regulations*. Washington, DC: US Department of State. Retrieved from https://www.pmddtc.state.gov/ regulations_laws/itar.html

Joint Defense Capabilities Study. (2004). *Improving DoD Strategic Planning, Resourcing and Execution to Satisfy Joint Capabilities*. Washington, DC: DoD.

Mitchell, S. W. (2010). *Model-Based System Development for Managing the Evolution of a Common Submarine Combat System*. Retrieved from http://www.omgsysml.org/Model_Based_ Approach_to_Manage_Evolution-SteveMitchell_CI_C4I_2010v6.pdf

Office of the Under Secretary of Defense for Acquisition, Technology, and Logistics (OUSD(AT&L)). (2014). *Guidelines for Creating and Maintaining a Competitive Environment for Supplies and Services in the Department of Defense*. Washington, DC: Author.

Thompson, M. (2013). The most expensive weapon ever built: The Pentagon's $400 billion F-35 is running into turbulence just as deeper budget cuts loom. *Time*. Retrieved from http://content. time.com/time/magazine/article/0,9171,2136312,00.html

USLegal. (2014). *Universal Commercial Code*. Retrieved from http://uniformcommercialcode.uslegal. com

US Navy Fact File. (2014). Destroyers–DDG. Retrieved from http://www.navy.mil/ navydata/fact_ display.asp?cid=4200&tid=900&ct=4

US Senate. (2014). *Defense Acquisition Reform: Where Do We Go from Here?* Testimony before the Permanent Subcommittee on Investigations, Committee on Homeland Security and Governmental Affairs. Retrieved from http://www.mccain.senate.gov/public/_cache/ files/7f54fe2e-9c26-4f66-b940-ebf8a9e9ef9c/psi-report---defense-acquisition-reform---a-compendium-of-views-10-2-14.pdf

Wydler, G., Chang, S., & Schultz, E. (2012). *The Limits of Competition in Defense Acquisition*. Retrieved from http://www.dau.mil/research/symposiumdocs/Wydler_Continuous%20Competition% 20slides.pdf

chapter seventeen

Strategy and military technology
The three offsets

Bud Baker

Contents

A case study in deterrence: The Strategic Air Command

The longstanding motto of the Strategic Air Command—"Peace is our Profession"—was often seen by critics as irony: "Yeah, war is just a hobby." But for those who served in that organization during the Cold War, there was no contradiction at all: In a nuclear world, where the two major adversaries had tens of thousands of nuclear warheads aimed at each other, the term "victory" had little meaning: To fight would be to lose. Thus came the strategy of nuclear deterrence, a preventative approach mirroring the wisdom of Sun Tzu in the *Art of War*: "The supreme art of war is to subdue the enemy without fighting."

The relationship between defense strategy and military technology leads to some principles that will underlie the remainder of this chapter, and those principles need to be acknowledged explicitly. First, military technology is constantly evolving, and that evolution is not random: Much technological innovation is built upon earlier technology breakthroughs, which themselves were designed to *counter* technological advances by a real or potential adversary. This is, of course, an endless cycle, and ensures that there is never an "ultimate weapon." So, to use an ancient example, bands of medieval marauders drove the creation of fortified towns, which in turn led to better siege engines, which in their turn produced even more elaborate fortifications, with their round towers, crenellations, moats, and drawbridges. Those defenses then led to the proliferation of high explosives, and so on. There is, to date, no apparent end to this pattern, no "weapon to end all weapons."

Another principle: Military technology need not be—indeed generally *is* not—symmetric in nature. That is, people will fight with whatever tools are available to them, and technological sophistication is not necessarily part of the mix. While in some cases technological innovation *can* move in parallel among potential combatants—consider the roughly similar triad of strategic nuclear forces maintained by the US and USSR/Russia since 1950—in many cases the opposite occurs: Opposing forces in the same conflict may operate from vastly different technological playbooks: Consider contemporary struggles, where one side relies on the technologically simple—airplanes flown into buildings, for example, or suicide bombings—while the other side relies on technologically sophisticated techniques like unmanned air vehicles, precision bombing, stealth technology, or cyberwarfare.

A third characteristic of modern military technology is the reality of ever-shortening product life cycles. Compare, for example, the Industrial Revolution, which played out over the better part of a century, to the rapidity of today's information revolution, in which giants like Google and Facebook are barely out of adolescence. Capabilities change rapidly in today's world, and anyone intending to counter those capabilities must be at least as agile. Even weapons technology that might seem on the surface to be relatively basic must be adapted and redeployed in a very short time: Consider, for example, the use of Improvised Explosive Devices (IEDs). American forces in Iraq learned that countermeasures that worked against an IED one day might be utterly ineffective against the next generation, just months—or even weeks, or days—later. If lethal weapons can change that fast, the associated countermeasures must be equally adaptive. Thus in 2006, the Joint Improvised Explosive Device Defeat Organization—JIEDDO—was established, to slash the countermeasure development response time from what was typically years to months, even weeks (Defense Threat Reduction Agency, (2016)).

So to summarize, military uses of technology can be offensive, defensive, or preventative. Such technologies are constantly evolving: They never stand still. Technologies need not be symmetrical: Adversaries may have different technological philosophies, based on distinctive competence, historical experience, or doctrinal tenets, which can be expected to drive variations in their technological investment. And whatever technologies are employed, the clear trend is toward ever more rapid change, driving a corresponding need for faster adaptation.

American defense strategies

Technology development as a core belief

For most of the last century, the American military has put technology at the forefront of military planning. There are many reasons for this, not all of them obvious. Nowhere is this focus on technology more obvious than in the development of what became the United States Air Force.

> The men in charge of the future Air Forces should always remember that problems never have final or universal solutions, and only a constant inquisitive attitude toward science and a ceaseless and swift adaptation to new developments can maintain the security of this nation through world air supremacy.

—Theodore von Karman, 1945 (as cited in Daso, 1997)

But it wasn't always so. The pioneering years of the US Army Air Service were marked by mere halting steps in technological development, as the United States lagged behind the nations of Europe. Of all the factors that led to the defeat of Germany in World War I, American aviation technology was clearly *not* one of them:

> ...by the time the Armistice came, we did have 2,768 completely trained pilots and observers on the Western Front. Out of 20,000 officers and 149,000 enlisted men of the Army Air Service at home and abroad, almost 40 percent of the officers and 50 percent of the enlisted men were in France or at advanced training bases in England. Many more would have been there if there were airplanes for them......*No American-designed combat planes flew in France or Italy during the entire war* (Italics added) (Arnold, 1949).

The author of those words, General Hap Arnold, went on to lead the US Army Air Forces from 1938 to 1946, and his enormous influence on aviation technology was felt for an even longer period than that. The only person ever designated as a 5-star "General of the Air Force," General Arnold devoted deep thought to the relationship between military forces and the larger society to which they belonged, and he came to believe that democracies would never be able to match up numerically with the armed forces of totalitarian states. To General Arnold, this insight meant that American Air Forces would always need to rely on technological advances, rather than superiority in numbers. And he understood that those technology breakthroughs were not likely to come solely from within the Air Corps, but also from partners in academia, science, engineering, and business. As he explained in a speech, shortly before he became US Army Air Corps Chief:

> Remember that the seed comes first: if you are to reap a harvest of aeronautical development, you must plant the seed called experimental research. Install aeronautical branches in your universities; encourage your young men to take up aeronautical engineering... Spend all the funds you can possibly make available on experimentation and research. Next, do not visualize aviation merely as a collection of airplanes. It is broad and far-reaching. It combines manufacture, schools, transportation, airdrome building and management, air munitions and armaments, metallurgy, mills and mines, finance and banking, and finally, public security—national defense (Daso, 1997).

Always a realist, General Arnold believed that his Air Force would not be able to attract or retain sufficient numbers of high quality scientists and technologists. For that reason, he stressed technology partnerships as an essential part of planning for future airpower capabilities. He reached out to an unprecedented consortium of strategic thinkers at leading universities, as well as to inventors, aviators, aeronautical designers, automotive manufacturers, and financiers. Through the National Academy of Sciences, he held meetings of top technology experts, gatherings which sometimes raised eyebrows in the more traditional senior officer ranks:

> Few high-ranking Army officers seemed aware of the close relationship developing between these specialists and the little Air Corps—a relationship that was to grow to such importance in World War II that civilian scientists would work side-by-side with staff officers in

our overseas operational commands, frequently flying on combat
missions to increase their data.

Once, after George Marshall became Chief of Staff, I asked him
to come to lunch with a group of these men. He was amazed that I
knew them. "What on earth are you doing with people like that!" he
exclaimed.

"Using them," I replied. "Using their brains to help us develop
gadgets and devices for our airplanes—gadgets and devices that are
far too difficult for the Air Force engineers to develop themselves."
(Arnold, 1949)

General Arnold commissioned the Army Air Force's Scientific Advisory Group (SAG,
today known as the Scientific Advisory Board) in 1944, to advise him and to guide the
technological strategies of the USAAF. Led by General Arnold's trusted colleague and
advisor Theodore von Karman, the SAG provided the foundation and blueprint for General
Arnold's vision of Air Force technological supremacy. Its prescient 1945 report *Toward New
Horizons* foresaw many of the scientific developments which would come to fruition over
the next seven decades, and which are taken for granted today.

While farsighted thinkers like General Arnold were focusing on technology, military
and political leaders elsewhere were relying on numbers: More planes. More tanks. Bigger
armies. The struggle between those competing views—in a sense, the question of quantity
vs. quality—is central to the three Offset strategies discussed in the coming pages.

The three offsets

The first offset strategy

Recall that one of our opening principles held that defense strategy need not be symmetrical. Grasping this idea is essential to understanding of what has since become known as
the First Offset Strategy.

Various definitions of offset strategy exist: One arguably too-simple definition is
"a technological response to a perceived military weakness" (Korb & Evans, 2017). Sadler
(2016) offer a more specific and nuanced definition, one which more fully captures the
technocentric essence of the "offset" concept: "An offset seeks to leverage emerging and
disruptive technologies in innovative ways in order to prevail in Great Power competition."

The Soviet Union had tested its first atomic bomb in 1949, and its first hydrogen bomb
just four years later. At that point, the USSR embarked on a major effort to build an offensive nuclear force, intending to match and indeed surpass the power of the US Air Force's
nascent Strategic Air Command. The Soviets launched their first ballistic missile in 1947
(Federation of American Scientists 2000), and by 1950, according to some sources, Soviet
military spending exceeded that of the United States. The Soviet Long Range Aviation arm
received similar attention, with well over two thousand long range bombers developed
and built beginning in the early 1950s, a number that was far beyond anything US air
defenses could be expected to intercept.

At this point the administration of US President Dwight D. Eisenhower faced a strategic decision: They could continue to invest in large numbers of manpower-intensive
conventional forces, aimed at fighting and winning conventional wars—like the Korean
Conflict just then winding down. This idea was found to be unappealing in several ways.
Most obvious was the enormous and unsustainable expense that would be involved.

President Eisenhower clearly believed that the USSR's massive forces were intended to force the US and its Western allies into a self-destructive and untenable spending surge. In a radio address to the American people just months after taking office, the new President explained:

> We must see, clearly and steadily, just exactly what is the danger before us. It is more than merely a military threat. It has been coldly calculated by the Soviet leaders, for by their military threat they have hoped to force upon America and the free world an unbearable security burden leading to economic disaster (Eisenhower, 1953).

There was another technological dimension that was just beginning to be understood, and that was the transformation created by nuclear weapons, which had been first employed just eight years before. Until Hiroshima and Nagasaki, wars were fought with the general understanding that one side would win, and the other would lose. But nuclear weapons changed all that: Both the size and the power of nuclear arsenals had forever altered the meaning of "winning": By 1953, it was dawning on decision makers that to even *fight* a nuclear war was to lose. Even if a country's defenses were *perfect*—a state that had never come close to being attained in any military confrontation—just the nuclear fallout from one's *own* attacks would likely bathe the earth in radioactive poisons for months, or longer.

What resulted from all this was what is known today as the First Offset Strategy. At the time it was called the "New Look," and it was classically adaptive in nature: It was designed to meet an offensive threat not with an appropriate *defense*, but with a countervailing *offense*. With the "New Look," American defense dollars were diverted from conventional defensive systems—especially land- and sea-based—to fund an enormous buildup of US offensive nuclear forces—the Strategic Air Command's thousands of manned bombers, and the creation of intercontinental and submarine-launched ballistic missile forces. The foremost aim of these forces was not to *fight* a nuclear war, but to *deter* one. The era of Nuclear Deterrence and "Massive Retaliation" had begun.

Deterrence, then, was not so much a choice between alternatives as it was a decision by default: There really *was* no alternative. Some called it a strategy of "Massive Retaliation," while others referred to it as "Mutual Assured Destruction." In the words of General Curtis E. LeMay, Commander-in-Chief of the Strategic Air Command (SAC) and later Air Force Chief of Staff, those terms were not apt:

> Massive retaliation was a term coined by either newspapermen or some public affairs guy someplace in the military. The idea was to have overwhelming strength so that nobody would dare attack us— at least that was my idea of it, and what I attempted to accomplish out at SAC—that we would have such strength that we would never have to do any fighting (Kohn & Harahan, 1988).

That objective cited by General LeMay—"that we would have such strength that we would never have to do any fighting"—proved elusive, of course. While nuclear deterrence proved an effective strategy for averting global war, its many shortcomings came to be well known. For one thing, nuclear deterrence strategy—the central tenet of the First Offset Strategy—did little to prevent lower-intensity conventional wars around the world, like the Vietnam Conflict or Arab-Israeli confrontations in the Middle East.

The second offset strategy

By the mid-1970s, a generation after the First Offset, history looked to be repeating itself. Just as after Korea, another Asian conflict had wound down, in Vietnam, and defense budgets were again falling. The end of the military draft in the US meant that the increased expense of military manpower—in the form of the "All Volunteer Force"—would put increased economic pressure on Department of Defense budgets. In Europe, the US and its NATO allies were facing Soviet and Warsaw Pact forces that outnumbered them by a factor of three-to-one. War games and simulations featuring a Warsaw Pact thrust through the Fulda Gap into West Germany predicted defeat for NATO Forces.

Meanwhile, in the Middle East, the Yom Kippur War in October 1973 had been short—just eighteen days from start to finish—but exceptionally costly to the Israeli military: Even with Israel's superbly trained aircrews flying the most modern western aircraft, the Soviet-provided air defenses exacted a huge price, downing over a hundred Israeli combat aircraft, about half of them in just the first three days of the war (Israel Defense Forces, (n.d.). (1973)).

Facing these military and economic challenges, President Carter's Secretary of Defense, Harold Brown, and his Undersecretary of Defense for Research and Engineering, William J. Perry, developed what became known as the Second Offset. Dr. Perry, its primary architect, identified areas in which American technological prowess could provide a significant competitive advantage. Consulting with the Defense Advanced Research Projects Agency (DARPA), he based the Second Offset on three emerging technologies: Battlefield awareness, through enhanced intelligence, surveillance, and reconnaissance; precision-guided munitions, and low-observables—stealth—technology (Perry, 2003).

In light of the technological successes of the Second Offset's weapon systems, the logic behind the Second Offset seems unassailable. But it was not that way at the time. Opponents like the Congressional Reform Caucus believed that the heavy reliance on technology was wrong-headed and unaffordable; Rather than smaller numbers of high technology weapons, the "defense reformers" argued that the real need was for large numbers of simple, low technology systems: As Secretary Perry saw it:

> The Caucus's view was that the offset strategy was a terrible idea, and what we ought to do instead was to focus on competing with the Soviets in numbers, setting aside the question of how we could persuade the public to support an army two or three times the existing size. They argued that the technology was a step backward and would introduce a complexity in weapon systems that the military personnel would be unable to operate or maintain. They didn't say it in so many words, but they implied that the military personnel were not capable. Instead, they would say things like, "It would take a Ph. D. to operate the equipment." I thought they were profoundly wrong (Goldberg & Trask, 1998).

The third offset strategy

In a November 15, 2014 memo in which he cited "eroding" American military dominance, Defense Secretary Chuck Hagel introduced his "Defense Innovation Initiative." The memo itself is oddly vague, but lurking on the second page, in lower-case letters, were the words "third offset strategy." Yet while the "offset" terminology was downplayed in the memo, the connection to the past offset strategies was clear (Hagel, 2014a).

Secretary Hagel made that connection even more clearly in a Reagan National Defense Forum speech that same night. He also outlined the specific technologies that he saw as the focus of the Third Offset, echoing the long-ago ideas of Hap Arnold, as he called for closer collaboration between the Pentagon and civilian technology experts:

> Our technology effort will establish a new Long-Range Research and Development Planning Program that will help identify, develop, and field breakthroughs in the most cutting-edge technologies and systems—especially from the fields of robotics, autonomous systems, miniaturization, big data, and advanced manufacturing, including 3D printing. This program will look toward the next decade and beyond.
>
> In the near-term, it will invite some of the brightest minds from inside and outside government to start with a clean sheet of paper, and assess what technologies and systems DoD ought to develop over the next three to five years and beyond (Hagel, 2014b).

Just as Secretary Perry had found thirty years previously, Secretary Hagel's Third Offset was met with both cautious cheers and caustic complaints. One critic called The Third Offset "fairy dust," and likened it to hitting the "Easy" button, *a la* the popular television commercials from Staples office supply. The same critic added that "basing a strategy on technological innovation that is not in hand is nothing more than wishful thinking" (Carafano, 2014).

Of course, that same critique could have been made about the Second Offset—stealth technology, precision guided munitions, and the other Second Offset innovations—which have now stood the test of time—and of combat operations—for three decades.

And next?

We saw at the beginning of this chapter some truths about the development of military technology, and the strategies related to it: It is often asymmetric, the rate of change is high and even increasing, and the cycle of weapon-and-counterweapon never ends. Considering Secretary Hagel's 2014 announcement of the Third Offset strategy, what are likely to be the impacts of the 2016 presidential election, and the resulting changes in the nation's political power structure?

Opinions abound, but all acknowledge the perilous situation facing any program when a new administration comes to town. Observers suggest that three options generally exist, for Third Offset or any similar initiative: Abandonment, neglect, or active support, even if that support is given under a different name. Given the present political atmosphere in Washington, the third option seems the least likely (Pomerleau, 2017).

Other defense experts doubt that the name of an initiative matters very much anyway. Much more important, they argue, is that the underlying ideas are recognized, understood, and supported. There is broad agreement on these underlying themes:

- Technology is most useful when focused on specific problems in need of solutions.
- Technology matters, and is an area in which the US has a proven distinctive competence.
- But technology is not the *only* thing that matters: The military must have in place the right workforce and the right processes to capitalize on technology as a competitive advantage.
- Asymmetry must be embraced. The US must look beyond fixing perceived weaknesses, to take advantage of its unique strengths (Johnson, 2016; Hicks, 2017).

Conclusion

Nearly a century has passed since a young Lieutenant Hap Arnold began to see the possibilities of advanced technology, as a means of offsetting an enemy's advantages in military strength. In those decades, General Arnold's ideas have been developed, tested, and refined, and yet his basic observations remain intact. Technology remains a distinctive American competence, and maintaining that technologic superiority will remain a critical challenge for the next century.

References

Arnold, H. H. (1949). *Global Mission*. New York, NY: Harper & Brothers.

Carafano, J. J. (2014). The third offset: The fairy dust strategy. Retrieved from http://www.heritage.org/defense/commentary/the-third-offset-the-fairy-dust-strategy

Daso, D. (1997). *Architects of American Air Supremacy: General Hap Arnold and Dr Theodore von Kármán*. Maxwell Air Force Base, AL: Air University Press.

Defense Threat Reduction Agency, (2016). *JIDO's History*. Retrieved from http://www.dtra.mil/Missions/Defending/JIDO/History/

Eisenhower, D. D. (1953). *Radio Address to the American People on the National Security and Its Costs*. Retrieved from http://www.presidency.ucsb.edu/ws/?pid=9854

Federation of American Scientists, (2000). *Strategic Missile Troops*. Retrieved from https://fas.org/nuke/guide/russia/agency/rvsn.htm

Goldberg, A., Trask, R., (Interviewers) & Perry, W. J. (Interviewee). (1998). *Interview with William J. Perry, October 6, 1998* [Interview transcript]. Retrieved from http://history.defense.gov/Portals/70/Documents/oral_history/OH_Trans_PERRYWilliam%20J10-06-1998.pdf?ver=2017-10-04-102345-740

Hagel, C. (2014a). The defense innovation initiative. Secretary of Defense Memorandum. Retrieved from http://archive.defense.gov/pubs/OSD013411-14.pdf

Hagel, C. (2014b). Reagan national defense forum keynote. [Transcript]. Retrieved from https://www.defense.gov/News/Speeches/Speech-View/Article/606635/

Hicks, K. (2017). What will replace the third offset? Lessons from past innovation strategies. Retrieved from http://www.defenseone.com/ideas/2017/03/what-will-replace-third-offset-lessons-past-innovation-strategies/136260/

Israel Defense Forces, (n.d.). (1973). History of the Yom Kippur War: Day-by-Day. Retrieved from https://www.idfblog.com/about-the-idf/history-of-the-idf/1973-yom-kippur-war-day-day/

Johnson, T. R. (2016). Will the Department of Defense invest in people or technology? *The Atlantic*. Retrieved from https://www.theatlantic.com/politics/archive/2016/11/trump-military-third-offset-strategy/508964/?utm_source=atlfb

Kohn, R. H., & Harahan, J. P. (Eds.). (1988). *Strategic Air Warfare: An Interview with Generals Curtis E. LeMay, Leon W. Johnson, David A. Burchinal, and Jack J Catton*. Washington, DC: Office of Air Force History.

Korb, L. J., & Evans, C. (2017). The third offset strategy: A misleading slogan. *Bulletin of the Atomic Scientists, 73*(2), 92–95.

Perry, W. J. (2003). Technology and national security: Risk and responsibilities. Conference on Risk and Responsibility in Contemporary Engineering and Science: French and US Perspectives. Retrieved from https://stanford.edu/dept/france-stanford/Conferences/Risk/Perry.pdf

Pomerleau, M. (2017). The fate of the third offset under President Trump. Retrieved from https://www.c4isrnet.com/it-networks/2017/01/18/the-fate-of-the-third-offset-under-president-trump/

Sadler, B. D. (2016). Fast followers, learning machines, and the third offset strategy. *Joint Force Quarterly, 83*, 13–18.

chapter eighteen

Prescription for an affordable full spectrum defense and innovation policy

Jan P. Muczyk

Contents

Introduction

Persons knowledgeable in international relations consider the US an indispensible nation. Hence, the US needs to pursue a full spectrum defense policy, according to [1], Muczyk (2017). However, a full spectrum defense policy is expensive indeed and must compete with pressing domestic priorities. Therefore, viable ways of making it more affordable have been presented. They include: total asset viability; looking in the right places; reducing federal bureaucracy; building weapons from low-hanging fruit; exploiting economies of scale; lesser reliance on military specifications, focused leadership education; and growing the technological fruit tree.

Economic limitations to the arms race

The belief by many of our civilian and military leaders based on outdated formulas developed by Frederick Lanchester at the height of WWI that technology will negate numerical superiority has led to a reliance on transformational technology which, in turn, has resulted in staggering product development costs and unprecedented product development life cycles. The cost of one B-2 bomber is $2 billion, which compelled Congress to limit its volume to 21 aircraft; and one has already been lost in an accident. The cost of one F-22A is $355 million ($420 million with retrofit items), and it took 22 years to field the F-22A. If it were being developed for WWII, it would not have seen service until the Vietnam conflict. The joke in the Pentagon has it that the 22 stands for the number of years it took to develop this plane. The F-35 is on the same glide path as the F-22A with respect to cost and product development time [2].

Since insurgencies, the existential and near-term threats, lack air forces and navies, the US can fight them without the so-called fifth generation platforms. However, insurgencies last a long time and are expensive, and the US cannot afford to bankrupt itself with prohibitively expensive high-tech weapon systems with dubious military advantages for fighting insurgencies. Former Congressman Barney Frank, D-Mass., speaks for many legislatures: "The math is compelling: If we do not make reductions approximating 25 percent of the military budget starting fairly soon, it will be impossible to continue to fund an adequate level of domestic activity ever, with a repeal of Bush's tax cuts for the very wealthy. American well-being is far more endangered by a proposal for substantial reductions in Medicare, Medicaid, Social Security or other important domestic areas then it would by cancelling weapon systems that have no justification from any threat we are likely to face."

Indeed, the opportunity costs of a large defense budget are considerable. Conservative historian Robert Kagan offers a rebuttal: "2009 is not the time to cut defense spending. A reduction in defense spending this year would unnerve American allies and undercut efforts to gain greater cooperation. There is already a sense around the world that the United States is in terminal decline. Many fear that the economic crisis will cause the United States to pull back from overseas commitments. The announcement of a defense cutback would be taken by the world as evidence that the American retreat has begun." What Robert Kagan overlooks is the fact that our allies have not paid their fair share of their own defense since the end of WWII, and it is about time that they become unnerved [2].

Historically, the US has contributed 50 percent of NATO's budget. Recently, the US share has jumped to 75 percent with Europeans using their economic woes as an excuse for not doing more. In light of the population size of the European Union and its combined GDP, this is inexcusable. Europe should heed the warning issued by former Secretary of Defense, Robert Gates, in his NATO valedictory address to contribute much more to its own defense because the US can easily lose the appetite to do so. A more recent Secretary of Defense, Ashton Carter, echoes Robert Gates. These gentlemen were not just crying "wolf." With the inauguration of Donald Trump as president, the time has actually arrived. There is some talk that the European Union should have its own unified military. This notion should receive full support from the United States [2].

Lessons learned from the arms race

Nations should learn lessons not only from their war experiences but from arms races as well. As the Soviets realized, quantity has its own quality advantages, even with superior equipment. Wonder weapons, with the exemption of nuclear warheads, are not a substitute for simpler but effective counterparts available in large numbers. When Soviet Field Marshal Gregory Zhukov, who knew more about large scale warfare than anyone, with the possible exception of Napoleon, was asked at the end of WWII what it took to win a large scale military conflict, he responded: "More—more troops, more tanks, more planes, more ships, more artillery, etc." The US WWII experience mirrors Marshal Zhukov's advice [2].

Does the US get good value for its huge expenditures?

There is an old British saying: "When you run out of money, you must begin thinking." It appears as though exotic weapon systems expand to exhaust the money available in the Defense Department (DoD) budget. As a result, fiscal austerity becomes the mother of an efficient and effective military. The size of the US defense budget shout not be confused with national security. It took a former general, President Eisenhower, to alert the nation to the military/industrial/congressional complex, but we did not listen. Eisenhower was convinced that the "Pentagon Boys" exaggerated threats in order to get larger military budgets. The politicians went along because jobs in my district get me elected and reelected, and that is what matters. Lockheed/Martin has subcontractors for the F-35 in 47 states to gain maximum political support. And this is not an isolated exception. The Navy F-18E/F has subcontractors in 44 states.

A report by the Government Accountability Office meticulously documented in 2012 that the Pentagon's 95 largest weapon systems were nearly $300 billion over budget. Deloitte Consulting LLP concluded that cost overruns have steadily worsened. Technical complexity accounts for an ever-increasing percentage of weapon's cost overruns. Complexity is also the enemy of reliability and meeting deadlines. The F-35 is so computer code dependent that writing and debugging the code has become the "long pole in the tent." The F-35 is not only over budget and behind schedule, but the critics of the F-35, the most expensive weapon system of all time, make a compelling case that the plane can't climb, and can't run, and is no match for the top of the line Russian fighters if it is thrust into aerial combat. Quite frankly, the US taxpayer and our allies who are counting on this place to be the backbone of their future air fleets deserve better. In time, the F-35 may become a viable platform since complex weapon systems experience lengthy teething problems. But that will not happen anytime soon [2].

Flawed funding processes based on unrealistic cost estimates are an integral part of the problem. Realistic cost estimates frequently are unavailable because most programs are funded and launched while there is still significant uncertainty about most everything. Hence, only fixed cost contracts should be negotiated by the DoD so that contractors also incur the risk associated with cost overruns.

How to make the arms race more affordable?

How much a nation spends on its national defense is a necessary condition, but the sufficient condition is how wisely the money is spent. We cannot risk unilateral disarmament because we no longer can count on two oceans for creating lead-time to rearm, as was the case in the past. Intercontinental ballistic missiles have seen to that. However, potential enemies continue to exist. Yet, we have pressing domestic priorities that compete with the defense budget. Hence, we must make a realistic defense policy more affordable. The ways exist. All we need is the will. First, we must guarantee that the books of the Pentagon and all the military branches are auditable. Until that is done, we cannot know what we need because we have no way of knowing what we have [3].

Relying on the intelligence community

The US has a robust Intelligence Community—both human intelligence as well as signals intelligence [4]. The information that it possesses should be the starting point with regard to identifying the assets needed to neutralize current and potential threats. Relying on government contractors may result in the procurement of inordinately expensive systems of dubious military value. Moreover, such systems could unnecessarily fuel the arms race.

Vital nature of total asset visibility

The United States sent twice as much material to the Persian Gulf as was required, and our troops did not know where half of it was at any given moment. Half of the 40,000 bulk containers shipped into the theater has to be opened in order to identify their contents, and most of it failed to contribute in any way to our success on the battlefield. If we recognize the coalition nature of present and future conflicts, then it becomes obvious that there is a big payoff associated with integrating our asset visibility system with those of our allies.

Look in the right places

The largest savings potential rests in the mission and roles category. For example, not only does the Navy have its own Air Force and an army (the Marines), the Navy's army has its own air force as well. Incidentally, the Army has its air force (and a large one at that when rotary aircraft are include) and a navy (Corps of Engineers) too. The Air Force is anxious to rid itself of the A-10 close air support aircraft, and the best one available; which leads the ground forces to question its commitment to close air support. Little wonder that the Marines insist on providing their own close air support. Perhaps, given the fact that Air Force generals appear to be ensorcelled by high tech wizardry, the close air support mission and the A-10 should be assigned to the Army [5].

Reducing the size of the federal defense bureaucracy

The US force structure and budget have declined by about one-third from their 1985 peak levels. The infrastructure, however, has declined about 18 percent [3]. Therefore, the two should be brought into balance before reducing the end strength of combat forces, and it should be done by proven re-engineering methods instead of for political reasons. After all, the WWII experience reveals that lean organizations produced the most impressive results [6].

Re-engineering means excising those activities that are either unrelated or marginally related to the central mission (occupational hobbies), removing redundancies, and creating or refining processes through which mission relevant goals and objectives are attained in an efficient and effective manner. Re-engineering requires evaluating the value chain and eliminating or reducing components that either add no value or very little, while retaining and even enhancing those that add considerable value.

A good place to begin re-engineering efforts is activity-based accounting (ABS), a systematic method for assigning costs to business activities. First, a reasonable number of business activities need to be defined, and all costs associated with each activity need to be assigned to the appropriate activity. Once this much has been accomplished, the activities with their associated costs can be allocated to products, processes, customers, or vendors. Next, activities need to be assigned priority on the basis of cost, with the most expensive activity receiving top priority for scrutiny with respect to redundancy, relevancy, and criticality. Last, whenever appropriate, the unnecessary or marginal activities are eliminated. Whenever practicable we must insist that all technology, processes, and procedures "buy" their way into the organization in terms of reducing the total cost of doing business [7].

We need to abandon practices that have been tried and found wanting. I have in mind trying to meet the needs of all the military branches with variants of one aircraft. That was tried in the past with the tactical fighter experimental (TFX) without success. Now the DoD is trying the same things with the F-35. To meet the Marine Corps requirements for Short Takeoff Vertical Landing (STOVL) aircraft, serious design compromises were made to the Air Force and Navy variants. The McDonnell Douglas F-4 Phantom II was first built for the US Navy and was later adopted by the US Air Force and the US Marine Corps with minor modifications. Also, a number of allied countries brought the aircraft. This airplane is considered among the best multi-mission aircraft ever to see service. This strategy, however, is not to be confused with building variants of a "joint strike fighter."

The concurrency doctrine of beginning production before testing is completed needs to be jettisoned as well. Testing reveals many problems that can only be fixed with redesign and major modification. Retrofitting is too time consuming, expensive, and often inadequate.

Economists agree that there are more cost efficient and socially beneficial job creation programs than building weapon systems. Military weapons should be justified on the basis of military necessities alone. While the author does not subscribe to the notion that national defense if too important to leave to generals (admirals), he is a strong supporter of vigilant oversight by Congressional committees and subcommittees.

Building weapon systems from low-hanging fruit

This effort demonstrates that being first with new technology provides a military advantage for a while. The length of time depends on how adversaries perceive the value of the weapon system in question. If considered critical, they will devote the necessary resources to minimize or eliminate the lead, providing they possess the economic and technical capacity to do so. Otherwise, they will either get around to it eventually or elect not to compete. The lead is important if a nation intends to start a war, and can serve as a deterrent for nations that wish to preserve the peace. Also, it is a military advantage if a nation is attacked. Simply getting the lead to demonstrate the political and economic superiority of the system a nation is committed to is of dubious military value.

Since a superpower needs to prepare for practically any contingency, and the US is indubitably such a superpower, it needs to design versatile weapon systems from low-hanging technological fruit with the capacity of being upgraded. Also, the reliance on military specifications should be restricted to areas where they are absolutely necessary. Modern weapon systems rely heavily on electronics, and electronic advances typically originate in consumer sectors of Information Technology such as computers and video games.

Also, in the interest of minimizing cost overruns change orders should be discouraged by setting "drop dead" deadlines for modifying requirements. Often, military leaders wish that a new defense system should do just about everything. Yet, typically it is the last 20 percent the accounts for a disproportionate amount of the cost. Hence, encouraging the 80 percent solution when viable should receive serious consideration from the defense acquisition community.

WWII examples

The Grumman F6F shared a heritage with the ineffective F4F. But evolutionary improvement, principally the Pratt and Whitney R-2800 double Wasp engine, made it the best Navy fighter plane during WWII, and is credited with destroying 5,163 Japanese planes.

The P-51; was an ordinary plane until it was upgraded with the Packard built Rolls-Royce Merlin engine and the bubble canopy, which made it the best fighter of WWII.

Cold War examples

The F-117 was constructed with off-the-shelf components with the exception of the foil and coating. As a result, its product development cycle and cost were uncommonly short and reasonable (schedule slippage of 13 months and cost overrun of merely 3 percent). The RQ-1A Predator is another example of matching maturing technologies with warfighter needs. The Air Force began taking deliveries of an upgraded RQ-1B less than 5 years from program inception. The best examples of upgrading weapon systems are the B-52 heavy bomber and the KC-135 aerial tanker. Both are still service. The GBU-28 Bunker Buster was developed from off-the-shelf parts, tested, and deployed in 28 days during Operation Desert Storm.

The F-18E/F Super Hornet is the evolutionary progeny of earlier F-18 models, which were designed to be upgraded. As a result of this approach, the Navy was able to field what it considers to be the most advanced multi-role strike fighter available today and for the foreseeable future. Other examples of the evolutionary approach are: The Trident II D-5, which is the sixth generation member of the navy Fleet Ballistic Missile Defense, and the Patriot Advanced Capability (PAC)-3, which was introduced during the first Gulf War [5]. The Soviet Union, now the Russian Federation, amplify the point with upgrades of the SU-27 and the MIG-29. The current US F-16s, F-15s, and F-18s are much superior platforms than the original versions as well, especially the F-15SE and F-16V. Ascertaining which upgrades provide the biggest bang for the buck is vital to this strategy. For example, while the F-22A and F-35B have limited thrust vectoring capability, providing robust thrust vectoring for all fighters and fighter bombers merits serious consideration. After all, if we accept the proposition that stealth is an asset of declining value, then eventually agility and speed will regain their historic preeminence. The US Air Force is getting ready to select a prime contractor for its next generation heavy bomber. Let us hope that it elects to upgrade the B-2 rather than rely on transformational technology to build a new one from a blank sheet. The DoD should learn from failed efforts to field weapon systems developed from transformational technology. Examples are: The Navy A-12 Avenger II; the Crusader mobile artillery; Comanche helicopter; the Army Future Combat Systems; and the Marine Corps Expeditionary Fighting Vehicle. Not only was a king's ransom spent developing these failed systems, but canceling them proved inordinately expensive as well.

Appreciating the significance of economies of scale

It is not unusual for the R&D phase of a complex weapon system to amount to as much as 50 percent of the production cost of the system. Ipso facto, purchasing such a system in small numbers drives up the cost to staggering proportions. Restricting the number of F-22A fighters to 187 was a serious blunder. The DoD could have purchased the F-22A, a superior plane to the F-35, at about the same price had it procured the required number. Now Congress has instructed the US Air Force to examine the feasibility of reopening the F-22A production line. Acquiring only 21 B2s was also a mistake that necessitated retaining three heavy bomber fleets, two of which are obsolete. Now the Air Force is compelled to launch a new heavy bomber program. Increasing joint ventures with allies and partners likewise will assist in securing the benefits of economies of scale [2].

Congress is also culpable when it comes to ignoring the benefits of economies of scale. When the DoD proposes a very expensive weapons system, rather than sending the DoD back to the drawing board to design a more affordable aircraft, it reduces the number of units, thereby driving up unit cost. Of course, producing an ineffective aircraft in large quantities is even a greater blunder.

The most meaningful force multiplier

Let us not forget that the most significant force multiplier is leadership. However, the most common degrees offered on military installations are business administration degrees, which prepare service members for post-retirement occupations. The military would get greater returns on its education dollars if it followed the example of the Air Force Institute of Technology (AFIT) and offered focused education. Approximately one-half of the AFIT faculty are civilians who see to it that best practices, even though they are derived from civilian organizations, are incorporated into the curriculum. "Little Israel" offers the best example of the multiplier effect of quality leadership with its repeated victories over the entire Arab world. In fairness, being supplied at first with modern French weapons and later with advanced US weapons helps the Israelis immensely [8].

Growing the technological fruit tree

When the Soviet Empire collapsed, the Russian Federation had to choose what parts of its defense establishment it would preserve. It elected to preserve its design bureaus rather than place orders for additional aircraft. That is to say, it chose the future over the present. Hence, the US should continue to grow the technological fruit tree by adequately funding basic as well as applied research. The Defense Advanced Research Projects Agency (DARPA), The Air Force Research Laboratory (AFRL), especially through its Air Force Office of Scientific Research Directorate (AFOSR), Air Force Institute of Technology (Graduate School of Engineering and Management), and the counterparts of the Navy, Army and Marine Corps should be funded in accordance with the high priority given pressing warfighter needs. Incentives should be provided to the private sector so that it would invest some of its capital to grow the technological fruit tree [5].

For example, Pratt and Whitney, the manufacturer of the F-135 engine that powers the Lockheed Martin F-35 fighter bomber, has upgraded the engine to produce a 6–10 percent thrust increase and a 5–6 percent fuel burn reduction by relying on the Navy sponsored Fuel Burn Reduction program and the Air Force Sponsored Component and Engine Structural Assessment Research Technology Maturation effort at no additional cost.

Conclusion

The Cold War left the US as the de facto leader of the free world with the obligation to create a defense policy capable of fighting regional conventional military engagements, counter insurgencies, as well as deterring major conflicts with the Russian Federation and China that could escalate into thermonuclear exchanges. All this created an unprecedented arms race between the US and the Soviet Union and their respective alliances, NATO and Warsaw Pact.

Since the US exited WWII with its economy unscathed by the war, it could afford guns and butter for the duration of the Cold War. Now, pressing domestic needs create serious competition for the federal dollar, and potential enemies, reverting to historical tendencies, refuse to go away. While arms limitation treaties have slowed the arms race, the US still needs to fashion an affordable defense policy. Toward that end recommendations have been made that include: Rationalizing missions and roles, streamlining the federal defense bureaucracy, discontinuing failed practices, exploiting economies of scale, lesser reliance on military specifications, setting "drop dead" deadlines on change orders, giving serious consideration to 80 percent solutions, integrating US asset visibility with that of our allies, increasing joint ventures with allies and partners, providing focused education, and building weapon systems through an evolutionary process rather than through transformational technology in case diplomatic strategies fail.

References

1. Muczyk, J. P. (2017), Prescription for an affordable spectrum defense policy, *Journal of Defense Management*, 7(1), 1–4.
2. Muczyk, J. P. (2013), Synchronizing foreign policy, trade policy, and defense budget in an era of fiscal austerity, *Journal of Defense Management*, 3(2), 1–7.
3. Muczyk, J. P. (1997), The changing nature of external threats, economic and political imperatives, and seamless logistics, *Airpower Journal*, 11, 81–92.
4. Hayden, M. V. (2016), *Playing to the Edge*, Penguin Press. New York.
5. Muczyk, J. P. (2007), On the road toward confirming Augustine's predictions and how to reverse course, *Defense Acquisitions Research Journal*, 14, 454–467.
6. Muczyk, J. P. (1999), The big L: American logistics in World War II Gropman A, *Air Power Journal*, 13, 119.
7. Muczyk, J. P. (1998), Generating needed modernization funds. Streamlining the bureaucracy—Not outsourcing and privatizing—Is the best solution, *Acquisition Review Quarterly*, 5, 317–332.
8. Muczyk, J. P., Kankey, R., Ely, N. (1997), Focused graduate education. An invisible but real competitive edge, *Acquisition Review Quarterly*, 4, 367–382.

chapter nineteen

Anatomy of arms races and technological innovation

Jan P. Muczyk

Contents

Introduction

The author proposes that arms races between nations can be explained by the history of nations, which in part produces their foreign policies; by the leadership of nations at a given point in time; by the wealth of a nation at a given historical juncture; and by the technological base available to a nation vis-a-vis its neighbors or competitors at a certain period.

Historians generally agree that the principal cause of military conflict between political entities is the "acquisitive motive" in the form of either territorial or influence expansion; which, in the extreme, results in empires. Perhaps the Lord should have included an eleventh commandment in the tablets given to Moses—"Nations Shalt Not Covet Their Neighbor's Territory," although wars can be initiated by accident as well, especially when countries are ready for military conflict. WWI is a case in point.

Typical reasons for territorial expansion are: Weak neighbors; population pressures; desire for more resources; religious differences; regaining former real estate (revanchism); and creating a military buffer in case of an invasion. Russia expanded its borders when its neighbors were weak. At the end of WWII, remembering recent history, the Soviet Union desired a military buffer which it created with the Soviet satellite block. Germany needed more living space (Lebensraum) and more resources. Japan required more resources to realize its ambitions (Greater East Asia Co-Prosperity Sphere), and a military buffer zone in the form of Pacific Ocean islands. The Middle East conflicts are religious in nature. Clearly, more than one of the aforementioned reasons is typically involved in military aggression. While arms races have existed for a long time, this effort will be restricted to the twentieth and twenty-first centuries.

The role of alliances

It is common for nation states to seek greater security by entering into alliances based on mutual interest. In most cases alliances obligate each member to come to the defense of other members if they are attacked. Article 5 of the NATO charter is a perfect example. WWI started in this manner, except it was the assassination of Archduke Ferdinand of Austria, the presumptive heir to the Austro-Hungarian throne, by a Serb nationalist, rather than an invasion that served as the trigger. Thus, Austria, a member of the "Triple Alliance" (Austria/Hungary, Germany and Italy), declared war on Serbia, which was supported by the "Triple Entente" (England, France and Russia).

Often, members of an alliance have an obligation to maintain a significant military presence under the questionable premise that "if you want peace, prepare for war." NATO requires members to spend a minimum of 2% of their GDP on defense, but the de facto percentage now is closer to 1.5% Hence, when a provocation occurs, nations are ready for war. Unfortunately, countries initially neutral are dragged into the military conflict, for example, the United States in both World Wars. Also, it should be kept in mind that the large members of an alliance supply smaller ones with military equipment, and as the pygmies begin fighting they invariably drag in the giants.

Saudi Arabia spent $80 billion on weaponry in 2014, more than either France or Britain, which put it in fourth place behind the US, China, and Russia. The US is supplying Middle East nations with most of their military needs as a counter to Iranian military threat. Thus, there is quite an arms race in the region. Israel is not objecting since it views Iran as the much greater threat than the Arab states receiving US weapons.

It is also common for nations to join alliances in order to create a "balance of power" condition if they believe that an adversary or potential adversary is getting too strong. Over the years, England frequently played that card. During the post WWII era, the role of preserving balance of power has perforce become a US responsibility. Some alliances have been notable successes. The North Atlantic Treaty Organization (NATO), whose purpose was to keep the Soviet Union out of Western Europe, the United States in, and Germany militarily down, is a classic case in point. US participation in the "Cold War" arms race certainly deterred Soviet aggression until the collapse of the Soviet Union in 1989 for economic reasons. It has also been argued that nuclear technology, which made the military reality of mutually assured destruction (MAD) possible, deserves some credit for the end of the Cold War, even though it did not prevent regional conventional military conflicts. That is likely why nations such as Sweden, Finland, and India renounced alliances, since too often they turned out to be gateways to war.

The balance of power doctrine has a mixed record with regard to preventing war. The 1870 war between Germany and France was about balance of power, but the reliance on this policy did not prevent WWI. On the other hand, the current balance of power has prevented an international conflagration, even though regional conflicts such as Korea and Vietnam did break out.

Brief histories of twentieth century powers

The history of a nation, to paraphrase Giuseppe Verdi, creates "a force of destiny." Of course, short-term exigencies can derail historic trends for a while (e.g., the disintegration of the Soviet Empire in 1989). However, the historic patterns have a tendency to resurface given propitious circumstances. A review of Russian history reveals a strong autocratic streak, and it will take generations before western democratic traditions take hold. Moreover, Russia expanded its territory whenever its neighbors were weak. It should be noted that the empire created by the czars was coterminous with the Soviet Empire.

It should not surprise anyone, therefore, that the Russian Federation should attempt to reassemble the empire that collapsed in 1989, and the weaker its neighbors, the more aggressive will be the effort. Unilateral disarmament by NATO members is tantamount to an invitation to the Russian Federation to pursue its irredentist instincts. While the Soviet Empire existed, ethnic Russians migrated to places such as Georgia, Ukraine, Estonia, and Moldova, thereby creating a pretext for the Russian Federation to intervene in the internal affairs of these countries. The Russian Federation is rich in natural resources, but with the exception of its defense sector, it is a manufacturing lightweight, and this in conjunction with population decline constitutes a serious constraint so far as Russian economic potential is concerned. Russia is also vulnerable to steep price declines in fossil fuels and trade sanctions.

China is a proud and ancient empire, and had been the dominant economy until the power shifted to Europe in the eighteenth century, and to North America in the twentieth. China hasn't forgotten any of this and will do whatever it can to restore its historic hegemonic and economic role, if not in the world, certainly in Asia. And much like the Russian Federation, the weaker its neighbors, the more determined will be the effort. China is serious about regaining Taiwan and control over the South China Sea. But given the size and growth rate of its economy, its primary goal is gaining access to natural resources that it lacks, principally oil.

Germany and Japan no longer harbor expansionist intentions. The Prussian militaristic ardor and the Samurai warrior ethos were extinguished by lessons learned in the most perspicuous manner—from devastating defeats during WWII. France is no longer an international power, and is primarily concerned with its own defense. Ditto for Great Britain and Italy. Simply put their economies can no longer support a significant international role. The United States escaped devastation because WWII was not waged on its territory, and thereby inherited the mantle of chief defender of democracy and liberty against authoritarian regimes led by the Soviet Union. While the German and Japanese economies recovered admirably from WWII, Europe and Asia are relieved that these two nations have demilitarized significantly, especially by eschewing nuclear weapons.

The role of technology in an arms race

During WWII the United States was not only a combatant in the European and the Pacific theaters, but also served as the "arsenal of democracy." The Soviet Union, though anything but a democracy, benefited from the US lend-lease program as well, since it was at war with Nazi Germany. This position gave the US a vital head-start in the Cold War arms race, which began September 2, 1945 and ended December 26, 1991. Of course, the technological base developed during WWII by all combatants was inherited by the victors, especially the US and the Soviet Union when they captured German rocket scientists. Moreover, much valuable German jet aircraft technology also fell into allied hands. European nuclear scientists escaped Nazi dominated Europe and emigrated to the US prior to the start of WWII.

The German lead in the arms race at the start of WWII

Hitler knew from the start that he would resort to war to realize his ambitions. Therefore, he launched a military buildup of unprecedented proportions in violation of the Versailles Treaty. Just as important, his generals developed a mobile war fighting strategy, "blitzkrieg," that gave the Germans an impressive advantage in the early stages of WWII, with the defeat of Poland and France in short order serving as object lessons. Blitzkrieg was heavily dependent on the use of tanks, and German industry supplied state of the art armor. Blitzkrieg worked quite well initially in the invasion of the Soviet Union, but was eventually neutralized by the immense size of the Soviet Union, its winter weather, and the T-34 tank, considered the best tank during most of the war.

As the war progressed, the allies, who had initial advantages as well, viz., radar and the Spitfire aircraft, not only caught up with the Germans but surpassed them. During the beginning of WWII in the Pacific, the Japanese Zero was the best plane, but its advantage was later negated by superior US aircraft in greater numbers.

Lessons learned from WWII

Perhaps the most important lesson is that quantity has its own quality advantage, both in military and economic terms. While the Germans pioneered the weaponization of intermediate range rockets (V-1 and V-2), their use on England did not prove decisive. In fact, the ubiquitous presence on the battlefield of the Soviet Katyusha tactical rockets caused much more critical damage. Ditto for the ship-borne rockets fired from US ships during the Pacific island invasions as well as Normandy. The capability of T-34 tank, as significant as it proved to be, was not decisive by itself. However, its availability in huge numbers was. The US Sherman tank was inferior to German armor in most respects, but the US prevailed because it possessed it in prodigious quantities. The German Me-262 jet fighter, even with its considerable speed advantage, had little bearing on the air war because of its limited numbers. The Japanese, with the assistance of their German allies, also spent a fortune on "wonder weapons." Expending that money on greater numbers of proven weapons and their upgrades probably would have made Germany and Japan more formidable adversaries.

Post WWII arms race

While WWII was being conducted, the Soviet Union and the US were allies, in spite of the different political and economic systems to which each was committed, because they were fighting a common enemy. Once the war ended, it didn't take long for the differences to fracture the alliance along ideological and geographic lines. Toward the end of WWII Soviet armies had overrun East Central Europe, and they imposed the Soviet version of communism on these countries, which perforce became satellites and created the Soviet buffer zone. The communist revolution in China provided the Soviet Union with a valuable ally in Asia, which it supplied with military technology, and to a lesser extent still does. Thus, the Cold War began and new alliances were formed.

Throughout the Cold War, US military strategy was predicated on the likelihood of confronting the USSR in a large scale force-on-force encounter in Europe and the possibility that such a conventional conflict would escalate into a thermonuclear exchange between the USSR and USA. Consequently, military strategy with its concomitant acquisition policies was designed around these two imperatives. While a conflagration between these two military superpowers never materialized, regional conventional conflicts did; creating a perceived need for a "full spectrum" defense policy.

NATO was unwilling to match the Warsaw Pact airplane for airplane. Therefore, the US Air Force followed the advice of its "think tank," the RAND corporation, to neutralize the Warsaw Pact's numerical advantage with "low-observable or stealth" technology. The US Navy is not as enamored with low-observable technology as is the US Air Force because advances in radar capable of detecting low-observable aircraft make stealth an asset of declining value.

Nuclear technology

So long as the US was the only country with nuclear weapons, it had a huge advantage which the Soviet Union was hellbent on overcoming by the development of its own atomic bomb by hook or crook. The US detonated an atomic device in New Mexico in July, 1945 and two atomic bombs on Japanese cities in August, 1945; while the Soviet Union detonated its first atomic bomb in August, 1949, much sooner than the experts predicted. Even with the relative technical complexity of the atomic bomb, being first bought the US a head start of only four years. The US exploded its hydrogen bomb in November, 1952, while the Soviet Union did the same in August of 1953. The US lead shrunk to one year. In no time at all, Britain, France, China, India, Pakistan, Israel, and probably North Korea followed with nuclear weapons and the means of their delivery. South Africa and the Ukraine are the only nations to give up nuclear weapons once they acquired them, and Ukraine probably regrets its decision.

The awesome destructive power of nuclear weapons and the technological lead that industrialized nations possess in conventional weapons, created a desire by lesser nations such as North Korea, Pakistan, and Iran to acquire nuclear weapons. Iran fought Iraq for eight years to a stalemate, while the US defeated Iraq in a matter of a few weeks in a strictly conventional war. Obviously, Iran was watching and learning. If it had nuclear weapons and effective means of delivering them, then a superior military power would be reluctant to wage war with it. Iran's opponents, especially Saudi Arabia and Israel, clearly understand this.

Aircraft technology

Since many different aircraft were introduced during the Cold War, out of necessity only the significant entrants in the arms race will be considered. Hence, planes such as the B-47, B-36, XB-70, F-100, F-101, F-102, F-4, F-104, F-105 and F-106 have been omitted from the analysis.

Fighters

Let us start with turbo jet fighter aircraft. The two dominant Korean War jet aircraft, Soviet MIG-15 and US F-86, took advantage of WWII German advances in turbo jet technology and were introduced about the same time. During the Cold War, the US introduced the F-15 in July, 1972 and the F-16 in January, 1974. The Soviets countered with the SU-27 in May, 1977 and the MIG-29 in October, 1977. Thus, the US had a 3- to 5-year advantage for a while.

The US introduced its subsonic low-observable fighter (stealth), F-117, in December, 1977 and the supersonic F-22A in September, 1990. Moreover, it introduced the multi-role/ multi-service supersonic stealth fighter, the F-35A and B for the Air Force and Marine Corps, and is close to introducing the F-35C naval version. The Russians (SU-T-50 PAK FA) and Chinese (Chengdu J-20) have their version of low observable fighter aircraft in the development stage and have begun test flights. Russia is still supplying engines for both. In the meantime, the Russian Federation is upgrading the SU-27, MIG-29, and MIG-25. Clearly, as China's economy continues to grow (it is second largest, but not on a per capita basis) and its technological base continues to expand, it will become a formidable contender in the arms race.

The Soviets fielded its vertical takeoff and landing (VTOL) aircraft in April, 1970 and retired it in 1991. The US bought the subsonic British Harrier (AV-8) in 1980 and improved on it. Now it declared the short takeoff and vertical landing (STOVL) F-35B operational in 2015. It is both low-observable and supersonic. The Russian Federation and China show no interest in competing in this type of aircraft, even though it is ideally suited for small aircraft carriers, called Amphibious Assault Ships by the US Navy, of which it has eleven. The export potential of this aircraft is significant to countries not interested in large aircraft carriers for a variety of reasons, if Lockheed Martin can fix the numerous problems that plague the development of this aircraft.

The US first flew a practical tiltrotor cargo plane, V-22, capable of VTOL and STOVL, in 1989 and fielded it in 2007, after an extensive and difficult development process. No other nation is interested in competing with this type of aircraft, but a few may purchase them, if they possess small aircraft carriers that lack catapults and arresting cables. Japan has already placed an order. The others will continue to rely on helicopters for the same mission.

Bombers

The US introduced its first medium range bomber, the B-58, in November, 1956, and the FB-111 in December, 1964. Both were supersonic and neither is currently in service. However, it was the Soviets who took the lead this time with the subsonic TU-16, which was introduced in April, 1952, and is no longer in service in Russia. However, the Chinese Air Force still flies an upgraded version of this plane. The Soviets introduced the supersonic TU-22M in August, 1969, a fine airplane that is still in service. With aerial refueling perfected and strategically located foreign bases, the US does not wish to compete in this type of aircraft.

So far as heavy bombers are concerned, the US possesses three fleets. The subsonic B-52 was introduced in 1955 and is the oldest, although it has been modified repeatedly. Next is the B1B, which is supersonic. The newest one is the B-2, which is low-observable, but subsonic. The US is also about to launch the development of the next generation heavy bomber very soon. The Soviets are still relying on the subsonic turboprop TU-95, introduced in 1952. The Soviets introduced the supersonic TU-160 in 1987, a plane comparable to the US B1B, but only 35 were produced, and fewer than 16 are still in service. The Russian Federation recently announced that it is re-opening the production line on this plane. They and the Chinese do not appear to be developing low-observable heavy bombers. With standoff weapons, such as air launched cruise missiles (ALCMs) and Joint Standoff Weapons (JSOWs), a bomber does not have to get near a target anymore. Ditto for fighter bombers such as the F-15E, F-16 and F-18.

Miscellaneous aircraft

Of course, many specialized aircraft are required in support of the war fighter. While reconnaissance satellites have replaced airplanes in many instances, cargo planes, aerial tankers, electronic counter measure aircraft, early warning and control platforms, airborne command posts, trainers, and marine patrol planes are important complements as well, and have a huge budgetary impact. Frequently, the aforementioned missions can be attained by modified older air frames fitted with appropriate electronics, but at considerable expense. The KC-46A aerial tanker and the Poseidon (P-8A) marine patrol aircraft are the latest examples. In any case, the issue will not be complicated by delving into the specifics other than to point out that the US has a substantial lead across the board regarding support aircraft.

The US enjoys a commanding lead in unmanned aerial vehicles (drones). However, Russia and China are determined to close the gap as soon as possible. Simply put, Russia and China are not about to cede uncontested hegemony to the US.

Naval warfare

No country is building or operating battleships. The US has 10 large aircraft carriers (with catapults and arresting cables) and the 11th is under construction—all nuclear powered. No other country has more than one. As has already been pointed out, the US also has 11 Amphibious Assault Ships (small carriers without catapults and arresting cables) with a well deck for amphibious landing craft. The F-35B and the V-22 are ideally suited for this kind of vessel, as are helicopters. The naval competition then is in smaller and faster ships equipped with missiles, and in submarines likewise equipped with missiles. Low-observable technology is also working its way into ship design.

Since the US is a naval power in need of protecting sea lanes and air routes to Europe, Asia, and the Middle-East, out of necessity it has developed considerable advantages in all areas of naval warfare. Russia is primarily a land and air power, and only secondarily a naval power. The same can be said for China at the moment, although its larger economy makes it possible for China to become a regional naval power in the near term. India, with its eye on China as a regional competitor, has regional naval aspirations as well, and may prove to be a customer for the F-35B; as the cost of large aircraft carriers for most nations is prohibitive.

Land warfare

The advantages of superior armor (including mine resistant, ambush protected vehicles; MRAPs), helicopters, unmanned aerial vehicles (drones), satellites, and the experience gained by US land forces in the protracted Iraqi and Afghanistan conflicts have provided the US with a lead in ground warfare for the time being. Russia and China have taken note and are committed to catching up as soon as their economies will permit.

Missiles

Intercontinental ballistic missiles

The intercontinental ballistic missile race started when the Soviet Union launched Sputnik in October, 1957. The US followed in October 1959, giving the Soviets a two-year temporary lead. The race was expanded when the US successfully launched a medium range missile from a submerged submarine in July, 1960. The soviets duplicated this feat in 1963, giving the US a temporary three-year lead. Now, both sides, and increasingly China, have just about every kind of missile.

The most intensive competition appears to be in the development of an effective anti-missile missile (ABM); perhaps the most daunting aspect of the arms race, since it is analogous to hitting a bullet with a bullet. So far the only realistic goal is to intercept a handful of relatively crude Intercontinental Ballistic Missiles (ICBMs) launched by a rogue state. The US has five programs underway at a staggering cost (Patriot—PAC 3; THAAD; Ground Based Mid-Course Interceptor; Aegis/SM3; and the Surface Launched AMRAAM with Norway). It is anyone's guess what the end result will be, but for the foreseeable future the offense will continue to hold a decisive advantage. In the end, satisfactory missile defense may turn out to be an elusive goal with the US simply wasting large sums of money trying to attain it.

In the meantime, the race continues. Russia just announced that it lifted its ban on the delivery to Iran of its anti-aircraft/anti-missile defense system—the S-3. Furthermore, it announced the sale of its most advanced anti-aircraft/anti-missile defense system—the S-4—to China. Even Turkey, a NATO member, has purchase the S-4. The US is supplying its allies all the anti-aircraft and anti-missile defenses they wish to purchase.

Also, the US continues to develop new technologies such as: High energy lasers, other directed energy weapons, hypersonic missiles, and electromagnetic rail guns, inter alia. The Russian Federation and China will try to keep up. The question is: Can their economies sustain such an economic burden? The Soviet Union disintegrated when it attempted to keep up with US defense spending during Ronald Reagan's first administration.

The self-perpetuating nature of arms races

Once parity is reached contestants do not cease the competition, even when treaties limit the development of certain types of weapons. There are a number of reasons for the unabated continuation of arms races. First of all, typically there are more than two actors at a given time engaged in arms races. At present we have the US, the Russian Federation, China, the EU (mostly through NATO), India, Pakistan, Saudi Arabia, Iran, North and South Korea, and Japan on the verge of entering. Hence, the sheer number of actors makes agreement on cessation or even moderation of the races quite difficult. The pressure by the US on NATO members to increase their spending to 2% of their GDP is a case in point. The parties to arms races are typically characterized by deep-seated and intractable differences based on historical factors such as: Predisposition to autocratic rule vs. Democratic governance, command economies vs. free market ones, and religious differences; which make reconciliation difficult.

Secondly, each contestant attempts to improve the effectiveness of its arsenal as well as those of its allies by improving the capability of existing weapons or creating new ones with greater capability. The others, to the extent of their economic and technological capacity, follow suite. Some of the contestants believe that they are entitled to catch up to the leaders who have a head start. Thus, it appears as though arms races will remain a part of the international landscape for the foreseeable future.

One way to slow down the arms race is to introduce only weapon systems that are needed to counter existential and likely threats. The US possesses a robust Intelligence Community (IC) that collects a plethora of human and signals intelligence that is useful in identifying existing and potential threats by our adversaries. Introducing weapons systems that are unrelated to threats merely adds fuel to the arms race.

Lastly, international organizations such as the United Nations at the moment are impotent to enforce decisions that they reach regarding disarmament because they were designed in such a way as not to infringe on sovereignty.

Conclusion

The Cold War left the US as the de facto leader of the free world with the obligation to create a defense policy capable of fighting regional conventional military engagements, counter insurgencies, as well as deterring major conflicts with the Soviet Empire and China that could escalate into thermonuclear exchanges. All this created an unprecedented arms race between the US and the Soviet Union and their respective alliances—NATO and Warsaw Pact. Since an arms race ineluctably becomes interwoven into a country's foreign policy, policy makers, civilian and military alike, are well served by being informed by a comprehensive anatomy of an arms race.

As an indispensable nation since World War II, the US needs a full spectrum defense policy because its existential and potential enemies just will not go away. But such a defense policy is very expensive and competes with high priority domestic exigencies. Consequently, the Defense Department and all the military branches must employ every practical means to make the full spectrum defense policy more affordable.

A prescription for an affordable full spectrum defense policy has been offered, which includes: Relying on the intelligence community; total asset visibility; looking in the right places; reducing federal bureaucracy; building weapons from low-hanging fruit; exploiting economies of scale; lesser reliance on military specifications; settling for the eighty percent solution; focused leadership education; and growing the technological fruit tree. But first we must guarantee that the books of the Pentagon and all the military branches are auditable. Until that is done, we cannot know what we need because we have no way of knowing what we have (Muczyk, 2017).

Reference

Muczyk, J. P. (2017), Prescription for an affordable full spectrum defense policy, *Journal of Defense Management*, 7(1), 1000163.

chapter twenty

Innovation dynamics in the defense space sector

Zoe Szajnfarber, Matthew Richards, and Annalisa Weigel

Contents

Introduction

DESPITE a rich legacy of delivering impressive technology, government space acquisitions are frequently characterized by schedule slips and cost overruns (e.g., AEHF, JWST, SBIRS-High, GPS III) (GAO, 2007, 2016). In recent years, in an effort to address these problems, multiple blue ribbon panels have been convened. Bringing to bear the members' vast experience, working in the current acquisition paradigm of large monolithic spacecraft, their recommendations, spanning a host of technology development, requirements management, and national space policy issues, emphasize a "back-to-basics" philosophy (e.g., maturing payload technologies outside of acquisition programs) summarized in Figure 20.1. However, the sharp contrast between the pace of change in the commercial space industry and the maturity which characterizes traditional satellite providers necessitates a more

		Rumsfeld (2001)	NDIA (2003)	Young (2003)	GAO (2006)	DAPA (2006)	NRC (2008)	Munson (2015)
technology	Restore funding for testing space technologies	X			X			
	Maintain U.S. technological lead in space	X						
	Keep R&D separate from systems acquisition				X		X	X
	Identify technology for rapid exploitation and control					X		
management	Establish Presidential and NSC space advisory groups	X						
	Integrate defense and intelligence space activities	X						
	Improve front-end systems engineering (req's=resources)		X	X	X	X	X	X
	Budget programs to most probable (80/20) cost				X			X
	Evaluate contractor cost credibility in source selections				X			X
	Conduct independent program assessments at MDA's				X			X
	Do not allow requirements creep			X	X	X	X	
	Match PM tenure with delivery of a product			X	X	X	X	X
	Pursue incremental increases in capability					X		X
	Withold contractor award fees when goals not met					X		X
	Establish a stable program funding account						X	X
	Structure development to achieve IOC within 3-7 years						X	
policy	Recognize space as top national security priority	X						
	Deter and defend against hostile acts in space	X						
	End practice of appointing only flight-rated space leadership	X						X
	Incentivize government career paths in acquisitions	X	X	X		X	X	X
	Improve workforce technical competence	X		X	X	X	X	X
	Compete acquisitions only when in best interest of gov't				X			X
	Develop integrated strategy for R&D and acquisitions					X	X	X
	Encourage LSI to compete major subsystems					X		

Figure 20.1 Key findings from recent studies. (From Rumsfeld, D. et al., Report of the commission to assess United States national security space management and organization, Washington, DC, 2001; NDIA, *Top Five Systems Engineering Issues in Defense Industry*, National Defense Industrial Association, Arlington, VA, 2003; Young, T. et al., Report of the defense science board/air force scientific advisory board joint task force on acquisition of national security space programs, Office of the Undersecretary of Defense for Acquisition, Technology, and Logistics, Washington, DC, 2003; GAO, *Defense Space Acquisitions: Too Early to Determine If Recent Changes Will Resolve Persistent Fragmentation in Management and Oversight*, GAO-16-529R, Washington, DC, 2016; DAPA, Defense acquisition performance assessment: Defense acquisition performance assessment project, Report for the Deputy Secretary of Defense, Washington, DC, 2006; NRC, *Pre-Milestone A and Early-Phase System Engineering: A Retrospective and Benefits for Future Air Force Acquisition*, National Academy Press, Washington, DC, 2008; Munson, A., *Why Can't We Get Acquisition Right?* Potomac Institute for Policy Studies, Arlington, VA, 2016.)

fundamental look at how the current government acquisition paradigm can be evolved to fully leverage these structural shifts in the industry.

The operationally responsive space (ORS) paradigm pursues a fundamentally different approach to spacecraft design and operation (Cebrowski & Raymond, 2005). Rather than emphasizing the delivery of long-lived, global, high-performance space capabilities, ORS missions envision pursuing short-term space capabilities tailored for specific operational scenarios. Broadly, this trend is enabled by the halving of launch costs, satellite component miniaturization, and distributed computing. Although ORS solutions will sacrifice performance on traditional measures of effectiveness with employment of smaller

Table 20.1 Distinguishing ORS from "Big Space"

Characteristic	"Big Space"	ORS
Historical Context	Cold War	Acquisitions crisis; fragilities inherent in integral, long-life designs
Original Beneficiary	White House	Theater combatant commander
Programmatic Drivers	Performance	Cost, schedule
Payloads	Customized, satisfy multiple missions	Off-the-shelf; single-mission focus
Design Life	10+ years	1+ year(s)
Risk Tolerance	Risk averse	Risk tolerant

satellites and commercial-off-the-shelf (COTS) technology, ORS offers large improvements in schedule performance as well as an opportunity to customize capability for emergent mission requirements.[1] Table 20.1 summarizes the characteristics which distinguish the ORS concept from the current "Big Space" paradigm.

In order to assess the ability of the ORS paradigm to complement and enhance the space acquisition enterprise, this paper proposes a framework of space sector innovation challenges, which serve as a common basis for comparison. Fundamentally, the goal of defense space acquisition is to facilitate the meeting of the Joint Combatant Commanders emerging needs. This requires technological innovation; be it by generating a wholly new capability, or reducing the resources required to achieve an existing capability (e.g., making the system cheaper or lighter). Encouraging innovation (i.e., generating new capabilities to meet these unmet needs) is a difficult problem in general and characteristics of the defense space sector make it harder still; the monopsony-oligopoly market structure and complexities of the product and associated operating environment limit the ways in which natural market dynamics can drive change. Nonetheless, given that (i) innovation is an implicit requirement of defense space acquisition and (ii) there is an extensive literature and theory on innovation in traditional markets, this paper seeks to answer the following question: What are the implications of the intrinsic characteristics of the space sector (i.e., monopsony-oligopoly market structure, extremely complex robust products) on how innovation can and should be encouraged in the defence space context?

Space sector innovation challenges: Nature, approach, and potential

In order to understand the implication of intrinsic characteristics of the space market for how spacecraft innovation can and should be encouraged, three steps were taken: First, strategic mechanisms proposed in the innovation dynamics and strategy literatures were

---

[1] The fundamental idea of ORS is to trade off the reliability, longevity, and performance achieved by satellites under the "Big Space" paradigm—the currently accepted way of conceptualizing, specifying, developing, and operating space systems—for the speed, responsiveness, and customization which may be achieved by architectures that incorporate elements such as small, modular spacecraft and low-cost, commercial launch vehicles (GAO, 2006). In addition to obtaining capability on-orbit quickly, ORS attributes include tactical control and assured access.

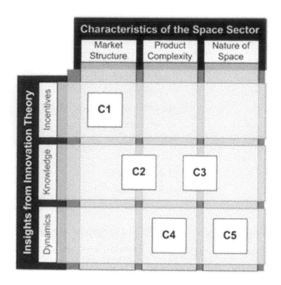

Figure 20.2 Conceptual outline of analysis approach.

reviewed and categorized; Second, unique characteristics of the space acquisition context which could potentially limit the applicability of theories developed in other contexts, were synthesized; Finally, the interactions of the first set of mechanisms with the second set of constraints were qualitatively assessed. The results of the analysis are captured in Figure 20.2.

The rows of the matrix capture the synthesized categories of insights from the innovation dynamics and strategy literature.[2] Starting from Schumpeter's basic supposition that long term economic growth can only be sustained through the entry of innovative entrepreneurs and the necessary value destruction of established (monopolistic) companies (Schumpeter, 1934), much of the business literature on innovation dynamics, developed over the subsequent eighty years, has addressed the question of why successful firms fail to traverse the discontinuity imposed by radical innovations. There are three complimentary ideas. One school of thought, epitomized by the Teece (1986) profit model, argues that innovation happens most effectively when the innovator profits from his efforts. It follows

---

[2] In order to focus this survey on the work that is most relevant to evolving management of innovation in government space, two related scoping decisions were made. First, literature focusing on commercialization and diffusion were de-emphasized. Although definitions of innovation typically combine the concepts of new and implemented, and in traditional markets, implemented is synonymous with commercialized (i.e., bringing an invention to market), in the space context, "implemented" means being integrated into, and flown on, a flight system. The fact that a flight system is often the only one of its kind, never to be mass produced or marketed, does not change the fact that the invention has been useful—the standard that differentiates an invention from an innovation. This is consistent with the way the term is used in the defense context (c.f., Sherwin and Isenson, 1967; Rosen, 1994; Grissom, 2004). Second, this framework relies strongly on innovation models developed pre-IT revolution and global networks of innovation. This is because the firms of that era are more representative of the defense space enterprise of today, than the modern innovation contexts analyzed in more modern studies. Given the national security context and the secrecy (i.e., open collaboration and information exchange across borders and industries is explicitly prohibited), their insights do not capture a realistic goal for the evolution of acquisition (this point is elaborated on in the section on space characteristics, in the body of this paper).

that established firms—who continue to profit from previous innovations if the status quo is maintained—will use their market power to resist competence destroying change (Stigler, 1971). We call this class of innovation mechanism "incentives." Another perspective is that incumbent firms don't fail to traverse discontinuous change because of a lack of capability; rather, it is because they remain focused on the needs of their core/mainstream customer until it is too late (Christensen, 2003). Further, even when a firm does recognize the need to address a new market base, there are multiple types of competence that can be destroyed by even seemingly small changes (Henderson & Clark, 1990). We call this class of mechanisms "knowledge." Finally, as articulated by Schumpeter, and supported by later empirical work (Utterback, 1994), the cycle of establishment and destruction is natural in a healthy market, and should be harnessed but not interfered with. We call this class of mechanism natural "dynamics."

Similarly, the columns of the matrix categorize intrinsic characteristics of the space sector as (a) market structure, (b) market complexity, and (c) nature of space. Firstly, the space "Market Structure" is relatively unique in that it is effectively a monopsony (single buyer) oligopoly (few sellers) contract market, which has implications for how transactions occur (Adams & Adams, 1972; Peck & Scherer, 1962). Secondly, spacecraft embody significant "Product Complexity." Each subsystem is itself a complex system; many disciplines, and many organizations are involved in each new acquisition; and multiple different levels of maturity exist simultaneously in any given system. While this characteristic is not unique to spacecraft, it has important implications for how maturity can be conceived (Sauser et al., 2008). Third, the "Nature of Space" has implications of its own. Space is a harsh, remote environment, with implications for system characteristics like survivability, serviceability etc. Space acquisitions represent an enormous public expenditure, bringing in questions of accountability and significant media attention. Further, as a strategic asset, space systems and their components are subject to stringent security protocols (e.g., ITAR, reduced communication across boundaries) and significant risk aversion.

After defining the rows and columns, the interactions of each of these innovation mechanisms and space sector characteristics were examined in detail, leading to the identification of five fundamental challenges for innovation in the space sector. Specifically: (1) generating bottom-up push in a predominantly top-down acquisition process, (2) representing the needs of a disaggregated buyer, (3) integrating fragmented sell-side knowledge from the top-down, (4) matching the innovation environment to the stage of development, and (5) balancing risk aversion and the need for experimentation. The discussion, in the sections that follow, is structured around these five challenges. In the remainder of the paper, the nature of the challenge is first explained in terms of the impact of characteristics of the space sector on the ability of mechanisms capable of encouraging innovation in *traditional* markets to function. Second, the current acquisition system is examined to determine how and to what extent it overcomes each of these challenges. Third, the ORS paradigm is evaluated to determine how, and the extent to which, its philosophies can be applied to improve the broader spacecraft innovation process.

The challenge of generating "bottom-up" push in a predominantly top-down acquisition process

Taking a classical economic view of innovation, market transactions are thought to be the fundamental driver of innovation. In a competitive market, both the consumer's needs and the supplier's capabilities are revealed through the mechanism of price (Adams & Adams,

1972). Innovation occurs (i.e., unmet needs are met) over time, through the continuous interaction of market pull and capabilities push (Rothwell & Zegveld, 1994). However, the market for spacecraft is neither competitive on the buy-, nor sell-side, and this holds important implications for innovation. Firstly, a monopsony market is discrete and specific since the market only exists when the buyer wants to buy, and as a result, user needs must be specified explicitly since there is no aggregate buyer behavior out on the open market from which they can be inferred. Further, in the stable oligopoly that exists on the space sector sell-side, there is little incentive for contractors to invest in innovation on their own; they tend to innovate in response to government requests.

In the traditional market conception (as illustrated in Figure 10.2), transactions can occur in one of two ways. Products are either sold by a third party retailer in a store, in which case prices are relatively standardized (i.e., every person will be charged the same amount for a given capability). Or, more customized products (i.e., a new roof for a home) and the labor associated with their installation are contracted directly with the supplier. Even in this case, enough sufficiently similar transactions occur to establish a market price. For the most part, buyers are limited to whatever is currently available on the market; however, the state-of-the-art is constantly changing to meet new needs. For example, if you decided to replace an iPod that was bought three years ago, in today's market you would expect to be faced with a different set of better model choices. Although you personally hadn't continued to reveal your preferences through ongoing purchases, the other millions of consumers had. Thus, as long as your values align with those of the general market, the continuous market feedback process will have driven innovation to create a next generation iPod model that better suits your needs.

Where in the traditional case, buyers are limited to what's available on the market, in the space sector, the government is the whole market; if the government doesn't buy it, it won't be sold. As a result, the buyer's needs and preferences must be revealed explicitly, thereby dictating what should be produced. Further, while a monopsony buy-side does not preclude price-competition among sellers (Adams & Adams, 1972), the earlier iPod example should give an intuitive understanding of why the incentives for such competition are weak. Specifically, the market growth potential is limited by the needs of the government, the profit margins are relatively small compared to commercial industries[3] and the barriers to entry are high as significant complementary assets and specialized knowledge are required for satellite manufacturing. These factors all contribute to the sell-side oligopoly that exists in the government space sector, resulting in the suppression of bottom-up capability development.

Extent to which the current Department of Defense structure resolves the issue
In order to generate the necessary technology "push," the Department of Defense (DoD) acquisition process employs a two-tiered organizational structure focused on (1) research and development and (2) formal acquisition programs. Initial technology development within the DoD is conducted by the Service Laboratories (e.g., Air Force Research Laboratory, Naval Research Laboratory, Army Research Laboratory) and several science and technology (S&T) organizations such as the Air Force Office of Scientific Research, the Office of Naval Research, and the Defense Advanced Research Projects Agency (DARPA). These latter S&T organizations are focused primarily[4] on a research-level investigation of

---

[3] The risk level for profit on government contracts varies with the type of contract vehicle used; more risk for firm-fixed-price contracts, less risk for cost-plus contracts. In the past, the government has often, but not exclusively, used cost-plus contracts for satellite procurement. But in the future, the government is expected to move to greater use of firm-fixed-price contracts for satellites.
[4] DARPA may also fund Advanced Technology Demonstrations.

basic physics and phenomenology. As these S&T organizations demonstrate concept fea-sibility, technologies are transferred to the Service Laboratories for further development, maturation, and demonstration of capability. Once these innovation organizations mature concepts to the point where they can be realistically assessed for cost, schedule, and per-formance contributions to a given set of program requirements, they may be considered as part of the Joint Capabilities and Integration Development System (JCIDS).

Figure 20.3 takes a highly simplified view of the acquisition process to illustrate how the two-tiered process generates the necessary push, despite its top-down structure. In Figure 20.3, capability-push is denoted with hashed arrows and need-pull with solid arrows. Nominally, the formal part of the acquisition process matches user needs with relatively mature tech-nologies to drive system-level innovation as dictated by the innovation theory. However, that pool of capabilities is not being created and marketed by the supply-side as would be the case in a traditional market. The lead on technology development efforts is still primarily the domain of the government, albeit a separate branch (both organizationally and culturally). The lab structure, as described earlier, contributes in two main ways: they conduct funda-mental research in areas that may one day be of use to the Joint Combatant Commanders, and they fund technology development contracts and studies. Although these contracts tend to be less specific than formal acquisitions, they follow the same general pattern; the customer identifies a need and puts out a request for comment, based on the response a more formal request for proposals is released, leading to a contractual relationship.

As a result, rather than the confluence of pull-push forces which drive innovation in a traditional market, the space drive is characterized by a coordinated pull-push-pull. One key disadvantage of this approach to top-down capability generation is that it creates a situation where much of the investment in product development for space applications originates from the government (Sherwin & Isenson, 1967). Also, while this system does technically create the required push, it is a fundamentally different push force than the independently supply-side initiated one described in the traditional market. The implica-tion of this difference will be discussed in the sections that follow.

The potential improvements offered by operationally responsive space
Operationally Responsive Space has been defined broadly by the Department of Defense as "assured space power focused on timely satisfaction of Joint Force Commanders'

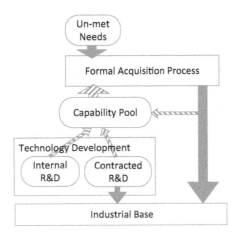

Figure 20.3 Capability generation in the acquisition process.

needs…while also maintaining the ability to address other users' needs for improving the responsiveness of space capabilities to meet national security requirements (DoD, 2007)." The purpose of ORS is to reduce the time constants associated with space system acquisition, design, and operation. ORS intends to enable rapid responses to changes in space capabilities by supplementing them quickly when they are lost, with lesser but still useful capabilities. In terms of the structure described in Figure 20.3, ORS is typically grouped into the category of technology development; however its functions really span both the roles of technology development and spacecraft acquisition. This has two key implications with respect to issue 1 (i.e., that the monopsony-oligopoly market structure enforces a top-down acquisition process).

First, by shifting to a greater reliance on standardized satellite buses and payload modules, ORS envisions lowering the barriers to entry for satellite suppliers by defining a potential future market for unarticulated products around common interfaces. If this *plug-and-play* market is successful it may generate more bottom-up initiative from the space industrial base and provide avenues for small, innovative companies to enter the DoD market. This process will be encouraged through a model of seed-funding rather than development contracts.[5] Where the historical lab structure, to a first order approximation, specifies a need and pays for the development required to meet it, the seed-funding model would allocate funding to firms in the early stages of a promising development. Conceptually, the difference between these two approaches is significant; the latter has the potential to reach non-traditional space firms and leverage bottom-up initiative, where the former perpetuates the traditional pull-push-pull. It remains to be seen whether the practical difference will be significant.

Secondly, the emphasis on rapid development cycles might create a more continuous innovation environment. One of the problems with the discrete nature of a monopsony market, as discussed earlier, is that it limits the opportunities for new capabilities to be "needed," while at the same time placing a high premium on major inter-generational improvements. Both of these factors serve to limit the incentives for bottom-up initiative. What the ORS paradigm may change (from the point of view of generating real push) is to create a more frequent market for incremental improvements. If there is a clear opportunity to capture the value of taking the functionality of a spacecraft beyond the specification, contractors may be more inclined to take the initiative.

The challenge of representing the needs of a disaggregated buyer

The necessity for a top-down process as described previously could theoretically foster ideal conditions for innovation because a knowledgeable buyer could: (1) decrease information asymmetries in the transactions by eliminating the need for suppliers to infer the future preferences of potential buyers; and (2) encourage investment in R&D by specifying sufficiently advanced needs that can only be solved through radical innovation. However in practice, the specialized knowledge required to drive change is fragmented across the space market structure, limiting the effectiveness of both 1 and 2 above. This section explains why knowledge fragmentation on the buy-side limits the efficiency of needs specification; the next section addresses how sell-side fragmentation exacerbates the challenge of integrating bottom-up push with top-down pull.

---

[5] As reported in an interview conducted by the authors with Dr. Adang, representing the recently stood up ORS office (2-26-08 1:00-2:30 EST).

The situation where sellers only make what buyers want, but can't currently get, could be an ideal environment for innovation. Where the business strategy literature emphasizes the importance of downstream aspects of the innovation process[6] (Bhide, 2008), in the space market, nearly everything that is developed is adopted. Similarly, on the upstream side, Von Hippel argues (as for e.g., in Ref. (Thomke & Hippel, 2004; von Hippel, 1988)) for increased emphasis on capturing lead user innovation. The idea is that people who actually use the product will be more likely to find its limits and potential extensions than engineers in a lab environment. In the DoD context, where the buying function nominally includes both use and needs specification, one might expect lead user innovation to be captured naturally. However, when the monopsonist buyer is as complex as the US government, incorporating multiple disaggregated interests, the assumption that the buyer knows what it needs is not always accurate.

As shown in Figure 20.4, in the traditional market conception, buyers are a relatively homogenous group of individuals or firms acting in their own interests. They determine what to buy based on an internal evaluation of their relative wants, budget and what is available on the market. However, in the government acquisition context, this evaluation is made among several independent organizations, based on presumed capabilities. Specifically, while the nature of the monopsony buyers' interests (i.e., those of the Department of Defense as a whole) is not dissimilar to that of the traditional buyer, in practice, having the interests disaggregated across organizational boundaries makes a

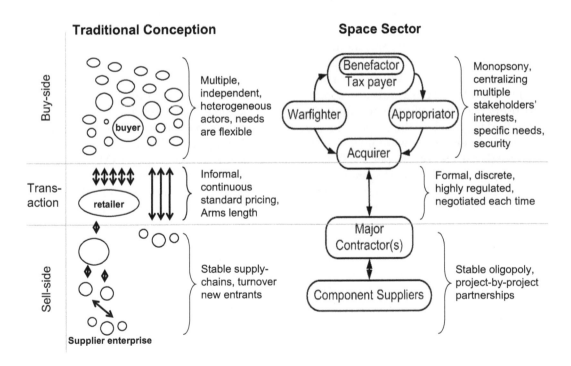

Figure 20.4 How the space market is different.

---

[6] For example, Bhidé argues that innovation depends as much on the user's willingness, and ability, to adopt new products and technologies, as the development of those products and technologies themselves.

significant difference. Rather than making an internally consistent determination of preferences, acquisition agents must integrate the inputs of needs (as expressed by the warfighters who use the system and possess the best understanding of how different systems will affect them operationally), budgetary constraints (as imposed by Congress which appropriates the funds and has the best appreciation for how spending in one sector will affect the overall national purse) and technical feasibility (as inferred from industry studies, in-house experts and through the contracting process). This creates a principal agent problem and complicates the needs representation process significantly. The key implication is that the ability of the space buyer to drive the radical innovation expected of spacecraft acquisitions is limited by incomplete architectural, component, operational and budgetary knowledge; all of which play an important role in driving change.

Extent to which the current Department of Defense structure resolves the issue
The short answer is that the acquisition structure does not address the knowledge disaggregation problem well. In fact, *challenge B* (i.e., needs representation) is in a sense a bi-product of the complex organizational structure that exists to resolve *challenge A* (i.e., insufficient bottom-up push). In the current DoD acquisition process, the intersection of what is possible and what is useful, is nominally identified through a series of "gap analyses" performed as part of the Joint Capabilities Integration and Development System (JCIDS). However, in practice the complexity of integrating the needs of such a disaggregated buyer as the US government leads to significant shortcomings in JCIDS' realization.

One critical aspect of prioritizing "next" acquisitions is in soliciting and integrating inputs from the operational arm of the DoD—the user warfighter. While it is relatively well accepted that a warfighter has a unique understanding of the impact of performance trade-offs on operational utility, and should thus be consulted to help refine needs; the extent to which a warfighter can contribute to the capability generation side of the innovation process is less well understood. When von Hippel's theory of lead-user innovation[7] (von Hippel, 1986) is extended to the acquisition context, warfighters are often identified as analogous to lead users because self-preservation is the highest possible incentive to innovate (Frisbee, 2003). To date, much of the empirical research on lead-user innovation has centered on systems that are either software intensive (e.g., personal computer—Computer Aided Design software (Urban & von Hippel, 1988)). Online public access library software (Morrison et al., 2000) or personal-use expert systems (e.g., canyoning, sailplaning, boarder cross and para-cyclists' equipment (Franke & Shah, 2003)). While some of the findings from these studies may generalize to government acquisition systems, key differences make the analogy suspect. Specifically, the cost of changing military systems is prohibitive,[8] warfighter culture emphasizes acceptance of the status quo and uniformity of equipment, and individual warfighters do not necessarily have the required knowledge to make substantial changes to the systems they use (particularly in the space context). Despite all these caveats, refining our understanding of lead-user innovation as it applies to complex products developed in government enterprises has the potential to lead to an improved approach to collecting and interpreting warfighter input into the acquisition process.

---

[7] Which asserts that users who (1) face needs in advance of the market at-large and (2) are positioned to benefit significantly by obtaining a solution to those needs, represent an important source of innovative product concepts.
[8] Developed over multiple years and designed to last for a decade or more, transition costs measure in the billions.

The potential for improvements offered by operationally responsive space

Given the extreme specialization required to develop and manage national security space assets, a large degree of organizational decomposition is inevitable. However, ORS does show some promise in reducing the magnitude of this principle agent problem. In addition to mitigating the complexity of traditional space assets by nature of a smaller, less-capable design paradigm, the ORS approach brings the warfighter closer to the acquisition process through its simplified concept of operations. On the front end, the tactical control provided to the warfighter by ORS assets may enable the lead-user innovation discussed in the previous section. On the back-end, the direct downlink of satellite data to the warfighter (removing traditional layers of analysis) may concretize the value of alternative satellite capabilities for improved needs representation in future satellite developments. This hypothesis will need to be tested over time.

The challenge of integrating fragmented sell-side knowledge from the top-down

The existence of a top-down acquisition structure presents a unique opportunity for the monopsony buyer to take a long-term, coherent perspective on driving innovation to their benefit. However for this to happen effectively, not only does the buyer need to know what they want, they also need to know what is possible. In practice, this proves extremely difficult since the required knowledge is fragmented across the space sell-side.

The innovation literature has historically differentiated between two types of innovation: incremental and radical (see e.g., ref. (Abernathy & Clark, 1993; Anderson & Tushman, 1990)). Incremental innovations are competence-enhancing; they generate a product that is better along dimensions that are familiar within the current paradigm. Radical innovations, on the other hand, are competence-destroying; they typically take a different approach to solving the same problem. For example, building bigger communication satellites that can carry more transponders would be an incremental innovation approach to increasing capacity, while developing a new method of performing on-board calculations (e.g., use of integrated circuits over core transistors on Apollo) is a way of addressing the same problem with a radical innovation. If established suppliers are driving change, not surprisingly, there is a tendency to avoid competence-destroying change (Christensen, 2003); on the other hand, if change is driven from the top, as in the space sector, by specifying sufficiently advanced needs, radical change may be the only option, thereby legitimating the risk.

However, in more complex product systems, Henderson & Clark (1990) observed that more than one type of knowledge is required to generate radical innovation. They differentiated between component level knowledge and architectural knowledge (i.e., knowledge of the linkages between components), which leads to a two dimensional spectrum of innovation types as shown in Figure 20.5. When these categories of knowledge are mapped onto the space market structure (see Figure 20.4), they are concentrated on the sell-side and divided between system integrators and component suppliers. This means that while acquisition agents may be in a position to drive radical innovation, they may not have all the knowledge required to do so. Further, unlike in the traditional market conception, where firms tend to establish stable supply-chain relationships (which enable them to integrate both component and architectural knowledge), in the space sector, stable relationships are effectively discouraged by the project-by-project acquisition structure. This further complicates the problem of determining the feasibility of future projects.

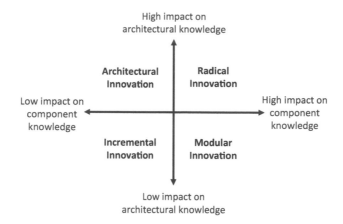

Figure 20.5 Knowledge required for innovation. (Adapted from Henderson, R. and Clark, K., *Adm. Sci. Q.*, 35, 9–30, 1990. With permission.)

Extent to which the current Department of Defense structure resolves the issue

Challenge C, the challenge of integrating fragmented sell-side knowledge, has been addressed differently over the history of the space age. Initially, significant in-house technical expertise was cultivated among government buyers and significant oversight spanning the entire sell-side supply-chain was common practice. The government buyer adopted the risk through cost-plus contracts, but retained design authority giving them the ability to intervene when contracts were not being executed as desired. More recently, as cost control became a primary focus, the role of system integrator has been delegated to industry contractors, with technical development subsequently delegated to subcontractors. The idea was that profit maximizing firms will allocate resources more efficiently. However in practice, the interests of industry do not always align with those of the government, limiting the effectiveness of the relationship. Coupled with the fact that the delegation of the oversight role has lead to a decrease in the technical competency of the acquisition core (NRC, 2008), this trend has exacerbated the challenge of integrating sell-side knowledge rather than helped. In fact, improving collaboration on requirements and increasing the technical competence of the acquisition core, are two of the nearly unanimously recommended remedies in the "blue ribbon" reports (see Figure 20.1).

The potential improvements offered by operationally responsive space

The ORS structure may lessen the knowledge disaggregation problem through its streamlined organization. Just as the rapid development cycles pursued by the ORS Office may strengthen the DoD industrial base by providing more opportunities for small, innovative firms, the approach might also improve the technical competency of space professionals by providing more opportunities for mid-career program managers to manage small-scale projects. As noted in a GAO report (2006), Navy and Air Force lab officials have found that the TacSat experiments have provided more hands-on and lifecycle management experience than would otherwise be possible on larger acquisition efforts.

The challenge of matching the innovation environment to the stage of development

The remaining two challenges relate to the nature of the "push" that must be generated. To this end, Utterback and Abernathy (1975, 1994) observed that the innovation process

Table 20.2 Phases of innovation process

	Fluid phase	Transitional phase	Specific phase
Innovation	Product changes/radical innovations	Major process changes, architectural innovation	Incremental innovations, improvements in quality
Product	Many different designs, customization	Less differentiation due to mass production	Heavy standardization in product design
Competitors	Many small firms, no direct competition	Many, but declining after the emergence of a dominant design	Few, classic oligopoly
Organization	Entrepreneurial, new entrants	More formal structure with task groups	Traditional hierarchical organization
Threats	Old technology, new entrants	Imitators and successful product breakthroughs	New technologies and firms bringing disrupting innovations
Process	Flexible and inefficient	More rigid, changes occur in large steps	Efficient, capital intensive and rigid

Source: Utterback, J.M., *Mastering the Dynamics of Innovation*, Harvard Business Press, Cambridge, MA, 1994.

proceeds in three phases—fluid, transitional, and specific, as shown in Table 20.2. During the fluid phase, the emphasis is on just getting the product to work. Lots of new, and very different, radical ideas are being tried. Some work, but many don't; it's a time of free experimentation. As a result of the high risk associated with this type of endeavor, the fluid phase is often carried out by entrepreneurial start-ups or single inventors working out of their garage. The goal is to prove-out the concept to the point that a larger company will buy-into the idea and facilitate its commercialization. Thus begins the transitional phase.

During the transitional phase, the emphasis is on making the invention mass producible. As a result, the product innovations tend to be architectural in nature. There may continue to be many players in the industry, but the number drops quickly after a dominant design emerges (e.g., the QWERTY keyboard set the standard for physical interfaces to computers). Where entrepreneurial organizations are best suited to the free experimentation required for the fluid phase, a more formal structure is required to standardize and commercialize the product in the transitional phase. And, once the standardization has occurred, this marks the beginning of the specific phase.

During the specific phase, the goal is to optimize the design within the framework of the dominant design. Changes tend to be incremental, reducing costs and increasing quality as the process improves. As the manufacturing process becomes increasingly specialized, investment in complementary assets leads to a market composed of few established firms with strong market positions. Unlike in the fluid and transitional phase, where the biggest threats are a lingering perception that the old way is better and losing out to competitors working in the same phase, in the specific phase, the biggest threat is complacency. There will always be entrepreneurial firms bringing in disruptive innovation; and so the cycle repeats itself.

There are very clear differences between the types of organizational environments that enable each phase of innovation. In the traditional market, different structures can easily be applied in each phase because the phases proceed relatively sequentially and distinctly. However, space products tend to integrate elements of each phase making the process harder to decouple. For example, many spacecraft are (1) fluid phase prototypes at the system level, in that they are accomplishing a task that has never previously been

Table 20.3 Phases of innovation in the acquisition system

	Technology development	Program acquisition
Innovation	Component/subsystem focus, some radical Innovation but mostly incremental	System level focus architectural innovation (major systems changes, minor component changes)
Product	Initially, as many different designs as funding allows (often few); later, heavy standardization	Two or three preliminary designs, but typically only one developed
Competitors	Few, classic oligopoly (even less for program acquisition)	
Organization	Siloed, hierarchical, but maintains research culture	Matrix organization
Threats	Resistance to change, funding cuts	Scope creep, funding cuts, political changes
Process	Less rigid, but still contract milestones	Inefficient, capital intensive, changes occur in large steps

accomplished; (2) built out of transitional components on the way to being standardized; and (3) machined using highly specialized specific phase equipment. As a result, the supporting organization incorporates elements of each phase, but is not optimized for any.

Extent to which the current Department of Defense process resolves the issue

In addition to generating the necessary capability push (as discussed in challenge 1), the "formal acquisition"/"technology development" separation has the effect of creating different innovation environments for different phases of development. These differences are highlighted in Table 20.3. For funding purposes, the capability development process is divided into seven categories; basic research (6.1), applied research (6.2), advanced technology development (6.3), demonstration and validation (6.4), engineering and manufacturing development (6.5), RDT&E management support (6.6) and operational systems development (6.7). Categories 6.1–6.3 are typically carried out in the research laboratories, while categories 6.4–6.7 are incorporated into the formal acquisition structure.

The cultures of the two tiers of acquisition are quite different, as desired. Especially for the 6.1–6.2 funds, the work is primarily contracted out to universities and research institutes or performed by in-house scientists. The nature is exploratory and the expected time frames for results relatively long (i.e., 15–20 years, although more emphasis has been put on near term focus, 5–10 years, of late (AFRL, 2009). As the concept matures (6.2–6.3) the emphasis on military usefulness increases. Projects are expected to show obviously useful areas of application; as a result, there is pressure to focus on near term development. For both the fundamental and applied research, projects may be siloed by discipline and collaboration across disciplines may be limited. On the formal acquisition side, the emphasis is obviously on immediate usefulness in the project and there is an expectation that the technology is mature and nearly ready to be implemented. The innovation at this stage primarily involves integrating the system components to accomplish a new task, although some emphasis on technical maturity persists, for example, through technology demonstrations.

While the acquisition system described earlier does nominally divide the spacecraft innovation process into phases, each of which has different expectations and cultures, the categories and strategies in each do not align with either the Utterback-Abernathy (1975) or Henderson-Clark (1990) model of innovation. Firstly, the research environment of the DoD technology development phase has a completely different effect on innovation than the entrepreneurial inventing environment of the fluid phase. Research is about exploration without a strong focus on how the results can be applied, while inventing is about taking what's known and making it useful. Research is collaborative, building on colleagues' insights; inventing is about competition—being the first to figure it out. Of course, inventors need research to be done, to generate new "knowns," but, it is inventors who generate the "creative destruction" at the heart of economic growth and innovation (Schumpeter, 1934). Thus, while 6.1–6.2 funds, and the laboratory structure, may generate breakthrough fundamental research, the applied value of those efforts won't be fully captured unless they get into the hands of inventors. This is the real implication of the lack of natural bottom-up push; an entrepreneurial environment is extremely difficult to manufacture and without it, innovation suffers.

The second limitation of the technology development and formal acquisition categorizations is that the former focuses primarily on developing component knowledge, leaving the dimension of architectural knowledge to the latter. This is an effective approach when the system level change is modular (Figure 20.5) as is the case with, for example, incorporating an advanced communication payload into a satellite with an otherwise standard architecture. However, when the system-level change is architecturally or radically innovative, mature component technology does not necessarily align with existing system program offices that are structured around legacy subsystem boundaries. Technology readiness needs to be defined along both the component and architectural dimensions.

The potential for improvements offered by operationally responsive space

The challenge of matching the innovation environment to the stage of product development identifies a fundamental limitation of the formal acquisition system. In the existing acquisition paradigm, the product development required to enable future missions is conceptualized as a linear progression from TRL 1-9. With this view in mind, the blue ribbon panels call for increased funding for technology testing. However, while increased funding for technology development is a needed step in the right direction, it only addresses part of the problem. It fails to appreciate the difference between architectural and component dimensions of knowledge and what that means for system level maturity. If the rest of the problem is to be addressed, there is a need for more than two organizational tiers: one for each of the three phases as well as the dimensions of component and architectural knowledge.

As discussed previously, ORS has the potential to incentivize (more) entrepreneurial contributions to space capability development and cut across the traditional tier system by emphasizing architectural innovation. By focusing on disruptive approaches to system-level integration of existing technologies (e.g., modular plug-and-play satellites), the ORS Program Office complements traditional technology development efforts focused on improving components and subsystems. The collective structure of an ORS office focused on architectural innovation and existing labs focused on technology development is well-aligned with the Henderson-Clark model of innovation in which product maturity is a function of both technology readiness and a readiness for system integration.

The independence of the ORS Program Office from existing organizations may also address other limitations of the formal acquisition system. As discussed in Frostman (2007), separation of entrepreneurial business units from mainstream business practices is important to prevent suppression of innovation. By focusing on operational experimentation for meeting existing requirements, ORS acquisitions are exempt from a traditional JCIDS approach. In addition, by emphasizing modular and flexible designs, ORS is able to rapidly integrate component-level technology innovation and may demonstrate the value of a more flexible design paradigm, for the current acquisition system. ORS platforms may also complement existing traditional space architectures by providing test beds for the on-orbit experimentation and maturation of emerging technologies outside of large scale acquisition programs—consistent with the "back to basics" philosophy.

The challenge of balancing risk aversion and the need for experimentation

Perhaps the biggest difference among the three phases is the extent to which innovation can be planned. Once a dominant design emerges (in the transitional phase) innovation can be achieved by systematically making incremental improvements along particular dimensions, but until that point, there is much less certainty about what will work. In the transitional and specific phase, increasingly formal organizational structures are put in place, and those structures facilitate the optimization aspect of the innovation process. Conversely, the fluid phase start-ups have very little in the way of formal organizational structure, in part because they are so transient. Many innovations fail to make it out of the fluid phase; in fact most successful entrepreneurs failed several times before they succeeded; and fail again many times afterwards. These are not risks that big companies typically take; it requires an undying belief in one's product that is often associated with entrepreneurs (Casson et al., 2006). As a result, society doesn't have a high expectation for the success of start-ups and it's not remarkable when they fail. This is not the case with space systems.

The critical mission areas fulfilled by government space programs and the drive for investor return in the commercial space industry, combined with the high cost of space systems, has led to an extremely risk-averse industry. Although decreasing launch costs are mitigating this to some extent, unlike in the fluid phase of traditional markets (where inventors get little attention until they succeed) space projects are highly visible (reinforcing the need to succeed the first time). However if radically different solutions are to arise, there is a need to shelter innovators from the constraining pressures of success.

Extent to which the current Department of Defense structure resolves the issue

The high cost of launching spacecraft combined with a focus on traditional strategic measures of effectiveness in the space industry (e.g., optimize cost-per-function) has driven US space architecture from an era of single-payload, short-lived spacecraft to the current state of multi-payload, long-lived systems. While this design philosophy is justified on the basis of economic arguments associated with the high initial cost of spacecraft and enabled by improvements in supporting subsystems, this design philosophy also has many negative implications. For example, noting that space system developments now take five to ten years, Brown (2007) describes how "complexity has bred fragility" in terms of unanticipated modes of failure. Such unanticipated modes of failure include an acquisitions crisis (Young et al., 2003) where development problems with an individual sensor can cripple the schedule and budget of multi-payload programs (e.g., the National Polar-Orbiting Environmental Satellite System), on-orbit failures that circumvent margin and redundancy (Leveson, 2004), and uncertain technological change.

The blue ribbon panels' recommendations (Figure 20.1) emphasize the need to conduct more technology development outside of programs of record. This will have the positive effect of sheltering high-risk developments from the public eye, but it only addresses the component level issue. As discussed earlier, system integration is, at a minimum, architectural innovation (but often radical innovation) and requires a fluid phase of free experimentation too. Without more system level technology demonstration missions, or a change in expectations regarding first-time mission success, innovation will be stifled.

The potential for improvements offered by operationally responsive space

To better balance risk aversion with the need for on-orbit experimentation, a major philosophical shift is needed in how space systems are designed and operated. In this case, "back to basics" might mean a return to the CORONA paradigm (e.g., recall that 12 launches of the revolutionary CORONA photoreconnaissance satellite were required before a successful demonstration of film capsule recovery on the 13th flight (Wheelon, 1995)). Advanced spacecraft must be sheltered from failure-is-not-an-option mentality, if the desired radical innovation is to be achieved.

ORS represents a major change in mentality, shifting from a performance oriented risk-averse paradigm, to a "good enough" approach; trading some failures for cost and schedule (Richards et al., 2008). While pursuing a "good enough" approach may increase risks of individual satellite failures, the approach may actually enhance the overall resilience of space architecture. For example, in addition to obtaining capability on-orbit quickly and providing the warfighter tactical control, a key attribute of ORS is assured access. Assured access refers to the potential ability of small, tactical spacecraft to be used to partially reconstitute Air Force space mission areas (i.e., Intelligence, Surveillance, and Reconnaissance; Position, Navigation, and Timing; Communications; Environmental Sensing; Missile Warning; and Space Control) should adversaries negate existing space capabilities (Cebrowski & Raymond, 2005). If this radical shift in organizational priorities can be achieved (and the ORS economic assumptions are validated), it may allow for risks associated with on-orbit failures or losses to be mitigated architecturally rather than the customary (and costly) approaches to reliability and survivability at the satellite-level. The Iridium constellation (Garrison et al., 1997) of 66 active satellites (with a reliability requirement of only 58% for a five-year mission) exemplifies this architectural approach to risk management.[9]

Way forward

This paper set out to answer the question: What are the implications of the intrinsic characteristics of the space sector (i.e., monopsony-oligopoly market structure, extremely complex, one-of-a-kind, robust products) for how innovation can and should be encouraged in the defense space context? By structuring the analysis around root-cause challenges, derived from innovation theory, rather than the existing acquisition norms as is typically done, this work contributes to the acquisition reform discussion by providing a baseline for identifying ways that the system could be different. To this end, it identified five core challenges of generating innovation in national security space: (A) generating bottom-up push in a predominantly top-down acquisition process, (B) representing the needs of a disaggregated buyer, (C) integrating fragmented sell-side knowledge from the top-down,

---

[9] These constellation level survivability features include dynamic control and routing of satellite crosslinks around unavailable nodes, on-orbit satellite spares, and the ability to control all 66 operational spacecraft from a single ground facility. For example, following the shattering of the satellite on February 10, 2009, Iridium was able to move one of its in-orbit spares into the network constellation within a month. As noted by Garrison, Pizzicaroli and Swan (1997), "...the design philosophy provides redundancy at the system level instead of the hardware configuration level. Autonomous operation and dynamic resource management and routing provide constellation failure mitigation. In effect, the traditional hardware redundancy is spread over many spacecraft".

(D) matching the innovation environment to the stage of development, and (E) balancing risk aversion and the need for experimentation.

From an innovation theory point of view, Challenge A (generating bottom-up innovation) the space market structure inhibits half of the natural competitive market innovation dynamic. As a result, until more buyers become involved in the space market,[10] any acquisition system will need a mechanism through which to ensure that new ideas continue to be infused into the acquisition system. Development contracts do accomplish this *capability development* to a certain extent, but as discussed earlier, they are limited in their ability to encourage *sell-side initiative* and the parallel and varied concept explorations it embodies. The emergence of "New Space" introduces other partial models for encouraging and leveraging sell-side initiative including COTS, seed-funding models such as Starburst Accelerator, prizes (e.g., Ansari X-Prize) and the market-independent funding associated with many of the New Space companies. The idea in each of these is to for the government to help sustain a market rather than subsidize the development of a particular technology (i.e., generate sell side initiative, not just capability development).

With regard to Challenge B (needs representation) and Challenge C (knowledge integration) the blue ribbon panels are almost unanimous in their recommendations to increase the technical competence of the acquisition core and emphasize the importance of front-end specification. However this only addresses half of the problem. No matter how many new capabilities are generated, their value will hinge on how well the original need was represented as a set of requirements. For the other half of the problem to be fully resolved, more emphasis must be given to the challenge of knowledge integration on both the buy- and sell-side. Specifically with respect to Challenge B, increased emphasis must be placed on flowing needs to requirements. This will involve a combined effort to educate users about their choices (what is possible) and help acquirers capture their needs more effectively. To this end, value-based system analysis methodologies to facilitate the process of capturing both articulated and unarticulated needs, early in the conceptual design phase, are currently being developed by researchers. Taking the value-centric perspective during conceptual design empowers stakeholders to rigorously evaluate and to compare different system requirements in the technical domain using a unifying set of attributes in the value domain (Mathieu & Weigel, 2005; Ross et al., 2004; Hanumanthrao et al., 2017). If deployed by System Program Offices, these emerging system analysis methodologies will contribute significantly to overcoming Challenge B.

Overcoming Challenge C will require more frequent interactions among contractors, integrators and the government through formal acquisitions. Where need-capability information is transferred continuously from buyers to sellers, and vice versa, in traditional markets, the transfer only happens during contracted hardware development in the space sector. As long as space acquisition continues to operate on a model of infrequent, extremely complex monoliths, the knowledge required to innovate will continue to be fragmented across the various players. Decreasing the acquisition cycle time will help both the knowledge integration problem identified in Challenge C, but also the risk aversion indentified in Challenge E.

Challenge D (matching) identifies a fundamental limitation of the current system. In the existing acquisition paradigm, the product development required to enable future missions is conceptualized as a linear progression from TRL 1-9. With this view in mind, the

---

[10] This has happened, to a certain extent, in the domain of communication satellites and earth imaging and may soon be the case if space tourism were to take off, but is arguably unrealistic in the near future for more advanced and military applications.

blue ribbon panels call for increased funding for technology development. However, while increased funding for technology development is a needed step in the right direction, it only addresses part of the problem. It fails to appreciate the difference between architectural and component dimensions of knowledge and what that means for system level maturity and infusion of new technologies. If the rest of the problem is to be addressed, there is a need for more than two organizational tiers—up to one for each of the three phases as well as the dimensions of component and architectural knowledge—and better coupling among them.

Similarly, the recommendations of the blue ribbon panels that pertain to Challenge E (risk shelter) emphasize a "back to basics" philosophy which keeps R&D separate from system acquisition. This would serve to shelter component development from political pressures, but do nothing at the spacecraft level. For spacecraft level development to achieve the risk shelter that is required, a major philosophical shift is needed. An ORS-like philosophy could serve that purpose.

The challenges identified in this paper are fundamental to generating innovation in the space sector; they will not be easy to overcome. The previous discussion provides some guidelines for how to approach solving the problems, but will require all stakeholders involved to come together to implement a solution.

Acknowledgments

The authors would like to thank Dr. Thomas Adang of the Aerospace Corporation and Maj. Luke Cropsey of the US Air Force for their valuable insights. Funding for this work was provided by the Center for Aerospace, Systems, Policy and Architecture Research (CASPAR); Cisco Systems Inc.; the Natural Science and Engineering Research Council of Canada (NSERC); and the Program on Emerging Technologies (PoET), an interdisciplinary research effort of the National Science Foundation at MIT.

References

Abernathy, W. J., & Clark, K. B. (1993). Innovation: Mapping the winds of creative destruction. *Research Policy*, 22(2), 102.

Adams, W., & Adams, W. J. (1972). The military-industrial complex: A market structure analysis. *The American Economic Review*, 62(1/2), 279–287.

AFRL. (2009). Air Force Research Laboratory website, http://www.wpafb.af.mil/afrl.aspx (last accessed 2009).

Anderson, P., & Tushman, M. L. (1990). Technological discontinuities and dominant designs: A cyclical model of technological change. *Administrative Science Quarterly*, 35(1), 604–633.

Bhide, A. (2008). *The Venturesome Economy: How Innovation Sustains Prosperity in a More Connected World*. Princeton, NJ: Princeton University Press.

Brown, O. (2007). *Speech by Dr. Owen Brown on Fractionated Spacecraft*. Anaheim, CA: DARPATech Symposium.

Casson, M., Yeung, B., Basu, A., & Wadeson, N. (2006). *The Oxford Handbook of Entrepreneurship*. Oxford, UK: Oxford University Press.

Cebrowski, A. K., & Raymond, J. W. (2005). Operationally responsive space: A new defense business model. *US Army War College Quarterly*, 35(2).

Christensen, C. (2003). *The Innovator's Dilemma*. New York: HarperCollins.

DAPA. (2006). Defense acquisition performance assessment: Defense acquisition performance assessment project. Washington, DC: Report for the Deputy Secretary of Defense.

DoD. (2007). Plan for operationally responsive space: A report to congressional defense committees. Washington, DC: National Security Space Office, Department of Defense.

Franke, N., & Shah, S. (2003). How communities support innovative activities: An exploration of assistance and sharing in end-user communities. *Research Policy*, 32(1), 157–178.

Frisbee, S. M. (2003). *Lead Warfighters and Innovations: Collaborative Methods of Filling Capability Gaps.* Montgomery, AL: Air University, Maxwell Air Force Bace.

Frostman, D. (2007). Leveraging the disruptive innovation of operationally responsive space. Paper Presented at the AIAA Space 2007. Long Beach, CA.

GAO. (2006). *Space Acquisitions: Improvement Needed in Space Systems Acquisitions and Keys to Achieving Them: Testimony before the Subcommittee on Strategic Forces, Senate Committee on Armed Services.*, US Government Accountability Office, GAO-06-626T.

GAO. (2007). *Space Acquisitions: Actions Needed to Expand and Sustain Use of Best Practices—Testimony before the Subcommittee on Strategic Forces, Senate Committee on Armed Services.* Washington, DC: US Government Accountability Office, GAO-07-730T.

GAO. (2016). *Defense Space Acquisitions: Too Early to Determine If Recent Changes Will Resolve Persistent Fragmentation in Management and Oversight.* Washington, DC: US Government Accountability Office, GAO-16-529R.

Garrison, T., Pizzicaroli, J., & Swan, P. (1997). Systems engineering trades for the IRIDIUM constellation. *Journal of Spacecraft and Rockets*, 34(5), 675–680.

Hanumanthrao, K., Mesmer, B. L., and Bloebaum, C. L. (2017). Increased system consistency through incorporation of coupling in value-based systems engineering. *Systems Engineering*, 20(1), 21–44.

Henderson, R., & Clark, K. (1990). Architectural innovation: The reconfiguration of existing product technologies and the failure of established firms. *Administrative Science Quarterly*, 35, 9–30.

Leveson, N. (2004). Role of software in spacecraft accidents. *Journal of Spacecraft and Rockets*, 41(4), 564–575.

Mathieu, C., & Weigel, A. L. (2005). Assessing the flexibility provided by fractionated spacecraft. Paper Presented at the AIAA Space 2005. Long Beach, CA.

Morrison, P. D., Roberts, J. H., & von Hippel, E. (2000). Determinants of user innovation sharing in a local market. *Management Science*, 46(12), 1513–1527.

Munson, A. (2016). *Why Can't We Get Acquisition Right?* Arlington, VA: Potomac Institute for Policy Studies.

NDIA. (2003). *Top Five Systems Engineering Issues in Defense Industry.* Arlington, VA: National Defense Industrial Association.

NRC. (2008). *Pre-Milestone A and Early-Phase System Engineering: A Retrospective and Benefits for Future Air Force Acquisition.* Washington, DC: National Academy Press.

Peck, M. J., & Scherer, F. M. (1962). *The Weapons Acquisition Process: An Economic Analysis.* Cambridge, UK: Harvard Business School Press.

Richards, M., Viscito, L., Ross, A., & Hastings, D. (2008). Distinguishing attributes for the operationally responsive space paradigm. *Paper Presented at the 6th Responsive Space Conference.* Los Angeles, CA.

Ross, A. M., Hastings, D. E., Warmkessel, J. M., & Diller, N. (2004). Multi-attribute tradespace exploration as front end for effective space system design. *Journal of Spacecraft and Rockets*, 41(1), 20–28.

Rothwell, R., & Zegveld, W. (1994). Reindustrialization and technology. In R. Rothwell (Ed.), *Towards the Fifth-generation Innovation Process* (Vol. 11, pp. 7–31). West Yorkshire, UK: International Marketing Review.

Rumsfeld, D., Andrews, D., Davis, R., Estes, R., Fogleman, R., Garner, J. et al. (2001). Report of the commission to assess United States national security space management and organization. Washington, DC.

Sauser, B., Ramirez-Marquez, J., Magnaye, R., & Tan, W. (2008). A systems approach to expanding the technology readiness level within defense acquisition. *International Journal of Defense Acquisition Management*, 1(3), 39–58.

Schumpeter, J. A. (1934). *The Theory of Economic Development.* New York: Harvard University Press.

Sherwin, C. W., & Isenson, R. S. (1967). Project hindsight. *Science*, 156(3782), 1571–1577.

Stigler, J. (1971). The theory of economic regulation. *Bell Journal of Economic Management Science*, 2, 3–21.

Teece, D. J. (1986). Profiting from technological innovation: Implications for integration, collaboration, licensing and public policy. *Research Policy*, 15(6), 285–305.

Thomke, S., & Hippel, E. V. (2004). *Customers as Innovators*. Boston, MA: Harvard Business Review.

Urban, G. L., & von Hippel, E. (1988). Lead use analyses for the development of new industrial products. *Management Science*, 34(5), 569–582.

Utterback, J. M. (1994). *Mastering the Dynamics of Innovation*. Cambridge, MA: Harvard Business Press.

von Hippel, E. (1986). Lead users: A source of novel product concepts. *Management Science*, 32(7), 791–806.

von Hippel, E. (1988). *The Sources of Innovation*. Oxford, UK: Oxford University Press.

Wheelon, A. (1995). Lifting the veil on CORONA. *Space Policy*, 11(4), 249–260.

Young, T., Hastings, D. E., & Schneider, W. (2003). Report of the defense science board/air force scientific advisory board joint task force on acquisition of national security space programs. Washington, DC: Office of the Undersecretary of Defense for Acquisition, Technology, and Logistics.

chapter twenty one

Innovative applications of polymer materials for 3D printing

Ibrahim Katampe

Contents

Classification of additive manufacturing processes

Three-dimensional (3D) printing, also known as additive manufacturing, is a layering process by which solid, 3D objects are created. The technology is gaining widespread interest in the aerospace, automotive, electrical, medical, and dental. The term "3D printing" in this context is used as a synonym for all types of Additive Manufacturing (AM) processes. Additive Manufacturing describes the technologies that build 3D objects by adding layer-upon-layer of material. Three types of materials are generally used: polymers, ceramics and metals; although polymers are the most commonly used. These polymer materials are often produced as liquid, powder form or in filament feedstock. In this chapter each manufacturing process is briefly described and the polymer types commonly used.

Individual processes differ depending on the material and machine technology used. Hence, in 2010, the American Society for Testing and Materials (ASTM) group "ASTM F42 – Additive Manufacturing" formulated a set of standards that classify the range of Additive Manufacturing processes into categories (Standard Terminology for Additive Manufacturing Technologies, 2012). This chapter briefly gives an overview of the different processes used in additive manufacturing practices.

Vat process – Stereolithography

This process [1–4] uses a vat of liquid photopolymer resin, out of which the model is constructed layer by layer. Resins are cured using a process of photo polymerization [5] or UV light, where the light is directed across the surface of the resin with the use of motor-controlled mirrors [6] (Figure 21.1).

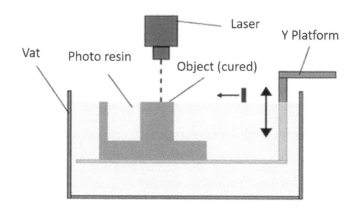

Figure 21.1

Step by Step – Photopolymerization process:

- The build platform is lowered from the top of the resin vat downwards by the layer thickness.
- A UV light cures the resin layer by layer. The platform continues to move downwards and additional layers are built on top of the previous.
- Some machines use a blade which moves between layers in order to provide a smooth resin base to build the next layer on.
- After completion, the vat is drained of resin and the object removed.

Material jetting process

This is an additive manufacturing process that uses inkjet printing technologies to manufacture 3D objects [7–9]. Material jetting creates objects in a similar method to a two-dimensional ink jet printer. Material is jetted onto a build platform using either a continuous or Drop on Demand (DOD) approach. Machines vary in complexity and in their methods of controlling the deposition of material. The material layers are then cured or hardened using ultraviolet (UV) light. As material must be deposited in drops, the number of materials available to use is limited. Polymers and waxes are suitable and commonly used materials, due to their viscous nature and ability to form drops (Figure 21.2).

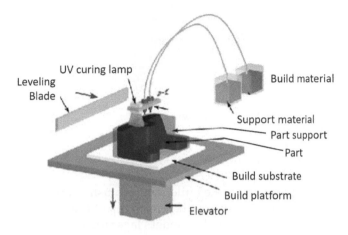

Figure 21.2

Step by Step – Material jetting process:

- The print head is positioned previously build platform.
- Droplets of material are deposited from the print head onto surface where required, using either thermal or piezoelectric method.
- Droplets of material solidify and make up the first layer.
- Further layers are built up as before on top of the previous.
- Layers are allowed to cool and harden or are cured by UV light. Post processing includes removal of support material.

Binder jetting process (3DP – MIT)

The binder jetting process [2,4] uses two materials; a powder-based material and a binder. The binder acts as an adhesive between powder layers. The binder is usually in liquid form and the build material in powder form. A print head moves horizontally along the x and y axes of the machine and deposits alternating layers of the build material and the binding material. After each layer, the object being printed is lowered on its build platform. Due to the method of binding, the material characteristics are not always suitable for structural parts and despite the relative speed of printing, additional post processing (see below) can add significant time to the overall process. As with other powder-based manufacturing methods, the object being printed is self-supported within the powder bed and is removed from the unbound powder once completed (Figure 21.3).

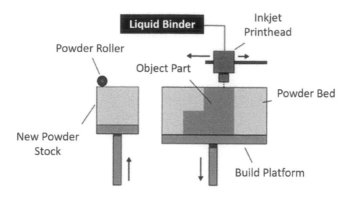

Figure 21.3

Step by Step – Binder jetting process:

- Powder material is spread over the build platform using a roller.
- The print head deposits the binder adhesive on top of the powder where required.
- The build platform is lowered by the model's layer thickness.
- Another layer of powder is spread over the previous layer. The object is formed where the powder is bound to the liquid.
- Unbound powder remains in position surrounding the object.
- The process is repeated until the entire object has been made.

Material extrusion process

Fused deposition modeling (FDM) is an additive manufacturing process [1,2] in which a thin filament of plastic feeds a typically of 0.25 mm. FDM is a common material extrusion process and is trademarked by the company Stratasys. Whilst FDM is similar to all other 3D printing processes, as it builds layer by layer, it varies in the fact that material is added through a nozzle under constant pressure and in a continuous stream (Figure 21.4).

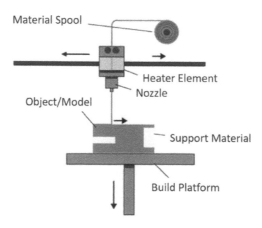

Figure 21.4

Step by Step – Material extrusion process:

- The first layer is built as the nozzle deposits material where required onto the cross-sectional area of first object slice.
- The following layers are added on top of previous layers.
- Layers are fused together upon deposition as the material is in a melted state.

Powder bed fusion process

Powder bed fusion (PBF) methods use either a laser or electron beam to melt and fuse material powder together [10–12]. The Powder Bed Fusion process includes the following commonly used printing techniques: Direct metal laser sintering (DMLS), Electron beam melting (EBM), Selective heat sintering (SHS), Selective laser melting (SLM) and Selective laser sintering (SLS). All PBF processes involve the spreading of the powder material over previous layers. Direct metal laser sintering (DMLS) is the same as SLS, but with the use of metals and not plastics. The process sinters the powder, layer by layer. Selective Heat Sintering differs from other processes by way of using a heated thermal print head to fuse powder material together (Figure 21.5).

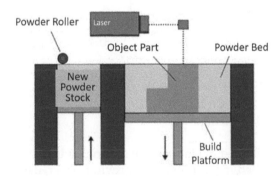

Figure 21.5

Step by Step – Powder bed fusion process:

- A layer, typically 0.1 mm thick, of material is spread over the build platform.
- A laser fuses the first layer or first cross section of the model.
- A new layer of powder is spread across the previous layer using a roller.
- Further layers or cross sections are fused and added.
- The process repeats until the entire model is created. Loose, unfused powder is remains in position but is removed during post processing.

Sheet lamination

Sheet lamination processes [13] include ultrasonic additive manufacturing (UAM) and laminated object manufacturing (LOM). The Ultrasonic Additive Manufacturing process uses sheets or ribbons of metal, which are bound together using ultrasonic welding. The process does require additional cnc machining and removal of the unbound metal, often during the welding process. Laminated object manufacturing (LOM) uses a similar layer by layer approach but uses paper as material and adhesive instead of welding. The LOM process uses a cross hatching method during the printing process to allow for easy removal post build. Laminated objects are often used for aesthetic and visual models and are not suitable for structural use. UAM uses metals and includes aluminum, copper, stainless steel and titanium (Ultrasonic Additive Manufacturing Overview, 2014). The process is low temperature and allows for internal geometries to be created. The process can bond different materials and requires relatively little energy, as the metal is not melted (Figure 21.6).

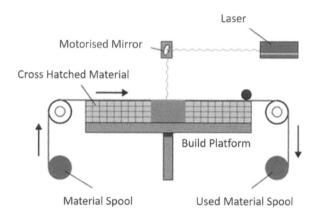

Figure 21.6

Step by Step – Sheet lamination:

- The material is positioned in place on the cutting bed.
- The material is bonded in place, over the previous layer, using the adhesive.
- The required shape is then cut from the layer, by laser or knife.
- The next layer is added.
- Steps two and three can be reversed and alternatively, the material can be cut before being positioned and bonded.

Directed energy deposition

Directed Energy Deposition [14] (DED) covers a range of terminology: "Laser engineered net shaping, directed light fabrication, direct metal deposition, 3D laser cladding" It is a more complex printing process commonly used to repair or add additional material to existing components [14]. A typical DED machine consists of a nozzle mounted on a multi axis arm, which deposits melted material onto the specified surface, where it solidifies. The process is similar in principle to material extrusion, but the nozzle can move in multiple directions and is not fixed to a specific axis. The material, which can be deposited from any angle due to 4 and 5 axis machines, is melted upon deposition with a laser or electron beam. The process can be used with polymers, ceramics but is typically used with metals, in the form of either powder or wire. Typical applications include repairing and maintaining structural parts (Figure 21.7).

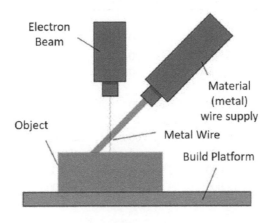

Figure 21.7

Step by Step – Direct energy deposition:

- A 4 or 5 axis arm with nozzle moves around a fixed object.
- Material is deposited from the nozzle onto existing surfaces of the object.
- Material is either provided in wire or powder form.
- Material is melted using a laser, electron beam or plasma arc upon deposition.
- Further material is added layer by layer and solidifies, creating or repairing new material features on the existing object.

Thermoplastic versus thermoset polymer materials

Thermoplastic

Most of the polymers described earlier are classified as **thermoplastic**. This reflects the fact that above T_g they may be shaped or pressed into molds, spun or cast from melts or dissolved in suitable solvents for later fashioning (Table 21.1 and Figure 21.8).

Thermoset

Thermoset is a group of polymers, characterized by a high degree of cross-linking, that resist deformation and solution once their final morphology is achieved. Partial formulas for four of these will be shown in the following (Figures 21.9 through 21.12).

Table 21.1 Specific thermoplastic properties

Properties of selected thermoplastics				
	Tensile strength (psi)	% Elongation	Elastic modulus (psi)	Density (g/cm³)	Izod impact (ft lb/in.)
Polyethylene (PE):					
Low-density	3,000	800	40,000	0.90	9.0
High-density	5,500	130	180,000	0.96	4.0
Ultrahigh molecular weight	7,000	350	100,000	0.934	30.0
Polyvinyl chloride (PVC)	9,000	100	600,000	1.40	
Polypropylene (PP)	6,000	700	220,000	0.90	1.0
Polystyrene (PS)	8,000	60	450,000	1.06	0.4
Polyacrylonitrile (PAN)	9,000	4	580,000	1.15	4.8
Polymethyl methacrylate (PMMA) (acrylic, plexiglas)	12,000	5	450,000	1.22	0.5
polychlortrifluroethylene	6,000	250	300,000	2.15	2.6
Polytetratrifluroethylene (PTFE, Teflon)	7,000	400	80,000	2.17	3.0
Polyoxymethylne (POM) (acetal)	12,000	75	520,000	1.42	2.3
Polymide (PA) (nylon)	12,000	300	500,000	1.14	2.1
Poltester (PET)	10,500	300	600,000	1.36	0.6
Polycarbonate (PC)	11,000	130	400,000	1.20	16.0
Polyimide (PI)	17,000	10	300,000	1.39	1.5
Polytheretherketone (PEEK)	10,200	150	550,000	1.31	1.6
Polyphenylene sulphide (PPS)	95,00	2	480,000	1.30	0.5
Polyether sulfone (PES)	12,200	80	350,000	1.37	1.6
Polyamide-imide (PAI)	27,000	15	730,000	139	4.0

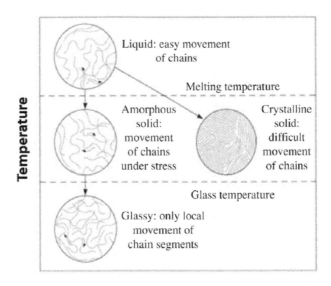

Figure 21.8 The effect of temperature on the structure and behavior of thermoplastics.

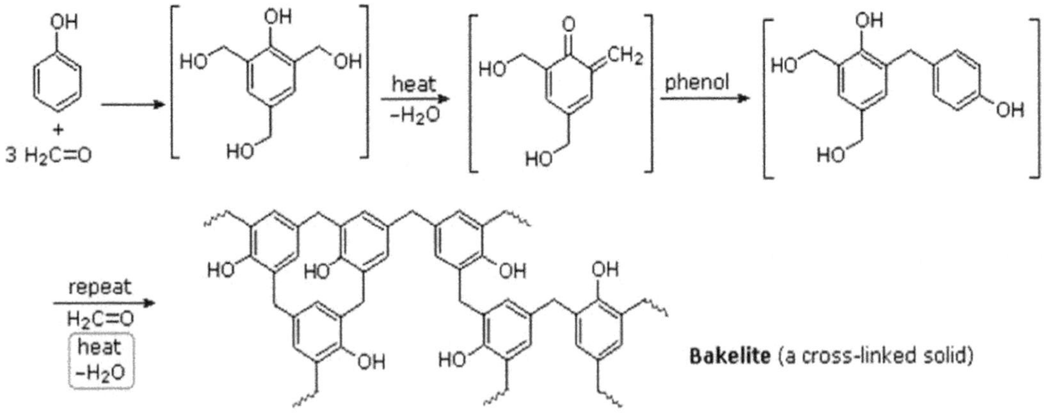

Figure 21.9 Phenol-Formaldehyde Resin.

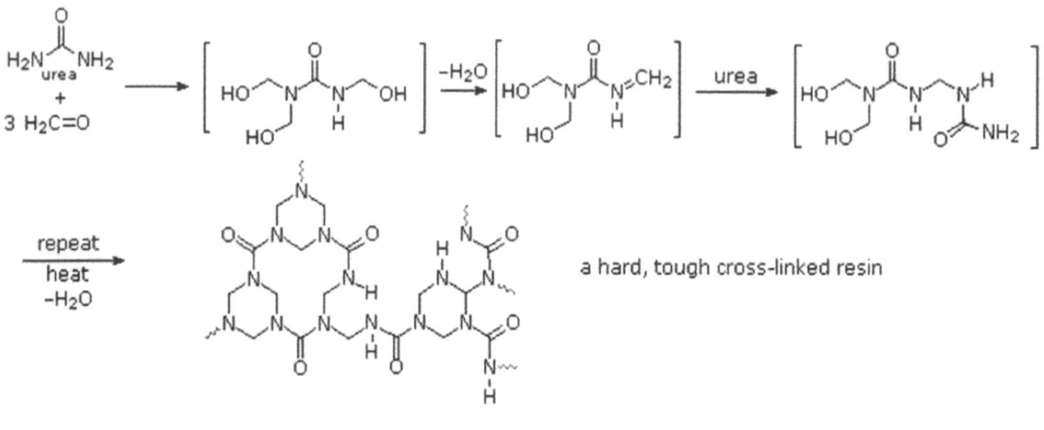

Figure 21.10 Urea-Formaldehyde Resin.

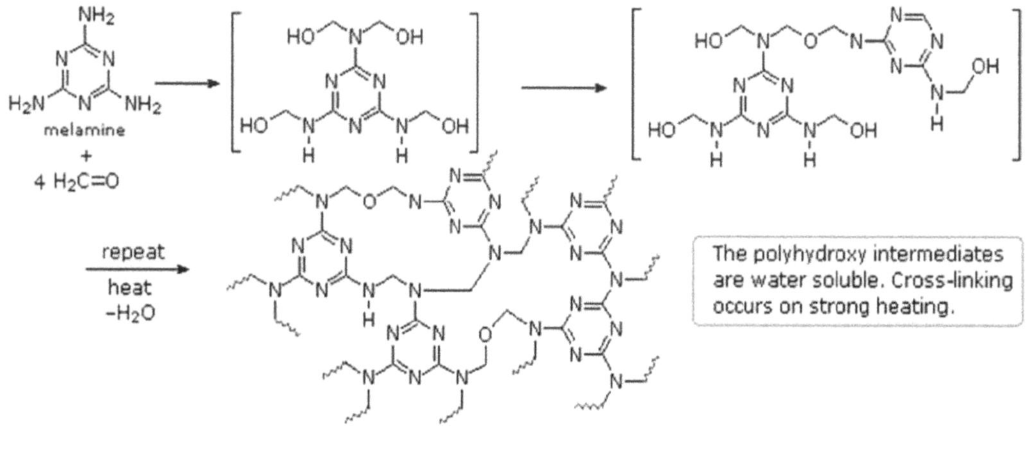

Figure 21.11 Melamine-Formaldehyde Resin.

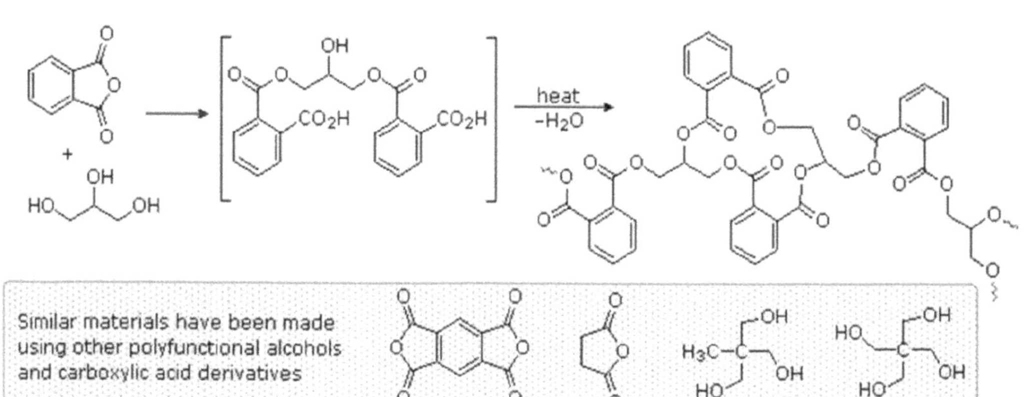

Figure 21.12 Glyptal Resins.

Main polymer materials for additive manufacturing

Plastic in 3D printing

In terms of raw materials that are put into the printing system to create the 3D objects, there seem to be relatively few limitations on what can be used.

Currently, plastics are the most widely used materials in additive manufacturing, and the important ones are listed below:

- ABS – acrylonitile butadiene styrene or "Lego" plastic – a very common choice for 3D printing.
- PLA – polylactic acid – available in soft and hard grades, is becoming very popular and may overtake ABS in the near future.
- PVA – polyvinyl alcohol – is used as a dissolvable support material or for special applications.
- PC – polycarbonate – requires high-temperature nozzle design and is in the proof-of-concept stage.
- SOFT PLA – polylactic acid – is rubbery and flexible, available in limited colors and sources; as 3D printing spreads, may get easy to find.

Photopolymer materials – for stereolithography (Vat) and material jetting

- Acrylics
- Acrylates
- Epoxies
- "ABS-like" (Material Jetting)

Materials for fused deposition modeling-amorphous thermoplastics

- ABS (Acryonitrile Butadiene Styrene)
- Polycarbonate
- PC/ABS Blend
- PLA (Polylactic Acid)
- Polyetherimide (PEI)
- Nylon Co-Polymer

Materials for laser sintering – crystalline/semi-crystalline thermoplastics

- Polyamide (Nylon) 11 and 12
- Polystyrene (Lost Wax Patterns)
- Polypropylene
- Polyester ("Flex")
- Polyether ether ketone (PEEK)
- Thermoplastic Polyurethane
- Nylon 6

Some innovative modification of polymers for 3D printing

With new materials, new methods, and new applications, the additive manufacturing printing field is revolutionizing prototyping and manufacturing, and changing the worlds of design, medicine, construction, and, of course, hobbying. This uptake in usage has been coupled with a demand for printing technology and materials able to print functional elements. Biomaterials are slowly gaining prominence as 3D printing materials in the healthcare industry for synthesizing artificial organs. Demand for lightweight and high strength materials is predicted to drive the development of reinforced composite filaments for use as 3D printing materials. Some examples include:

Conductive thermoplastic composite materials

For use in electronic sensor development [15]. A significant advantage in using 3D printing to create electronic components such as these is that sockets for connecting to standard equipment such as interface boards and multimeters can be printed as part of the printed structure. This approach will open up many new applications for 3DP where fully interactive devices can be printed, for instance, designers could understand how people tactilely interact with their products by monitoring sensors embedded inside. Although arbitrary polymeric microfluidic systems can be readily constructed, the 3D printing process has not been widely applied to the general field of Microelectromechanical Systems (MEMS). One reason is the difficulty in producing good 3D conductive layers, which are essential in most functional devices, as special equipment and techniques are required [16].

Tissue engineering by biodegradable polymers

Bone and cartilage generation by autogenous cell/tissue transplantation is one of the most promising techniques in orthopedic surgery and biomedical engineering [16]. Rapid prototyping technologies, such as 3-D printing (3-DP) and FDM, allow the development of manufacturing processes to create porous scaffolds that mimic the microstructure of living tissue [17].

Soft-tissue engineering – hydrogel-forming polymers

Hydrogel-forming polymers that are suitable for soft tissue engineering with a focus on materials that can be fabricated using additive manufacturing (3D-printing) is an innovative trend [18].

Bioprinting

Bioprinting has emerged as a flexible tool in regenerative medicine with potential in a variety of applications. Bioprinting is a relatively new field within biotechnology that can be described as robotic additive biofabrication that has the potential to build or pattern viable organ-like or tissue structures in 3 dimensions. Generally, bioprinting devices have the ability to print cell aggregates, cells encapsulated in hydrogels or viscous fluids, or cell-seeded microcarriers – all of which can be referred to as "bioink" – as well as cell-free polymers that provide mechanical structure [19].

Creation of light weight, high strength material structures called "Gyroids"

3D printed "Gyroid" shaped objects are the latest structural material that could compete with plastic foams and light weight composite. "Gyroids" may have the potential use for structural parts or components for airplanes, automobiles and the construction building sector of Industry. The new structural materials could withstand various compression and deformation forces when subjected to stress and load [20].

References

1. R. Noorani, *Rapid Prototyping—Principles and Applications*, John Wiley & Sons, Hoboken, NJ, 2006.
2. K. Cooper, *Rapid Prototyping Technology*, Marcel Dekker, New York, 2001.
3. P. P. Kruth, Material incress manufacturing by rapid prototyping techniques, *CIRP Annals—Manufacturing Technology*, 40(2), 603–614, 1991.
4. J. W. Halloran et al., Photopolymerization of powder suspensions for shaping ceramics, *Journalof the European Ceramic Society*, 31(14), 2613–2619, 2011.
5. I. Gibson, D. Rosen, and B. Stucker, *Additive Manufacturing Technologies—3D Printing, Rapid Prototyping and Direct Digital Manufacturing*, 2nd ed. Springer, New York, 2015.
6. E. Grenda, *Printing the Future—The 3D Printing and Rapid Prototyping Source Book*, 3rd ed. Castle Island, Arlington, MA, 2009.
7. Wohlers Report, 3D printing and additive manufacturing state of the industry annual worldwide progress report, Wohlers Associates, Fort Collins, CO, 2010.
8. R. Singh, Process capability study of PolyJet printing for plastic components, *Journal of Mechanical Science and Technology*, 25(4), 1011–1015, 2011.
9. V. Petrovic et al., Additive layered manufacturing: Sectors of industrial application shown through case studies, *International Journal of Production Research*, 49(4), 1061–1079, 2011.
10. T. Hwa-Hsing et al., Slurry-based selective laser sintering of polymer-coated ceramic powders to fabricate high strength alumina parts, *Journal of the European Ceramic Society*, 31(8), 1383–1388, 2011.
11. G. V. Salmoria et al., Microstructural and mechanical characterization of PA12/MWCNTs nanocomposite manufactured by selective laser sintering, *Polymer Testing*, 30(6), 611–615, 2011.
12. D. Slavko and K. Matic, Selective laser sintering of composite materials technologies, *Annals of DAAAM & Proceedings*, 21, 1527–1528, 2010.
13. B. Vaupotic et al., Use of PolyJet technology in manufacture of new product, *Journal of Achievements in Materials and Manufacturing Engineering*, 18(1–2), 319–322, 2006.
14. I. Gibson et al., *Additive Manufacturing Technologies—3D Printing, Rapid Prototyping, and Direct Digital Manufacturing*, 245–268, 2010.
15. J. L. Simon et al., A simple, low-cost conductive composite material for 3D printing of electronic sensors. doi:10.1371/journal.pone.0049365, 2012.
16. S.-Y. Wu, C. Yang, W. Hsu, and L. Lin, 3D-printed microelectronics for integrated circuitry and passive wireless sensors, *Microsystems & Nanoengineering*, 1, 15013, 2015. doi:10.1038/micronano.2015.
17. C. W. Patrick Jr, A. G. Mikos, and L. V. McIntire. Prospectus of tissue engineering, *Frontiers in Tissue Engineering*, 3–14, 1998.
18. D. M. Kirchmajer, R. Gorkin III, and M. In Het Panhuis, An overview of the suitability of hydrogel-forming polymers for extrusion-based 3D-printing, *Journal of Materials Chemistry B*, 3(20), 4105–4117, 2015.
19. R. P. Visconti et al., Towards organ printing: Engineering an intra-organ branched vascular tree, *Expert Opinion on Biological Therapy*, 10, 409–420, 2010.
20. J. B. Markus et al., The mechanics and design of a lightweight three-dimensional graphene assembly, *Science Advances*, 3(1), e1601536, 2017.

chapter twenty two

Innovation project management

Adedeji B. Badiru

Contents

Badiru (2012) defines project management as the process of managing, allocating, and timing resources in order to achieve a given objective in an expeditious manner. The objective may be in terms of time, monetary, or technical results. Project management is the process of achieving objectives by utilizing the combined capabilities of available resources. It represents a systematic execution of tasks needed to achieve project objectives. In a new technology environment, the basic functions of project management cover the following:

1. Planning
2. Organizing
3. Scheduling
4. Control

Because of the complexity often encountered when installing new high-tech equipment, the steps of the design process require thinking outside the conventional project box. It has been shown again and again that the majority of technology failures can be traced to communication failures at the initial stages of a project. Thus, communication constitutes an important foundation for achieving success of innovation projects. When embarking on new innovation projects, particularly those involving technology products, some of the issues of crucial consideration include the following:

- Purchasing process and contracting requirement
- Delivery timeline
- Safety concerns
- Training requirements

- Maintenance
- Skilled operators
- Service contract
- Space requirements (equipment footprint and supporting infrastructure)
- Power supply
- Water needs
- HVAC needs
- Operational requirements
- Occupational Safety and Health Administration (OSHA) requirements
- Sustained utilization
- Funding (initial and subsequent)
- Vibration control
- Facilities upkeep (housekeeping around equipment)
- Production level requirements
- Minimum acceptable quality

All of these, and some more not listed here, require a whole lot of coordinated project management. Essentially, a comprehensive project management is required.

Basics of the Triple C model for innovation project management

The Triple C model introduced by Badiru (2008) is an effective project planning tool that has been successfully utilized for projects of all types. It can be particularly effective for a distributed product development environment, such as interdisciplinary innovation, where personnel coordination is very crucial. The model states that project management can be enhanced by implementing it within the integrated functions of

- Communication
- Cooperation
- Coordination

The Triple C model facilitates a systematic approach to planning, organizing, scheduling, and control. The model is shown graphically in Figure 22.1. It highlights what must be done and when. It can also help to identify the resources, such as personnel, peripheral

Figure 22.1 Triple C project management framework for innovation.

equipment, facilities, power supply, space requirements, and so on, associated with the products of new innovation.

Typical questions to be addressed in innovation project management for new high-tech equipment installation include the following:

Who: Who is the point of contact for the new equipment? Who made the selection? Who else is involved? Who has been informed? Who will run the equipment? Who are the users? Who will maintain the equipment? Who is proving the funding for all the needs affiliated with the equipment?

What: What is being purchased? What will the equipment be used for? What are the options? What will be equipment replace or supplement? What peripheral installation needs are involved? Safety concerns? Security concerns? Power supply needs? Fire suppressant? Water supply needs? Lighting needs? HVAC needs? Vibrant concerns? Emission concerns? Stability concerns?

Which: Which functional and/or administrative units are responsible for the equipment?

When: When will the equipment be purchased? When is the delivery timeline? When is the contracting timeline, if applicable?

Where: Where will be equipment be placed? Is co-location with other organization facilities possible?

How: How will be equipment be used? How will be equipment be maintained? How will the equipment utilization be sustained? How will the equipment be de-commissioned, when applicable?

Why: Why is the equipment needed at all?

Communication

Communication facilitates team work. The communication function of project management involves making all those concerned become aware of project requirements and progress. Those who will be affected by the project directly or indirectly, as direct participants or as beneficiaries, should be informed regarding the following:

- Scope of the product
- Personnel contribution required
- Expected cost and merits of the project
- Project organization and implementation plan
- Potential adverse effects if the project should fail
- Alternatives, if any, for achieving the project goal
- Potential direct and indirect benefits of the product development project

The communication channel must be kept open throughout the project life cycle. In addition to internal communication, appropriate external sources should also be consulted. This is particularly essential for a distributed product design environment where design participants may be geographically dispersed over large distances. Figure 22.2 presents a specific application to inter-module communication in innovation product development. Using Triple C helps to clarify the following questions, particularly when the modules are designed at geographically dispersed locations:

- Does each product development participant know what the objective is?
- Does each product development participant know his or her role in achieving the objective?
- What obstacles may prevent a participant from playing his or her role effectively?

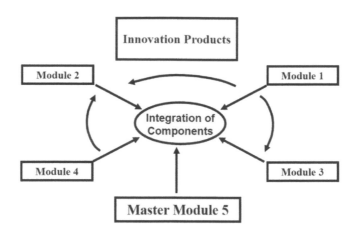

Figure 22.2 Innovation product inter-module communication channels.

Some of the sources of communication problems for high-tech technology project management are summarized in the following:

Social environment: Communication problems sometimes arise because people have been conditioned by their prevailing social environment to interpret certain issues in unique ways, particularly when new pieces of technological equipment are being contemplated. Vocabulary, idioms, organizational status, social stereotypes, and economic situation are among the social factors that can impede effective communication in advanced manufacturing organizations. Innovation is not immune to these adverse scenarios.

Cultural background: Cultural differences are among the most pervasive barriers to technological project communications, especially in today's multinational organizations. Language and cultural idiosyncrasies often determine how communication is approached, received, and interpreted.

Semantic and syntactic factors: Semantic and syntactic barriers to communications usually occur in written documents. Semantic factors are those that relate to the intrinsic knowledge of the subject of the communication. Syntactic factors are those that relate to the form in which the communication is presented. The problems created by these factors become acute in situations where response, feedback, or reaction to the communication cannot be observed directly or face-to-face. Explicit efforts must be made to bring everybody on board for new innovation undertakings.

Organizational structure: Frequently, the organization structure within which a technical project is housed has a direct influence on the flow of information and, consequently, on the effectiveness of communication. Organization hierarchy may determine how different personnel levels perceive specific information. One key aspect to keep in mind is the proverbial guide of "the higher the level of management, the lower the level of details needed." An overly technical presentation of an innovation project can quickly lose the interest of management. This is particularly important where funding decisions are involved.

Communication medium: The method of transmitting a message may also affect the value ascribed to the message and, consequently, how it is interpreted or used. With the excessive prevalent of email communications nowadays, it is essential to determine where and when direct face-to-face communication is better than email transmission of critical information about a proposed technological innovation.

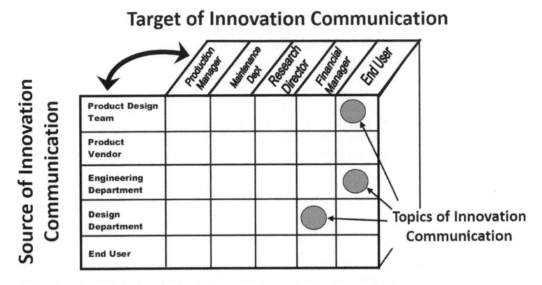

Figure 22.3 A template for communication matrix for innovation project management.

Figure 22.3 shows a condensed sample of multi-dimensional communication matrix for innovation implementation environment. Actual users will include all the pertinent elements for their specific operating environment. Communication across various functional lines is important to bring everyone on board for a cohesive innovation effort. Of particular importance is the need to keep end-user requirements in mind throughout the development process. The cells in the communication matrix indicate the source-to-target communication linkages as well as specific topic of communication. This helps to identify not only who is communication with whom, but also what is expected to be communicated.

Cooperation

Cooperation of the personnel involved in innovation must be elicited using explicit means. Merely voicing consent for a project is not enough assurance of full cooperation. Participants and beneficiaries of the project must be convinced of the merits of the project. The pros and cons should be addressed. Never shy away from the "cons" of a project. Rather than being a source of ire for team members, a specification of the "cons" may be vital for garnering support, as long as individuals know what to expect and what not to expect. Some of the factors that influence cooperation in a project environment include personnel requirements, resource requirements, budget limitations, past experiences, conflicting priorities, space limitation, resource sharing constraints, and lack of uniform organizational support. A structured approach to seeking cooperation for innovation should clarify the following:

- The level and type of cooperative efforts required
- Precedents for collaborative projects
- The possible implication of lack of cooperation
- The criticality of cooperation to project success
- The expected organizational impact of cooperation
- The time frame involved in the project
- The organizational benefits of cooperation
- The personal benefits or rewards of cooperation

The types of cooperation required for a successful product development include functional cooperation, social cooperation, legal cooperation, administrative cooperation, proximity cooperation, dependency cooperation, lateral cooperation, vertical cooperation, and imposed cooperation. Some of these are possible only in certain types of project scenarios. Following are some guidelines for securing cooperation for innovation:

- Establish achievable goals for the project.
- Clearly outline individual commitments required.
- Integrate project priorities with existing priorities.
- Allay the fear of job loss due to innovation products compared to traditional manufacturing.
- Anticipate and preempt potential sources of resource conflicts.
- Remove skepticism by referring to earlier communication of the merits of the project.

Coordination

After communication and cooperation functions have been initiated successfully, the efforts of the project personnel must be coordinated. Many projects fail because the project team anxiously jumps to the coordination stage. But where there has not been sufficient communication and there is a lack of cooperation, coordination cannot be accomplished effectively. Coordination facilitates congruent organization of efforts. The construction of a responsibility chart can be very helpful at this stage. A responsibility chart is a matrix consisting of columns of individual or functional departments and rows of required actions. Cells within the matrix are filled with relationship codes that indicate who is responsible for what. The matrix helps avoid neglecting crucial communication requirements and obligations. It helps resolve questions such as:

- Who is to do what?
- How long will it take?
- Who is to inform whom of what?
- Whose approval is needed for what?
- Who is responsible for which results?
- What personnel interfaces are required?
- What support is needed from whom and when?

When implemented as an integrated process, the Triple C model can help avoid conflicts in new high-end equipment installation. When conflicts do develop, it can help in resolving the conflicts. Several sources of conflicts can exist in complex technical projects, including the following:

Schedule conflict: Conflicts can develop because of improper timing or sequencing of project tasks. This is particularly common in large multiple projects spread over multiple locations. Procrastination can lead to having too much to do at once, thereby creating a clash of project functions and discord among team members. Inaccurate estimates of time requirements may also lead to infeasible activity schedules.

Cost conflict: Product development cost may not be generally acceptable to the clients of a project. This will lead to project conflicts. Even if the initial cost of the product development is acceptable, a lack of cost control during implementation can lead to conflicts. Poor budget allocation approaches and the lack of a financial feasibility study will cause cost conflicts later on in the product development process. One area of concern for innovation is the cost of supplies to sustain the operation of the innovation equipment. Adequately funding the purchase of innovation equipment is one thing, but funding the recurring purchase of supplies is an entirely different things.

Performance conflict: If clear performance requirements are not established, innovation product performance conflicts will develop. Lack of clearly defined quality standards and expectations can lead each person to evaluate his or her own performance based on personal value judgments. In order to uniformly evaluate quality of innovation outputs and monitor project progress, performance standards should be established based on the intended scope of the innovation project.

Management conflict: There must be a two-way alliance between management and the innovation team. The views of management should be understood by the team. The views of the team should be appreciated by management. If this does not happen, management conflicts will develop.

Technical conflict: If the technical basis of a project is not sound, technical conflicts will develop. New manufacturing projects are particularly prone to technical conflicts because of their significant dependence on technology. Lack of a comprehensive technical feasibility study will lead to technical conflicts. Clear communication, solid cooperation, and tight coordination can help defuse the adverse impacts of reluctance to embrace new innovation.

Priority conflict: Priority conflicts can develop if project objectives are not defined properly and applied uniformly across a project. A lack of a direct project definition can lead each project member to define his or her own goals which may be in conflict with the intended goal of the project. A lack of consistency of the project mission is another potential source of priority conflicts. Over-assignment of responsibilities with no guidelines for relative significance levels can also lead to priority conflicts. One person taking on the task of what should be a team effort is a sure basis for priority conflict. Again, using the Triple C model can help preempt or resolve priority conflicts.

Resource conflict: Resource allocation problems are a major source of conflicts in any project management. Competition for resources, including personnel, tools, hardware, software, space, and so on, can lead to disruptive conflicts.

Power conflict: Project politics lead to a power play which can adversely affect the progress of a project. Project authority and project power should be clearly delineated. Project authority is the control that a person has by virtue of his or her functional position. Project power relates to the clout and influence, which a person can exercise due to connections within the administrative structure of an organization. People with popular personalities can often wield a lot of project power in spite of low or nonexistent project authority.

Personality conflict: Personality conflict is a common problem in projects involving a large group of people. The larger the project, the larger the size of the management team needed to keep things running. Unfortunately, the larger management team creates an opportunity for personality conflicts. Communication and cooperation can help defuse personality conflicts.

Distributed innovation product development

This section covers the fundamentals of distributed product development in innovation projects. Figure 22.4 presents the product development process in a distributed environment across functional areas. The inputs are in terms of capital, raw material, and labor.

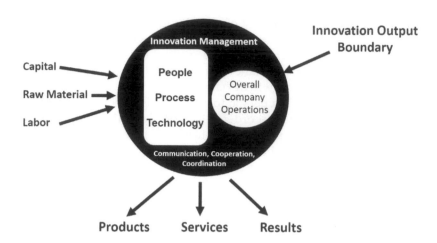

Figure 22.4 Input-Output framework for distributed innovation product development.

At the output end, the physical products are complemented by organizational services and a metric of market share. The project management approach embodies technology, people, and work process. In this environment, the Triple C model serves as the tool to integrate the various project management efforts.

Analysis of innovation project requirements

A typical project is undertaken to create a unique **product**, **service**, or **result**. In the case of innovation, the project output is a certain **product**, hopefully of high quality, that meets the market needs of the organization. The key to getting everyone on board with the innovation process is to ensure that product objectives are clear and comply with the principle of **SMART** as outlined in the following:

- *Specific:* Task objective must be specific. Project objectives must be specific, explicit, and unambiguous. Objectives that are not specific are subject to misinterpretations and misuse.
- *Measurable:* Task objective must be measurable. Project objectives should be designed to be measurable. Any factor that cannot be measured cannot be tracked, evaluated, or controlled.
- *Aligned:* Task objective must be achievable and aligned with overall project goal.
- *Realistic:* Task objective must be realistic and relevant to the organization. A project's goals and objectives must be aligned with the core strategy of an organization and relevant to prevailing needs. If not aligned, an objective will have misplaced impacts. A project and its essential elements must be realistic and achievable. It is good to "dream" and have lofty ideas of what can be achieved. But if those pursuits are not realistic, a project will just end up "spinning wheels" without any significant achievements.
- *Timed:* Task objective must have a time basis. Timing is the standardized basis for work accomplishment. If project expectations are not normalized against time, there will be no basis for an accurate assessment of performance.

If a task has the aforementioned intrinsic characteristics, then the function of communicating the task will more likely lead to personnel cooperation. A SMART approach to developing and communicating innovation objectives can ensure the cooperation of everyone. Specific means that an observable action, behavior, or achievement is described. It also means that the work links to a rate of performance, frequency, percentage, or other quantifiable measure. For some jobs, being specific can, itself, be nebulous. However, to whatever extent possible and reasonable, we should try to achieve specificity. That is exactly what project management seeks to achieve. This ensures that the leadership team, operators, staff, and customers all share the same expectations.

The word "measurable" means observable or verifiable, which implies that a method or procedure must be in place to track and assess the behavior or action on which the objective focuses and the quality of the outcome. As not all work lends itself to measurability, objectives can be written in a way that focuses on observable or verifiable behavior or results, rather than on measurable results. If no measurement system exists, the project manager must be able to monitor performance to ensure it complies with the specified objective.

An aligned objective provides a conceptual basis to draw a linkage line from the objective to other factors throughout the project. It means that the objectives throughout the organization pull in the same direction. In this way the performance of the project team and whole organization is improved.

Project managers must have a clear understanding of their own objectives before they can work with project team members to establish their job objectives. This is one of the key building blocks of performance assessment in project management. If managers know the functions on which people actually are spending time, they can make meaningful improvements in organizational performance by ensuring effort is focused on work that the organization values and by eliminating inefficient processes. Job objectives align work with organizational goals and the mission, drawing the line of sight between the employee's work, the work unit's goals, the project functions, and the organization's success. The letter "R" in SMART has two meanings that are both important: Realistic and Relevant.

Realistic has two meanings:

- The achievement of an objective is something an employee or a team can do that will support a work unit's goal. The objective should be sufficiently complex to challenge the individual or team, but not so complex that it cannot be accomplished. At the same time, it should not be so easy that it does not bring value to the individual or the team.
- The objective should be achievable within the time and resources available to the project, which is usually expressed as Triple Constraints of Time, Cost, and Quality.

Relevant implies it is important for the advancement of the employee and the organization.

Figure 22.5 illustrates the application of the Triple C approach of project management in the context of using the SMART principle of project performance assessment.

Project management implementation flowchart

This chapter has presented general principles of project management and applicability to innovation project management. The coverage in this chapter barely scratches the surface of the tools and techniques of project management. A vast array of references is available on the topic. The knowledge areas compiled by the Project Management

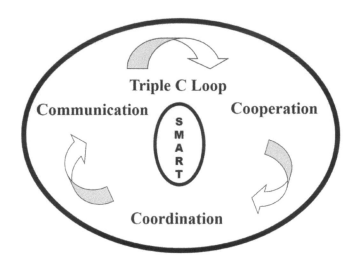

Figure 22.5 Application of Triple C in a SMART loop.

Institute (PMI) are generally applicable to the theme of this chapter. Readers are encouraged to seek more in-depth techniques of project management within the specific knowledge areas listed in the following, based on PMI's Project Management Body of Knowledge (PMBOK®):

1. Project **Integration** Management
2. Project **Scope** Management
3. Project **Time** Management
4. Project **Cost** Management
5. Project **Quality** Management
6. Project **Human** Resource Management
7. Project **Communications** Management
8. Project **Risk** Management
9. Project **Procurement** Management

The previous segments of the body of knowledge of project management cover the range of functions associated with any project, particularly complex ones, such as untested innovation pursuits. Multinational projects particularly pose unique challenges pertaining to reliable power supply, efficient communication systems, credible government support, dependable procurement processes, consistent availability of technology, progressive industrial climate, trustworthy risk mitigation infrastructure, regular supply of skilled labor, uniform focus on quality of work, global consciousness, hassle-free bureaucratic processes, coherent safety and security system, steady law and order, unflinching focus on customer satisfaction, and fair labor relations. Assessing and resolving concerns about these issues in a step-by-step fashion will create a foundation of success for a large project. While no system can be perfect and satisfactory in all aspects, a tolerable trade-off on the factors is essential for project success. That is what this chapter advocates for new endeavors along the lines of innovation. Figure 22.6 presents a generic flowchart applicable for project scheduling, which is a cornerstone for any project execution.

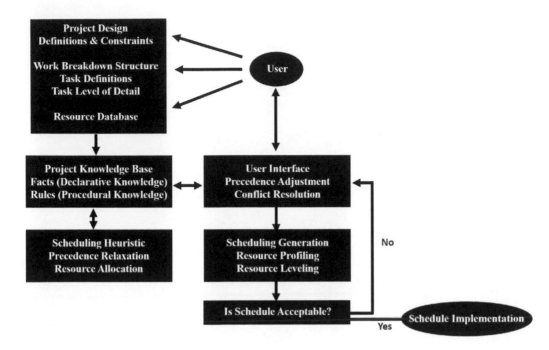

Figure 22.6 Decision support model for project scheduling.

References

Badiru, A. B. (2008), *Triple C Model of Project Management: Communication, Cooperation, and Coordination*, CRC Press/Taylor & Francis Group, Boca Raton, FL.

Badiru, A. B. (2012), *Project Management: Systems, Principles, and Applications*, CRC Press/Taylor & Francis Group, Boca Raton, FL.

chapter twenty three

Innovation in systems framework for intelligence operations

Adedeji B. Badiru and Anna E. Maloney

Contents

Introduction

Since September 11, 2001 (9/11) homeland security concerns have dominated the national agenda. Amidst unprecedented tragedy, many government agencies rapidly initiated a vast array of security improvements as stop-gap measures to protect critical facilities, transportation systems, and other infrastructure. As the country and the Intelligence Community (IC) adjust to the "new normalcy," it is logical to question whether

- The right things are being done to deter, detect, prevent, or mitigate future terrorist attacks.
- Too much or too little effort, and too many or too few resources, are being applied.
- New technologies will help deter attacks or simply cause the attacker to use a different, but equally effective, means to achieve the same result.
- All the factors that impinge on national security are being factored into intelligence decisions.

These are logical questions. No one can state definitively what should or should not be done to detect and prevent terrorist attacks, but some level of consensus decision-making will have to occur. It has been shown that effective homeland security is not just a matter of technological gadgetry; but also of an effective high-fidelity decision process. It is the position of this paper that a systems-based adaptive risk-based approach can help improve decisions, particularly decisions that must be based on uncertain information. Systems Risk-based Decision-Making (SRBDM) can help decision makers (users) address these questions and a wide range of other related questions.

For example, the US Navy (Navy) is using a risk-based approach to evaluate and implement an interdependent suite of antiterrorism (AT) capabilities aimed at increasing the Navy's ability to deter, detect, and respond to terrorist threats. While many capabilities already exist and others are being developed, the Navy must make decisions on how to allocate resources in a resource-constrained environment to best manage the risks associated with security threats. The Commander, Navy Installations (CNI) has sponsored efforts to link the results of a risk-based model with classic operations research modeling to help optimize the allocation of limited resources among many Anti-Terrorist (AT) capabilities. The prototype Navy model (at that time) was not comprehensive and did not adequately cover all decision factors; but it did demonstrate the feasibility of developing an adaptive risk-based resource-allocation decision tool for homeland security application. Although the AT capabilities were considered individually, significant dependencies exist between some capabilities due to the integrated performance expected from the systems that will be implemented to reduce the overall threat of a terrorist attack. For example, certain robust command and control actions rely heavily on information management and display infrastructure, and on communication capabilities between various Naval and civilian agencies. Therefore, the benefits associated with improved command and control systems cannot be fully realized without also addressing infrastructure needs related to information management and communication. This presents the following challenges:

1. How to directly incorporate interdependencies into a risk model?
2. How to determine the benefit (reduction in risk) that might be derived if a particular capability is only partially implemented?
3. How to change the assessment of overall risks and adjust strategies given the two-player nature of intentional attacks?
4. How to incorporate intelligence (both HUMINT and SIGINT) into AT decision-making?
5. How to ensure an integrative continuity of the AT strategy?

Background

Researchers all over the nation are developing, testing, and integrating advanced signal-processing, image-processing, and data-processing technologies for high-fidelity sensing systems. Improved technologies will increase reliability and accelerate the speed at which data is transmitted from sensing systems to humans, who monitor and analyze the data. Badiru and Maloney (2017) present a conceptual framework for an innovative application of a systems engineering model to intelligence operations. Intelligence on terrorist activities will always be incomplete and imperfect. Intentional planned attacks involve two opponents with competing objectives. Each opponent will consider the options and objectives of his adversary in formulating a strategy, and will change his strategy based on what his opponent does or what he expects him to do. Although it is clear that some terrorists are willing to die while attacking a target, there is no evidence to suggest that a terrorist will deliberately attack a target if he perceives that he will fail in the attempt. Therefore, making a target invulnerable to a particular mode of attack will not necessarily put the terrorist out of business; rather, it may cause him to consider other targets or modes of attack. Therefore, taking action to reduce a specific threat may increase the likelihood that an alternative target will be attacked. These are all issues that must be addressed from a systems perspective within the realm of effective human decision-making; not just from a technical assets point of view, as embodied in the DEJI model (Badiru 2012, 2014).

Decision-makers must be aware that terrorists will adjust attack modes and adopt strategies that will exploit vulnerabilities in a dynamic manner. Actions designed to reduce vulnerability to a specific attack mode must not only be considered to ensure that they are effective; their impact on other scenarios must also be considered. Myopic focus on safeguards for a specific target or attack mode could simply shift the risk to other targets or attack modes, and could actually increase overall risk. The research question is posed as follows:

> Can an effective systems-based methodology be devised that will effectively **Design**, **Evaluate**, **Justify**, and **Integrate** the trade-offs between risk and costs, acknowledge the interdependencies in resource-allocation decisions and the integrated nature of systems execution, respond in a timely manner to counter strategies perceived to have been made by the terrorist adversaries, take into account the imperfect and incomplete nature of intelligence on terrorist activities, and effectively guide resource allocation decisions so as to minimize the overall threat of an intentional attack?

In the approach of this paper, we recommend using the systems-based DEJI model, which has been applied to a variety of practical problems (Badiru 2012, 2014) dealing with designing, evaluating, justifying, and integrating problem parameters and factors. Figure 23.1 illustrates the basic structure of the DEJI model.

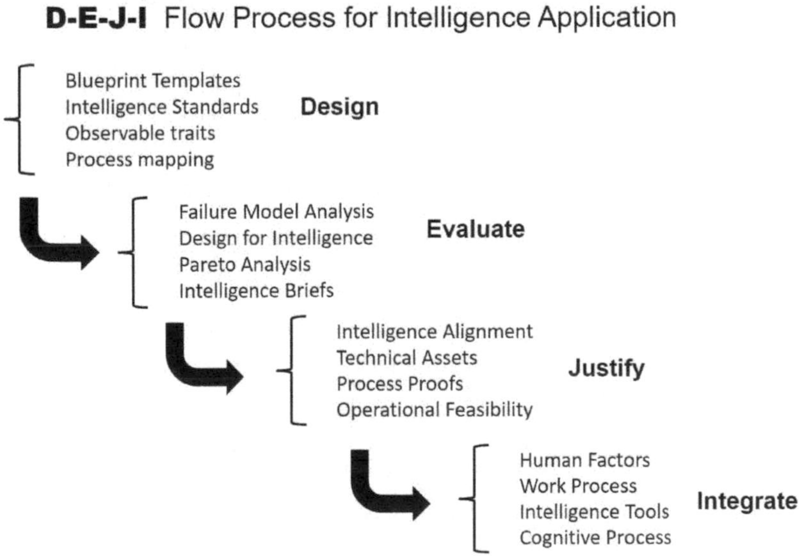

Figure 23.1 DEJI model for intelligence analysis.

The research problem has two components:

- The *risk assessment* part in which benefits (defined as reduction in the risk of attack) are moving targets that must be continuously re-evaluated based on current intelligence and dynamically fed into a capital budgeting model for strategy development and adjustment.
- The *resource allocation* part in which decisions are generated as to which capabilities should be implemented, and at what level, in order to minimize the overall threat based on the most recent systems risk assessment.

Proposed methodology

Given the wide range of attack scenarios (i.e., possible terrorist targets and attack modes), and the uncertainty associated with each scenario, the benefit provided by a particular resource allocation can be difficult to predict. However, we can establish a structured methodology to evaluate the risk benefit for each capability based on how each capability might change the threat, target vulnerability, or consequences (TVC) for each scenario.

Design section

Under the DEJI model approach, the Design aspect relates to the resource allocation model as formulated in the following large-scale nonlinear mixed binary integer programming problem:

I = set of counterterrorism capabilities

T = set of funding cycles (e.g., fiscal years)

J = set of funding sources = $\{1,2,3,4\}$

  e.g., funding source 1 = capital works; 2 = research/development;

  3 = procurements; 4 = operations/maintenance

x_{ijt} = proportion of full funding allocated to capability i from

  funding source j in time period t

$\phi(x_{i4t})$ = utility function; expressed as the % benefit derived as a function of

  funding level (% of total funding) in time t – applied only to

  operations/maintenance funding source

$y_{it} = \begin{cases} 1, \text{ if capability } i \text{ is deployed at time } t \\ 0, \text{ otherwise} \end{cases}$

  $i \in I \quad j \in J \quad t \in T$

c_{it} = reduction in risk (threat) achieved by fully implementing capability i in time t.

Objective: Maximize the reduction in the risk of a terrorist attack.

$$Max \quad \xi = \sum_i \sum_t c_{it} y_{it} \phi(x_{i4t})$$

1. Appropriation Limits (Budgets from each funding source cannot be exceeded).

2. Dependencies Across Funding Sources (e.g., sequencing of spending).

3. Dependencies Across Time Periods (e.g., sustaining funding commitments).

4. Dependencies Across Capabilities (e.g., funding prevention before detection).

5. Contingencies (either/or, if a then b).

6. Deployment (only allocate funds to deployed capabilities).

7. Bounding (force minimum funding levels if appropriate).

8. Structural (e.g., linearization, binary).

The proposed model consists of the modules shown in Figure 23.2.

Evaluation section

The capital budgeting class of problems, including uncertainty and risk, has been investigated from numerous perspectives. Recent work includes the use of fuzzy numbers to estimate uncertain returns, a modified weighted average cost of capital methodology to project returns, pooling of risks across multiple projects, analytic hierarch process (AHP) as a decision framework, zero-one integer programming to accommodate sequencing decisions in a dynamic environment, an integrated approach that combines risk management with capital budgeting, a methodology for selecting projects in high risk R&D environments, goal programming for decision making in an uncertain environment, a generalized dynamic capital allocation methodology using distortion risk measures, sensitivity analysis as a tool applied to capital budgeting under uncertainty and risk, and game theory combined with Monte Carlo simulation for timing resource allocations in a homogeneous commodity market.

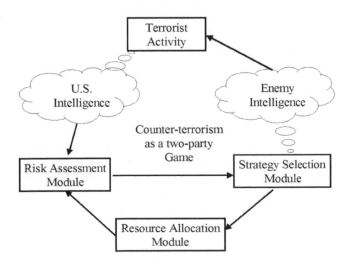

Figure 23.2 DEJI systems modeling for intelligence problem abstraction.

The desired methodology will allocate resources sequentially across competing highly interdependent projects, where each project may be partially funded, with non-linear utility functions that are dependent on the counter-moves of an intelligent adversary. A model to allocate resources to combat terrorism must address all these issues from an overall systems perspective. Developing such a model requires the integration of mathematical programming, stochastic processes, risk assessment, and classical game theory.

1. There is an urgent need for decision tools that will reduce the threat, help decision makers to spend public money more wisely, and rapidly respond to changes in the threat due to actions by the terrorists.
2. The integrative nature of the proposed model represents an application of established research to new areas of expertise.
3. The use of a decision model that directly links resource allocation strategy dynamically with counter-moves by a terrorist adversary, factoring in the imperfect nature of intelligence information and the political process, has largely been untested.

Because acts of terrorism can be vastly different, widespread, and involve ever-changing methods, analysis of information needs to be performed every step of the way. Those who gather intelligence need to realize that some information is more vital to national security and the analysts themselves need to take into account the very human aspect of terrorism. Humans are irrational at best, especially when angry and trying to get a point across. Analysts in today's world need to be trained in the art of analysis, not just specific topics or regions. Analysts nowadays must have a global perspective that was not needed during the predictable days of the Cold War. Additional funding to analysis is vital. This funding would allow additional analysts to keep an eye on small threads linking bits of information to potential terrorist attacks. While the large amount of information possessed by the IC is impressive, national security interests are only protected if that information is transformed into intelligence.

We are no longer in a Cold War world. In our world, terrorist organizations are proud of their blatant disregard of humanity. The IC can no longer fund specialists like they could in the mid to late twentieth century. When the United States had one large, known enemy, it was possible for analysts to pick a specialty and stick with it their entire career. Now it is nearly impossible to have specialists on every threat to the US. From Russia, China, and North Korea, we face states that would like to see the United States taken down a peg; from the Islamic State of Iraq and the Levant, Hezbollah, and Boko Haram we face non-state actors with no limitations on their cruelty and barbarism.

Justification section

Assuming a set of predetermined strategic scenarios, risk-based payoff values will be developed and evaluated using a modified risk-assessment procedure and appropriate statistical procedures. The best strategy can be found by solving a mixed (randomized) strategy game problem based on the payoff matrix obtained from the Risk-assessment Module. The optimal solution of a Linear Programming (LP) formulation will represent the probabilities for each strategy to return the best expected payoff value. Based on the optimal solution from the Strategy Selection Module, the predetermined strategies can be prioritized. The objective of the proposed resource allocation model is to maximize the reduction in the risk of a terrorist attack, and can be formulated as the large-scale nonlinear mixed binary integer programming problem.

The money it takes to build a new satellite is astronomical compared to the money it takes to hire on additional analysts to different intelligence agencies. As stated by Best (2015), "Unfortunately, sophisticated political and social analysis is often not emphasized in intelligence agencies, especially within the Defense Department, that are focused on technical collection and direct support to operational commanders." The US government is setting the IC up to fail should policy-makers not fund what the IC deems necessary. The cost is very low when it comes to linking intelligence agencies and allowing them to share their information and analysis. The way the IC had grown to be so bureaucratic and have immense tangles of red tape can be reversed only if policy makers decide that information sharing between agencies is as vital as many IC members claim.

Integration section

This section is presented as a hypothetical case example of the importance of integration in intelligence analyses.

Hypothetical case example 1: More attention to human intelligence in order to augment signals intelligence

The intelligence community, including the armed forces, has focused the majority of their research, time, and money on Signals Intelligence (SIGINT) while leaving Human Intelligence (HUMINT) without sufficient attention and funding. Though SIGINT worked wonderfully during the Cold War to decode and decrypt Soviet messages, the developing world has seen a rise in non-state actors threatening the United States. These non-state actors utilize the exponential growth of social media and the internet. Because there are entirely too many pieces of information circulating every day, vital information can fall through the cracks. SIGINT cannot, and should not, be responsible for keeping track of all signals. Therefore, HUMINT should come back into the spotlight. The intelligence community should put a renewed focus on quality over quantity.

Many countries around the world have developed high-performing and gainful human intelligence agencies. Though these nations tend to use tactics considered inhumane or illegal, places like Russia and Israel have found a way to gather relatively secure sources in places where SIGINT is no longer the best option. The United States needs human intelligence in order to be better prepared to deal with small terrorist cells and lone-wolf attacks. While the United States values human rights and should not go to cruel measures to secure information, the US needs to understand the importance of intelligence "boots on the ground."

In many ways, Signals Intelligence intrigues people. Congress is fascinated by new technology and likes to physically see what they are funding. This emphasis on SIGINT, however, is not always the most cost effective. Human intelligence, though occasionally subject to manipulation and deception, can provide data and information that signals intelligence cannot. A human being can see if a person looks worried while talking to different individuals or takes extra caution crossing a certain stretch of the road. A human being can detect when a voice seems weary or cautious. A human being can begin to bond and build relationships with people who hold vital information. While reading communications and listening to recordings is wonderful for quantitative information, the United States is in dire need of qualitative information if there is any hope in staying two steps ahead of her adversaries. As the Head of Intelligence Collation Management at the North Atlantic Treaty Organization's mission in Sarajevo,

Palfy (2015) said it best when he stated, "... increased collection does not necessarily or automatically lead to better intelligence outcomes."

We can look to history to compare and contrast SIGINT and HUMINT. During the Cold War, missions like the Bay of Pigs had many pictures and technical information about Soviet intervention in Cuba, but there was very little focus on HUMINT. Had the United States noticed that the Cuban people did not want to overthrow Castro, they may have been spared the embarrassing blemish on the Kennedy administration.

The implementation of this shift in concentration would not likely be difficult. The most time consuming and difficult step in the process would likely be funding approval from Congress. The Congress people would likely be unhappy to cut funding to Signals Intelligence because, in many cases, creating technology used for SIGINT brings in money and jobs to their constituents back home. As long as Congress approves the reassignment of funds, integrating this new policy should be smooth. SIGINT should continue to be funded, of course, because of its vital role in the information age.

Conclusion

This paper has presented a conceptual framework for the application of a systems approach to addressing systems challenges. Although no specific problem is tackled in the paper, the framework can give readers in the intelligence community an expanded idea of how to design, evaluate, justify, and integrate intelligence strategies. In recent years, sensor-based systems have increased in development and applications. The variety and diversity of HUMINT and SIGINT systems necessitate the application of a systems approach. This paper has introduced the application of the DEJI model by proposing the use of HUMINT as an integrated augmentation of SIGINT.

References

Badiru, A. B. (2012), Application of the DEJI model for aerospace product integration, *Journal of Aviation and Aerospace Perspectives (JAAP)*, 2(2), 20–34, 2012.

Badiru, A. B. (2014), Quality insights: The DEJI model for quality design, evaluation, justification, and integration, *International Journal of Quality Engineering and Technology*, 4(4), 369–378.

Badiru, A. B. and A. E. Maloney (2016), A conceptual framework for the application of systems approach to intelligence operations: Using HUMINT to augment SIGINT, *American Intelligence Journal*, 33(2), 41–46, 2016.

Best, R. A. (2015), Intelligence and US national security policy, *International Journal of Intelligence and CounterIntelligence*, 28(3), 449–467. doi:10.1080/08850607.2015.1022460.

Palfy, A. (2015), Bridging the gap between collection and analysis: Intelligence information processing and data governance, *International Journal of Intelligence and CounterIntelligence*, 28(2), 365–376. doi:10.1080/08850607.2015.992761.

Appendix 1: Conversion factors for innovation & logistics

Numbers and Prefixes

yotta (10^{24}):	1 000 000 000 000 000 000 000 000
zetta (10^{21}):	1 000 000 000 000 000 000 000
exa (10^{18}):	1 000 000 000 000 000 000
peta (10^{15}):	1 000 000 000 000 000
tera (10^{12}):	1 000 000 000 000
giga (10^{9}):	1 000 000 000
mega (10^{6}):	1 000 000
kilo (10^{3}):	1 000
hecto (10^{2}):	100
deca (10^{1}):	10
deci (10^{-1}):	0.1
centi (10^{-2}):	0.01
milli (10^{-3}):	0.001
micro (10^{-6}):	0.000 001
nano (10^{-9}):	0.000 000 001
pico (10^{-12}):	0.000 000 000 001
femto (10^{-15}):	0.000 000 000 000 001
atto (10^{-18}):	0.000 000 000 000 000 001
zepto (10^{-21}):	0.000 000 000 000 000 000 001
yacto (10^{-24}):	0.000 000 000 000 000 000 000 001
Stringo (10^{-35}):	0.000 000 000 000 000 000 000 000 000 000 000 01

Constants

speed of light	$2.997,925 \times 10^{10}$ cm/sec
	983.6×10^{6} ft/sec
	186,284 miles/sec
velocity of sound	340.3 meters/sec
	1116 ft/sec
gravity	9.80665 m/sec square
(acceleration)	32.174 ft/sec square
	386.089 inches/sec square

Area

Multiply	by	to obtain
acres	43,560	sq feet
	4,047	sq meters
	4,840	sq yards
	0.405	hectare
sq cm	0.155	sq inches
sq feet	144	sq inches
	0.09290	sq meters
	0.1111	sq yards
sq inches	645.16	sq millimeters
sq kilometers	0.3861	sq miles
sq meters	10.764	sq feet
	1.196	sq yards
sq miles	640	acres
	2.590	sq kilometers

Volume

Multiply	by	to obtain
acre-foot	1233.5	cubic meters
cubic cm	0.06102	cubic inches
cubic feet	1728	cubic inches
	7.480	gallons (US)
	0.02832	cubic meters
	0.03704	cubic yards
liter	1.057	liquid quarts
	0.908	dry quarts
	61.024	cubic inches
gallons (US)	231	cubic inches
	3.7854	liters
	4	quarts
	0.833	British gallons
	128	US fluid ounces
quarts (US)	0.9463	liters

Energy, Heat Power

Multiply	by	to obtain
BTU	1055.9	joules
	0.2520	kg-calories
watt-hour	3600	joules
	3.409	BTU
HP (electric)	746	watts
BTU/second	1055.9	watts
watt-second	1.00	joules

Mass

Multiply	by	to obtain
carat	0.200	cubic grams
grams	0.03527	ounces
kilograms	2.2046	pounds
ounces	28.350	grams
pound	16	ounces
	453.6	grams
stone (UK)	6.35	kilograms
	14	pounds
ton (net)	907.2	kilograms
	2000	pounds
	0.893	gross ton
	0.907	metric ton
ton (gross)	2240	pounds
	1.12	net tons
	1.016	metric tons
tonne (metric)	2,204.623	pounds
	0.984	gross pound
	1000	kilograms

Temperature

Conversion formulas	
Celsius to Kelvin	$K = C + 273.15$
Celsius to Fahrenheit	$F = (9/5)C + 32$
Fahrenheit to Celsius	$C = (5/9)(F - 32)$
Fahrenheit to Kelvin	$K = (5/9)(F + 459.67)$
Fahrenheit to Rankin	$R = F + 459.67$
Rankin to Kelvin	$K = (5/9)R$

Velocity

Multiply	by	to obtain
feet/minute	5.080	mm/second
feet/second	0.3048	meters/second
inches/second	0.0254	meters/second
km/hour	0.6214	miles/hour
meters/second	3.2808	feet/second
	2.237	miles/hour
miles/hour	88.0	feet/minute
	0.44704	meters/second
	1.6093	km/hour
	0.8684	knots
knot	1.151	miles/hour

Pressure

Multiply	by	to obtain
atmospheres	1.01325	bars
	33.90	feet of water
	29.92	inches of mercury
	760.0	mm of mercury
bar	75.01	cm of mercury
	14.50	pounds/sq inch
dyne/sq cm	0.1	N/sq meter
newtons/sq cm	1.450	pounds/sq inch
pounds/sq inch	0.06805	atmospheres
	2.036	inches of mercury
	27.708	inches of water
	68.948	millibars
	51.72	mm of mercury

Distance

Multiply	by	to obtain
angstrom	10^{-10}	meters
feet	0.30480	meters
	12	inches
inches	25.40	millimeters
	0.02540	meters
	0.08333	feet

(Continued)

Multiply	by	to obtain
kilometers	3280.8	feet
	0.6214	miles
	1094	yards
meters	39.370	inches
	3.2808	feet
	1.094	yards
miles	5280	feet
	1.6093	kilometers
	0.8694	nautical miles
millimeters	0.03937	inches
nautical miles	6076	feet
	1.852	kilometers
yards	0.9144	meters
	3	feet
	36	inches

Physical Relationships

$$D = \frac{m}{V}$$

D density
m mass
V volume

$$\left(\frac{g}{cm^3} = \frac{kg}{m^3} \right)$$

$$d = v \cdot t$$

d distance m
v velocity m/s
t time s

$$a = \frac{vf - vi}{t}$$

a acceleration m/s^2
vf final velocity m/s
vi initial velocity m/s
t time s

$$P = \frac{W}{t}$$

P power W(=watts)
W work J
t time s

$$K.E. = \frac{1}{2} \cdot m \cdot v^2$$

K.E. kinetic energy
m mass kg
v velocity m/s

$$Fe = \frac{kQ_1Q_2}{d^2}$$

Fe electrical force N
k Coulomb's constant

$$\left(k = 9 \times 10^9 \frac{N \cdot m^2}{c^2} \right)$$

$Q_1 \cdot Q_2$ are electrical charges C
d separation distance m

(Continued)

$$d = vi \cdot t + \frac{1}{2} \cdot a \cdot t^2$$

d distance m
vi initial velocity m/s
t time s
a acceleration m/s²

$$F = m \cdot a$$

F net force N(=newtons)
m mass kg
a acceleration m/s²

$$Fg = \frac{G \cdot m_1 \cdot m_2}{d^2}$$

Fg force of gravity N
G universal
  gravitational constant

$$\left(G = 6.67 \times 10^{-11} \frac{N - m^2}{kg^2} \right)$$

m₁, m₂ masses of the two
  objects kg
d separation distance m

$$p = m \cdot v$$

p momentum kg·m/s
m mass
v velocity

$$W = F \cdot d$$

W work J(=joules)
F force N
d distance m

$$V = \frac{W}{Q}$$

V electrical potential difference V(=volts)
W work done J
Q electric charge moving C

$$I = \frac{Q}{t}$$

I electric current ampères
Q electric charge flowing C
t time s

$$W = V.I.t$$

W electrical energy J
V voltage V
I current A
t time s

$$P = V \cdot I$$

P power W
V voltage V
I current A

$$H = c \cdot m \cdot \Delta T$$

H heat energy J
m mass kg
ΔT change in temperature °C
c specific heat J/Kg·°C

Units of measurement

English system			Metric system		
1 foot (ft)	= 12 inches (in) 1′=12″	mm	millimeter	.001 m
1 yard (yd)	= 3 feet	cm	centimeter	.01 m
1 mile (mi)	= 1760 yards	dm	decimeter	.1 m
1 sq. foot	= 144 sq. inches	m	meter	1 m
1 sq. yard	= 9 sq. feet	dam	decameter	10 m
1 acre	= 4840 sq. yards = 43560 ft²	hm	hectometer	100 m
1 sq. mile	= 640 acres	km	kilometer	1000 m

Note: Prefixes also apply to l (liter) and g (gram).

Common Notations

Units of meas.	Abbrev.	Relation	Units of meas.	Abbrev.	Relation
meter	m	length	degree Celsius	°C	temperature
hectare	ha	area	Kelvin	K	thermodynamic temp.
tonne	t	mass	pascal	Pa	pressure, stress
kilogram	kg	mass	joule	J	energy, work
nautical mile	M	distance (navigation)	Newton	N	force
			watt	W	power, radiant flux
knot	kn	speed (navigation)	ampere	A	electric current
liter	L	volume or capacity	volt	V	electric potential
second	s	time	ohm	Ω	electric resistance
hertz	Hz	frequency	coulomb	C	electric charge
candela	cd	luminous intensity			

Measurement Conversion Units

A pinch	1/8 tsp. or less
3 tsp	1 tbsp.
2 tbsp	1/8 c.
4 tbsp	1/4 c.
16 tbsp	1 c.
5 tbsp. + 1 tsp	1/3 c.
4 oz	1/2 c.
8 oz	1 c.
16 oz	1 lbs.
1 oz	2 tbsp. fat or liquid
1 c. of liquid	1/2 pt.
2 c	1 pt.
2 pt	1 qt.
4 c. of liquid	1 qt.
4 qts	1 gallon
8 qts	1 peck (such as apples, pears, etc.)
1 jigger	1 ½ fl.oz.
1 jigger	3 tbsp.

Appendix 2: Glossary of innovation project management terms

- ABC. Activity Based Costing. Bottom up estimating and summation based on material and labor required for activities making up a project.
- Activity. A component of work performed during the course of a project. See also schedule activity.
- Activity Duration. The time in calendar units between the start and finish of a schedule activity. See also actual duration, original duration, and remaining duration.
- Activity Resource Estimating. The process of estimating the types and quantities of resources required to perform each schedule activity.
- Activity Sequencing. The process of identifying and documenting dependencies among schedule activities.
- Authority. The right to apply project resources, expend funds, make decisions, or give approvals.
- Bar Chart. A graphic display of schedule-related information. In the typical bar chart, schedule activities or work breakdown structure components are listed down the left side of the chart, dates are shown across the top, and activity durations are shown as date-placed horizontal bars. Also called a Gantt chart.
- Baseline. The approved time phased plan (for a project, a work breakdown structure component, a work package, or a schedule activity), plus or minus approved project scope, cost, schedule, and technical changes. Generally refers to the current baseline, but may refer to the original or some other baseline. Usually used with a modifier (e.g., cost baseline, schedule baseline, performance measurement baseline, technical baseline). See also performance measurement baseline.
- Baseline Start Date. The start date of a schedule activity in the approved schedule baseline. See also scheduled start date.
- Best Practices. Processes, procedures, and techniques that have consistently demonstrated achievement of expectations and that are documented for the purposes of sharing, repetition, replication, adaptation, and refinement.
- Change Control. Identifying, documenting, approving or rejecting, and controlling changes to the project baselines.
- Close Project. The process of finalizing all activities across all of the project process groups to formally close the project or phase.
- Common Cause. A source of variation that is inherent in the system and predictable. On a control chart, it appears as part of the random process variation (i.e., variation from a process that would be considered normal or not unusual), and is indicated

by a random pattern of points within the control limits. Also referred to as random cause. Contrast with special cause.

- Configuration Management System. A subsystem of the overall project management system. It is a collection of formal documented procedures used to apply technical and administrative direction and surveillance to: identify and document the functional and physical characteristics of a product, result, service, or component; control any changes to such characteristics; record and report each change and its implementation status; and support the audit of the products, results, or components to verify conformance to requirements. It includes the documentation, tracking systems, and defined approval levels necessary for authorizing and controlling changes. In most application areas, the configuration management system includes the change control system.

- Constraint. The state, quality, or sense of being restricted to a given course of action or inaction. An applicable restriction or limitation, either internal or external to the project, that will affect the performance of the project or a process. For example, a schedule constraint is any limitation or restraint placed on the project schedule that affects when a schedule activity can be scheduled and is usually in the form of fixed imposed dates. A cost constraint is any limitation or restraint placed on the project budget such as funds available over time. A project resource constraint is any limitation or restraint placed on resource usage, such as what resource skills or disciplines are available and the amount of a given resource available during a specified time frame.

- Contingency Reserve. The amount of funds, budget, or time needed above the estimate to reduce the risk of overruns of project objectives to a level acceptable to the organization.

- Control. Comparing actual performance with planned performance, analyzing variances, assessing trends to effect process improvements, evaluating possible alternatives, and recommending appropriate corrective action as needed.

- Control Chart. A graphic display of process data over time and against established control limits, and that has a centerline that assists in detecting a trend of plotted values toward either control limit.

- Control Limits. The area composed of three standard deviations on either side of the centerline, or mean, of a normal distribution of data plotted on a control chart that reflects the expected variation in the data. See also specification limits.

- Cost Control. The process of influencing the factors that create variances, and controlling changes to the project budget.

- Cost of Quality (COQ). Determining the costs incurred to ensure quality. Prevention and appraisal costs (cost of conformance) include costs for quality planning, quality control (QC), and quality assurance to ensure compliance to requirements (i.e., training, QC systems, etc.). Failure costs (cost of non-conformance) include costs to rework products, components, or processes that are non-compliant, costs of warranty work and waste, and loss of reputation.

- Cost Performance Index (CPI). A measure of cost efficiency on ^project. It is the ratio of earned value (EV) to actual costs (AC). CPI = EV divided by AC. A CPI value equal to or greater than one indicates a favorable condition and a value less than one indicates an unfavorable condition.

- Cost-Plus-Fee (CPF). A type of cost reimbursable contract where the buyer reimburses the seller for seller's allowable costs for performing the contract work and seller also receives a fee calculated as an agreed upon percentage of the costs. The fee varies with the actual cost.

- Cost-Plus-Fixed-Fee (CPFF) Contract. A type of cost-reimbursable contract where the buyer reimburses the seller for the seller's allowable costs (allowable costs are defined by the contract) plus a fixed amount of profit (fee).
- Cost-Plus-Incentive-Fee (CPIF) Contract. A type of cost-reimbursable contract where the buyer reimburses the seller for the seller's allowable costs (allowable costs are defined by the contract), and the seller earns its profit if it meets defined performance criteria.
- Cost-Plus-Percentage of Cost (CPPC). See cost-plus-fee.
- Cost-Reimbursable Contract. A type of contract involving payment (reimbursement) by the buyer to the seller for the seller's actual costs, plus a fee typically representing seller's profit. Costs are usually classified as direct costs or indirect costs. Direct costs are costs incurred for the exclusive benefit of the project, such as salaries of full-time project staff. Indirect costs, also called overhead and general and administrative cost, are costs allocated to the project by the performing organization as a cost of doing business, such as salaries of management indirectly involved in the project, and cost of electric utilities for the office. Indirect costs are usually calculated as a percentage of direct costs. Cost-reimbursable contracts often include incentive clauses where, if the seller meets or exceeds selected project objectives, such as schedule targets or total cost, then the seller receives from the buyer an incentive or bonus payment.
- Cost Variance (CV). A measure of cost performance on a project. It is the algebraic difference between earned value (EV) and actual cost (AC). CV = EV minus AC. A positive value indicates a favorable condition and a negative value indicates an unfavorable condition.
- Crashing. A specific type of project schedule compression technique performed by taking action to decrease the total project schedule duration after analyzing a number of alternatives to determine how to get the maximum schedule duration compression for the least additional cost. Typical approaches for crashing a schedule include reducing schedule activity durations and increasing the assignment of resources on schedule activities. See schedule compression and see also fast tracking.
- Create WBS (Work Breakdown Structure). The process of subdividing the major project deliverables and project work into smaller, more manageable components.
- Critical Activity. Any schedule activity on a critical path in a project schedule. Most commonly determined by using the critical path method. Although some activities are "critical," in the dictionary sense, without being on the critical path, this meaning is seldom used in the project context.
- Critical Chain Method. A schedule network, analysis technique that modifies the project schedule to account for limited resources. The critical chain method mixes deterministic and probabilistic approaches to schedule network analysis.
- Critical Path. Generally, but not always, the sequence of schedule activities that determines the duration of the project. Generally, it is the longest path through the project. However, a critical path can end, as an example, on a schedule milestone that is in the middle of the project schedule and that has a finish-no-later-than imposed date schedule constraint. See also critical path method.
- Critical Path Method (CPM). A schedule network analysis technique used to determine the amount of scheduling flexibility (the amount of float) on various logical network paths in the project schedule network, and to determine the minimum total project duration. Early start and finish dates are calculated by means of and forward pass, using a specified start date. Late start and finish dates are calculated by means of a backward pass, starting from a specified completion date, which sometimes is the project early finish date determined during the forward pass calculation.

- Decision Tree Analysis. The decision tree is a diagram that describes a decision under consideration and the implications of choosing one or another of the available alternatives. It is used when some future scenarios or outcomes of actions are uncertain. It incorporates probabilities and the costs or rewards of each logical path of events and future decisions, and uses expected monetary value analysis to help the organization identify the relative values of alternate actions. See also expected monetary value analysis.
- Decomposition. A planning technique that subdivides the project scope and project deliverables into smaller, more manageable components, until the project work associated with accomplishing the project scope and providing the deliverables is defined in sufficient detail to support executing, monitoring, and controlling the work.
- Defect. An imperfection or deficiency in a project component where that component does not meet its requirements or specifications and needs to be either repaired or replaced.
- Defect Repair. Formally documented identification of a defect in a project component with a recommendation to either repair the defect or completely replace the component.
- Deliverable. Any unique and verifiable product, result, or capability to perform a service that must be produced to complete a process, phase, or project. Often used more narrowly in reference to an external deliverable, which is a deliverable that is subject to approval by the project sponsor or customer. See also product, service, and result.
- Delphi Technique. An information gathering technique used as a way to reach a consensus of experts on a subject. Experts on the subject participate in this technique anonymously. A facilitator uses a questionnaire to solicit ideas about the important project points related to the subject. The responses are summarized and are then re-circulated to the experts for further comment. Consensus may be reached in a few rounds of this process. The Delphi technique helps reduce bias in the data and keeps any one person from having undue influence on the outcome.
- Develop Project Charter. The process of developing the project charter that formally authorizes a project.
- Discrete Effort. Work effort that is directly identifiable to the completion of specific work breakdown structure components and deliverables, and that can be directly planned and measured. Contrast with apportioned effort.
- Dummy Activity. A schedule activity of zero duration used to show a logical relationship in the arrow diagramming method. Dummy activities are used when logical relationships cannot be completely or correctly described with schedule activity arrows. Dummy activities are generally shown graphically as a dashed line headed by an arrow.
- Early Finish Date (EF). In the critical path method, the earliest possible point in time on which the uncompleted portions of a schedule activity (or the project) can finish, based on the schedule network, logic, the data date, and any schedule constraints. Early finish dates can change as the project progresses and as changes are made to the project management plan.
- Early Start Date (ES). In the critical path method, the earliest possible point in time on which the uncompleted portions of a schedule activity (or the project) can start, based on the schedule network logic, the data date, and any schedule constraints. Early start dates can change as the project progresses and as changes are made to the project management plan.

- Earned Value (EV). The value of completed work expressed in terms of the approved budget assigned to that work for a schedule activity or work breakdown structure component. Also referred to as the budgeted cost of work performed (BCWP).
- Earned Value Management (EVM). A management methodology for integrating scope, schedule, and resources, and for objectively measuring project performance and progress. Performance is measured by determining the budgeted cost of work performed (i.e., earned value) and comparing it to the actual cost of work performed (i.e., actual cost). Progress is measured by comparing the earned value to the planned value.
- Earned Value Technique (EVT). A specific technique for measuring the performance of work for a work breakdown structure component, control account, or project. Also referred to as the earning rules and crediting method.
- Effort. The number of labor units required to complete a schedule activity or work breakdown structure component. Usually expressed as staff hours, staff days, or staff weeks. Contrast with duration.
- Enterprise. A company, business, firm, partnership, corporation, or governmental agency.
- Enterprise Environmental Factors. Any or all external environmental factors and internal organizational environmental factors that surround or influence the project's success. These factors are from any or all of the enterprises involved in the project, and include organizational culture and structure, infrastructure, existing resources, commercial databases, market conditions, and project management software.
- Execute. Directing, managing, performing, and accomplishing the project work, providing the deliverables, and providing work performance information.
- Expected Monetary Value (EMV) Analysis. A statistical technique that calculates the average outcome when the future includes scenarios that may or may not happen. A common use of this technique is within decision tree analysis. Modeling and simulation are recommended for cost and schedule risk analysis because it is more powerful and less subject to misapplication than expected monetary value analysis.
- Expert Judgment. Judgment provided based upon expertise in an application area, knowledge area, discipline, industry, and so on. as appropriate for the activity being performed. Such expertise may be provided by any group or person with specialized education, knowledge, skill, experience, or training, and is available from many sources, including: other units within the performing organization; consultants; stakeholders, including customers, professional and technical associations; and industry groups.
- Failure Mode and Effect Analysis (FMEA). An analytical procedure, in which each potential failure mode in every component of a product is analyzed to determine its effect on the reliability of that component and, by itself or in combination with other possible failure modes, on the reliability of the product or system and on the required function of the component; or the examination of a product (at the system and/or lower levels) for all ways that a failure may occur. For each potential failure, an estimate is made of its effect on the total system and of its impact. In addition, a review is undertaken of the action planned to minimize the probability of failure and to minimize its effects.
- Fast Tracking. A specific project schedule compression technique that changes network logic to overlap phases that would normally be done in sequence, such as the design phase and construction phase, or to perform schedule activities in parallel. See schedule compression and see also crashing.
- Finish-to-Finish (FF). The logical relationship where completion of work of the successor activity cannot finish until the completion of work of the predecessor activity. See also logical relationship.

- Finish-to-Start (FS). The logical relationship where initiation of work of the successor activity depends upon the completion of work of the predecessor activity. See also logical relationship.
- Firm-Fixed-Price (FFP) Contract. A type of fixed price contract where the buyer pays the seller a set amount (as defined by the contract), regardless of the seller's costs.
- Fixed-Price-Incentive-Fee (FPIF) Contract. A type of contract where the buyer pays the seller a set amount (as defined by the contract), and the seller can earn an additional amount if the seller meets defined performance criteria.
- Fixed-Price or Lump-Sum Contract. A type of contract involving a fixed total price for a well-defined product. Fixed-price contracts may also include incentives for meeting or exceeding selected project objectives, such as schedule targets. The simplest form of a fixed price contract is a purchase order.
- Float. Also called slack. See total float and see also free float.
- Flowcharting. The depiction in a diagram format of the inputs, process actions, and outputs of one or more processes within a system.
- Free Float (FF). The amount of time that a schedule activity can be delayed without delaying the early start of any immediately following schedule activities. See also total float.
- Gantt Chart. See bar chart.
- Imposed Date. A fixed date imposed on a schedule activity or schedule milestone, usually in the form of a "start no earlier than" and "finish no later than" date.
- Influence Diagram. Graphical representation of situations showing causal influences, time ordering of events, and other relationships among variables and outcomes.
- Integrated Change Control. The process of reviewing all change requests, approving changes and controlling changes to deliverables and organizational process assets.
- Invitation for Bid (IFB). Generally, this term is equivalent to request for proposal. However, in some application areas, it may have a narrower or more specific meaning.
- Lag. A modification of a logical relationship that directs a delay in the successor activity. For example, in a finish-to-start dependency with a ten-day lag, the successor activity cannot start until ten days after the predecessor activity has finished. See also lead.
- Late Finish Date (LF). In the critical path method, the latest possible point in time that a schedule activity may be completed based upon the schedule network logic, the project completion date, and any constraints assigned to the schedule activities without violating a schedule constraint or delaying the project completion date. The late finish dates are determined during the backward pass calculation of the project schedule network.
- Late Start Date (LS). In the critical path method, the latest possible point in time that a schedule activity may begin based upon the schedule network logic, the project completion date, and any constraints assigned to the schedule activities without violating a schedule constraint or delaying the project completion date. The late start dates are determined during the backward pass calculation of the project schedule network.
- Latest Revised Estimate. See estimate at completion.
- Lead. A modification of a logical relationship that allows an acceleration of the successor activity. For example, in a finish-to-start dependency with a ten-day lead, the successor activity can start ten days before the predecessor activity has finished. See also lag. A negative lead is equivalent to a positive lag.
- Life Cycle. See project life cycle.
- Materiel. The aggregate of things used by an organization in any undertaking, such as equipment, apparatus, tools, machinery, gear, material, and supplies.

- Matrix Organization. Any organizational structure in which the project manager shares responsibility with the functional managers for assigning priorities and for directing the work of persons assigned to the project.
- Milestone. A significant point or event in the project. See also schedule milestone.
- Monte Carlo Analysis. A technique that computes, or iterates, the project cost or project schedule many times using input values selected at random from probability distributions of possible costs or durations, to calculate a distribution of possible total project cost or completion dates.
- Opportunity. A condition or situation favorable to the project, a positive set of circumstances, a positive set of events, a risk that will have a positive impact on project objectives, or a possibility for positive changes. Contrast with threat.
- Organizational Breakdown Structure (OBS). A hierarchically organized depiction of the project organization arranged so as to relate the work packages to the performing organizational units. (Sometimes OBS is written as Organization Breakdown Structure with the same definition.)
- Parametric Estimating. An estimating *technique* that uses a statistical relationship between historical data and other variables (e.g., square footage in construction, lines of code in software development) to calculate an *estimate* for activity parameters, such as *scope, cost, budget,* and *duration.* This technique can produce higher levels of accuracy depending upon the sophistication and the underlying data built into the model. An example for the cost parameter is multiplying the planned quantity of work to be performed by the historical cost per unit to obtain the estimated cost.
- Pareto Chart. A histogram, ordered by frequency of occurrence, that shows how many results were generated by each identified cause.
- Position Description. An explanation of a project team member's roles and responsibilities.
- Precedence Relationship. The term used in the precedence diagramming method for a logical relationship. In current usage, however, precedence relationship, logical relationship, and dependency are widely used interchangeably, regardless of the diagramming method used.
- Predecessor Activity. The schedule activity that determines when the logical successor activity can begin or end.
- Product Life Cycle. A collection of generally sequential, non-overlapping product phases whose name and number are determined by the manufacturing and control needs of the organization. The last product life cycle phase for a product is generally the product's deterioration and death. Generally, a project life cycle is contained within one or more product life cycles.
- Product Scope. The features and functions that characterize a product, service, or result.
- Product Scope Description. The documented narrative description of the product scope.
- Program. A group of related projects managed in a coordinated way to obtain benefits and control not available from managing them individually. Programs may include elements of related work outside of the scope of the discrete projects in the program.
- Program Management. The centralized coordinated management of a program to achieve the program's strategic objectives and benefits.
- Program Management Office (PMO). The centralized management of a particular program or programs such that corporate benefit is realized by the sharing of

resources, methodologies, tools, and techniques, and related high-level project management focus. See also project management office.

- Project. A temporary endeavor undertaken to create a unique product, service, or result.
- Project Charter. A document issued by the project initiator or sponsor that formally authorizes the existence of a project, and provides the project manager with the authority to apply organizational resources to project activities.
- Project Life Cycle. A collection of generally sequential project phases whose name and number are determined by the control needs of the organization or organizations involved in the project. A life cycle can be documented with a methodology.
- Project Organization Chart. A document that graphically depicts the project team members and their interrelationships for a specific project.
- Project Scope Statement. The narrative description of the project scope, including major deliverables, project objectives, project assumptions, project constraints, and a statement of work, that provides a documented basis for making future project decisions and for confirming or developing a common understanding of project scope among the stakeholders. A statement of what needs to be accomplished.
- Resource Leveling. Any form of schedule network analysis in which scheduling decisions (start and finish dates) are driven by resource constraints (e.g., limited resource availability or difficult-to-manage changes in resource availability levels).
- Responsibility Matrix. A structure that relates the project organizational breakdown structure to the work breakdown structure to help ensure that each component of the project's scope of work is assigned to a responsible person.
- Risk. An uncertain event or condition that, if it occurs, has a positive or negative effect on a project's objectives. See also risk category and risk breakdown structure.
- Risk Acceptance. A risk response planning technique that indicates that the project team has decided not to change the project management plan to deal with a risk, or is unable to identify any other suitable response strategy.
- Risk Avoidance. A risk response planning technique for a threat that creates changes to the project management plan that are meant to either eliminate the risk or to protect the project objectives from its impact. Generally, risk avoidance involves relaxing the time, cost, scope, or quality objectives.
- Risk Breakdown Structure (RBS). A hierarchically organized depiction of the identified project risks arranged by risk category and subcategory that identifies the various areas and causes of potential risks. The risk breakdown structure is often tailored to specific project types.
- Rolling Wave Planning. A form of progressive elaboration planning where the work to be accomplished in the near term is planned in detail at a low level of the work breakdown structure, while the work far in the future is planned at a relatively high level of the work breakdown structure, but the detailed planning of the work to be performed within another one or two periods in the near future is done as work is being completed during the current period.
- Root Cause Analysis. An analytical technique used to determine the basic underlying reason that causes a variance or a defect or a risk. A root cause may underlie more than one variance or defect or risk.
- Schedule Milestone. A significant event in the project schedule, such as an event restraining future work or marking the completion of a major deliverable. A schedule milestone has zero duration. Sometimes called a milestone activity. See also milestone.
- Scope. The sum of the products, services, and results to be provided as a project. See also project scope and product scope.

- S-Curve. Graphic display of cumulative costs, labor hours, percentage of work, or other quantities, plotted against time. The name derives from the S-like shape of the curve (flatter at the beginning and end, steeper in the middle) produced on & project that starts slowly, accelerates, and then tails off. Also a term for the cumulative likelihood distribution that is a result of a simulation, a tool of quantitative risk analysis.
- Statement of Work (SOW). A narrative description of products, services, or results to be supplied.
- SWOT Analysis (Strengths, Weaknesses, Opportunities, and Threats Analysis). This information gathering technique examines the project from the perspective of each project's strengths, weaknesses, opportunities, and threats to increase the breadth of the risks considered by risk management.
- Triple Constraint. A framework for evaluating competing demands. The triple constraint is often depicted as a triangle where one of the sides or one of the corners represent one of the parameters being managed by the project team.
- Value Engineering (VE). A creative approach used to optimize project life cycle costs, save time, increase profits, improve quality, expand market share, solve problems, and/or use resources more effectively.
- Work Breakdown Structure (WBS). A deliverable-oriented hierarchical decomposition of the work, to be executed by the project team to accomplish the project objectives and create the required deliverables. It organizes and defines the total scope of the project. Each descending level represents an increasingly detailed definition of the project work. The WBS is decomposed into work packages. The deliverable orientation of the hierarchy includes both internal and external deliverables. See also work package, control account, contract work, breakdown structure, and project summary work, breakdown structure.

Index

Note: Page numbers in italic and bold refer to figures and tables respectively.